HANDBOOK OF NONRESPONSE IN HOUSEHOLD SURVEYS

WILEY SERIES IN SURVEY METHODOLOGY

Established in Part by WALTER A. SHEWHART AND SAMUEL S. WILKS

A complete list of the titles in this series appears at the end of this volume.

HANDBOOK OF NONRESPONSE IN HOUSEHOLD SURVEYS

JELKE BETHLEHEM
FANNIE COBBEN
BARRY SCHOUTEN

WILEY
A JOHN WILEY & SONS, INC., PUBLICATION

Published by John Wiley & Sons, Inc., Hoboken, New Jersey.
Published simultaneously in Canada.

Library of Congress Cataloging-in-Publication Data:

Bethlehem, Jelke G.
Handbook of nonresponse in household surveys / Jelke Bethlehem, Fannie Cobben, Barry Schouten.
p. cm.
Includes index.
ISBN 978-0-470-54279-8 (cloth)
1. Household surveys–Response rate. 2. Nonresponse (Statistics) I. Cobben, Fannie, 1979–
II. Schouten, Barry, 1971– III. Title.
HB849.49.B48 2011
001.4'33–dc22

2010016636

Printed in Singapore

10 9 8 7 6 5 4 3 2 1

Contents

PREFACE

This is a book about nonresponse in household surveys. When persons selected for a survey do not provide the requested information, researchers lose control of the sample selection mechanism of the survey. Some groups in the population will be underrepresented in the survey, while other groups will be overrepresented. The immediate consequence of nonresponse is that, without taking special measures, it is not possible to compute reliable estimates of population characteristics. Validity of inference about the population is at stake.

Nonresponse is a worldwide problem. Survey researchers everywhere are confronted by it, and the problem seems to be rising over time. Survey response rates are decreasing in many countries. This book discusses many theoretical and practical aspects of nonresponse. It can be used as a handbook by survey researchers working in official statistics (e.g. in national statistical institutes), by academics, and by commercial market researches.

The book is the result of many years of research in official statistics at Statistics Netherlands. Already since the 1970s survey researchers in the Netherlands are confronted with decreasing response rates. This has forced Statistics Netherlands to conduct research aimed at reducing nonresponse in the field and correcting for the remaining nonresponse.

The long years of experience with attempts at lessening the nonresponse problem have been used in a European course that is given several times for researchers of European national statistical institutes. This course is part of the ESTP, the European Statistical Training Program. The objective behind the training course in the ESTP is to improve the quality of national and regional statistics by transmitting methodological know-how and by fostering the exchange of good practices among EU countries. The material of the training course on nonresponse in household surveys forms the basis of this book.

The obvious way to approach the nonresponse problem is to prevent it from happening in the field. However, experience shows that regardless of effort, a substantial amount of nonresponse always remains. So, to achieve reliable estimates, effort needs to be made to correct survey responses for the possible negative effects of nonresponses. This book contains a systematic overview of the main existing correction techniques. It also presents some new methods and techniques.

The first two chapters of the book are an introduction into the nonresponse problem. Basic concepts are described and definitions are given. For example, it is shown that it is not always possible to compute something as simple as a response rate. Two models are presented that can be used to include nonresponse in the theory of survey sampling: the fixed response model and the random response model.

Chapter 3 is about reducing the nonresponse in the field. It focuses on the psychological aspects of nonresponse reduction research. Reduction of nonresponse is considered in terms of the behavior of potential respondents by addressing what causes persons not to participate in a survey and how this can be prevented. The chapter shows how survey organizations can translate this knowledge into fieldwork strategies that aim at obtaining representative and high response rates.

Different modes can be used for survey data collection. There are face-to-face surveys, telephone surveys, mail surveys, Web surveys, and so forth. The magnitude and effects of nonresponse can be different in each mode. For example, it matters whether or not interviewers are involved in the data collection. This is the topic of Chapter 4.

Chapters 5 and 6 are about the analysis of nonresponse. It is important to carry out an analysis, because this helps to determine to what extent the survey outcomes are affected by nonresponse. Chapter 6 focuses on comparing nonresponse of surveys in different countries. It shows that this is not easy to do.

Nonresponse is closely related to the concept of representativity. If, due to nonresponse, the survey response is not representative, estimates of population characteristics will be biased. Chapter 7 is about representativity. It introduces a new indicator for the quality of survey response. This so-called R-indicator was recently developed in the RISQ project. This research project was financed by the 7th Framework Program of the European Union. Three national statistical institutes (in the Netherlands, Norway, and Slovenia) and two universities (in Southampton and Leuven) cooperated in the project.

Weighting adjustment is the most important family of techniques used to correct for a possible nonresponse bias. Several weighting adjustment techniques are described in Chapter 8: simple poststratification, linear weighting (as a form of generalized regression estimation), and multiplicative weighting (raking ratio estimation).

Auxiliary information is required to correct for a possible bias due to nonresponse. In some countries auxiliary variables are scarce, but in other countries, such as the Netherlands and the Scandinavian countries, researchers have access to public registers and other governmental information. This gives rise to the question which auxiliary variables should be used for nonresponse correction. Chapter 9 presents old and new techniques for the selection of such variables.

One way to obtain information about nonrespondents is to contact them again. Two such approaches are discussed in Chapter 10. One approach is the callback approach. This means another attempt is made to get all answers to the questions in the questionnaire form. Another approach is the basic-question approach, whereby the nonrespondents are re-approached with a very short questionnaire containing only a few basic questions.

According to the random response model, all persons in the population have unknown and different probabilities to respond in a survey. Chapter 11 is about estimating these response probabilities. The estimated response probabilities (the response propensities) can also be used to correct a possible nonresponse bias.

Experience has shown that analysis and correction of nonresponse can be more effective if different causes of nonresponse can be distinguished. Chapter 12 gives an overview of a number of techniques that take into account different types of nonresponse.

Chapter 13 is about a new development. Traditionally survey designs are fixed before the fieldwork starts. If it becomes clear in the course of the fieldwork, however, that nonresponse will seriously affect the survey results, different measures should be taken. One way to do this is by implementing an adaptive survey design. Adaptive survey designs assign different survey design features to different sample units based on their characteristics and response behavior. Characteristics may be available beforehand or may be observed during data collection.

A large part of the book is about unit nonresponse. Unit nonresponse occurs when the questionnaire remains completely empty. Not one question is answered. There is also item nonresponse. This occurs when only some (possible sensitive) questions are left unanswered. Item nonresponse is the topic of Chapter 14.

Chapter 15, the final chapter, contains some miscellaneous topics. Among these are the combined treatment of unit and item nonresponse, nonresponse in panels, and the use of paradata (information that is collected during data collection about the data collection process).

The accompanying website www.survey-nonresponse.com contains the survey data set of the General Population Survey (GPS). This data set has been used for many examples and applications in the book. The data set is available in both SPSS and Stata format.

CHAPTER ONE

The Nonresponse Problem

1.1 Introduction

We live in an information society. There is an ever growing demand for statistical information about the economic, social, political, and cultural shape of the country. Such information enables policy makers and others to make informed decisions for a better future.

Sometimes it is possible to retrieve such statistical information from existing administrative sources such as public registers. More often there is no such sources. Then a survey is the best instrument to use for collecting new statistical information.

A *survey* collects information about a specific population. This population need not necessarily consist of persons. For example, the elements of the population can be households, farms, companies, or schools. Typically information is collected by asking questions about the elements in the population. To do this in a uniform and consistent way, a questionnaire is developed.

One way to carry out a survey is to collect information on all elements in the population. Such a survey is called a *census* or a *complete enumeration*. This approach has a number of disadvantages. In the first place, it is very expensive. Surveying a large population requires a lot of people (e.g., interviewers) and a lot of computer resources. In the second place, it is very time-consuming. Collecting and processing a large amount of data takes time. This affects the timeliness of the results of the survey, and less timely information is less useful. In the third place, large surveys increase the response burden more than small surveys. As more and more people are asked to participate in surveys, they are less inclined to cooperate.

Handbook of Nonresponse in Household Surveys, By Jelke Bethlehem, Fannie Cobben, and Barry Schouten

A sample survey is a solution to many of the problems of complete enumeration because it collects information on only a small part of the population. This small part is called the *sample*. In principle, the sample only provides information on the sampled elements of the population. There is no information on the nonsampled elements. Still, if the sample is selected in a "clever" way, it is possible to make inference about the population as a whole. In this context, "clever" means that the sample is selected by means of probability sampling. A random selection procedure determines which elements are selected, and which not. If the survey researcher knows how the selection mechanism works and if it is possible to compute the probabilities of being selected in the sample, the results can be used to draw reliable conclusions about the nonsampled elements.

At first sight, the idea of introducing an element of uncertainty in the investigation seems odd. How can a survey researcher say something about a population as a whole by investigating only a small randomly selected part of it? The theory of statistical inference shows that this indeed is possible. Many books about the mathematical and statistical background of survey sampling have been published. Examples are Deming (1950) and Hansen et al. (1953), Cochran (1977), and Bethlehem (2009). The basic principles and concepts of survey sampling are summarized in chapter 2.

The first ideas on survey sampling emerged around the year 1895. See Bethlehem (2009) for an historic overview. The principles of probability sampling have been successfully applied on a regular basis in official and academic statistics since the 1940s, and to a much lesser extent also in commercial market research. Nevertheless, the survey organization does not have full control over the survey process. Practical problems may occur while collecting survey data. One of these problems is nonresponse. *Nonresponse* occurs when elements in the population that are sampled, and that are eligible for the survey, do not provide the requested information, or provide information that is not usable.

When confronted with nonresponse in a survey, a researcher loses control over the sample selection mechanism of the survey. Some groups in the population will be underrepresented in the survey, while other groups will be overrepresented. The immediate consequence of nonresponse is that, without taking special measures, it is not possible to compute reliable estimates of population characteristics. Validity of inference about the population is at stake. Both the composition and size of the sample are affected.

This handbook is about the nonresponse problem. It shows what the effects of incomplete sampling on the outcomes of surveys can be. It also suggests what can be done about the problem. An obvious consideration would be to prevent nonresponse from happening in the first place. This is discussed in more fully in Chapter 3. Practical experience, however, shows that it is impossible to eliminate nonresponse completely. Therefore some corrective action always has to be taken.

This chapter provides a general introduction on the phenomenon of nonresponse and its effect on the usefulness of survey-based estimates. As is shown, nonresponse has become a serious problem.

1.2 Theory

1.2.1 CAUSES AND EFFECTS OF NONRESPONSE

Surveys are often compromised by nonresponse. If the sampled population does not provide the requested information on selected items the collected information is unusable. Two types of nonresponse can be distinguished:

- *Unit nonresponse.* The selected person does not provide any information at all, meaning the questionnaire form remains completely empty.
- *Item nonresponse.* Some questions have been answered but not all questions, especially sensitive questions. So the questionnaire form has been partially completed.

A consequence of unit nonresponse is that the realized sample size is smaller than planned. If nonresponse is random, it will result in increased variances of the estimates, and thus in a lower precision of estimates. Valid estimates can still be obtained, however, because the computed confidence intervals will have the proper confidence level.

If a specific sample size is required, it is important to take into account that nonresponse will occur. For example, if a researcher wants to have at least 1000 completed questionnaires, and the nonresponse rate is expected to be in the order of 60%, the initial sample size should be approximately equal to $1000/0.6 = 1667$.

The main problem of nonresponse is that estimates of population characteristics may be biased. This situation occurs if some groups in the population are over- or underrepresented in the sample, and these groups behave differently with respect to the characteristics to be investigated. This is called *selective nonresponse.*

Indeed estimates must be assumed to be biased unless very convincing evidence to the contrary is provided. Bethlehem and Kersten (1985) mention a number of Dutch surveys where nonresponse is selective:

- A follow-up study of the Dutch Victimization Survey showed that people who are afraid to be home alone at night are less inclined to participate in the survey.
- In the Dutch Housing Demand Survey, it turned out that people who refused to participate have fewer housing demands than people who responded.
- For the Survey of Mobility of the Dutch Population it was obvious that the more mobile people were underrepresented among the respondents.

It will be shown in Chapter 2 that the amount of nonresponse is one of the factors determining magnitude of the bias of estimates. The higher the nonresponse rate, the larger the bias will be.

EXAMPLE 1.1 Nonresponse in the Dutch Housing Demand Survey

The effect of nonresponse is shown in a somewhat simplified example using data from the Dutch Housing Demand Survey. Statistics Netherlands carried out this survey in 1981. The initial sample size was 82,849. The number of respondents was 58,972, which comes down to a response rate of 71.2%.

To obtain more insight in the nonresponse, a follow-up survey was carried out among the nonrespondents. They were also asked whether they intended to move within two years. The results are summarized in the table below:

Do you intend to move within 2 years?	Response	Nonresponse	Total
Yes	17,515	3,056	20,571
No	41,457	20,821	62,278
Total	58,972	23,877	82,849

Based on the response, the percentage of people with the intention to move within two years is 29.7%. However, for the complete sample (response and nonresponse) a much lower percentage of 24.8% is obtained. The reason is clear: there is a substantial difference between respondents and nonrespondents with respect to the intention to move within two years. For nonrespondents this percentage is only 12.8%

Nonresponse can have many causes, and it is important to distinguish these causes. To reduce nonresponse in the field, one needs to know what the underlying reasons and motives are. Moreover different types of nonresponse can have different effects on estimates, and therefore may require different treatment. (For a model of survey participation, see Groves and Couper, 1998.)

The many ways to classify nonresponse according to cause make it difficult to compare nonresponse for different surveys. Unfortunately, no internationally accepted standardized classification exists. There have been some attempts. The American Association for Public Opinion Research (AAPOR) has published a report with a comprehensive list of definitions of possible survey outcomes (see AAPOR, 2000). However, these definitions only apply to household surveys with one respondent per household, and samples selected by means of Random Digit Dialing (RDD). Lynn et al. (2002) have proposed a more general classification. This is the classification used here. The classification follows probable event outcomes when selected population members are approached in an attempt to obtain cooperation in a survey; see Figure 1.1.

First, contact must be established with the selected person. If this is not successful, there are two explanations. If the selected contact belongs to the target population of the survey (i.e., *eligible* population) and should rightly be

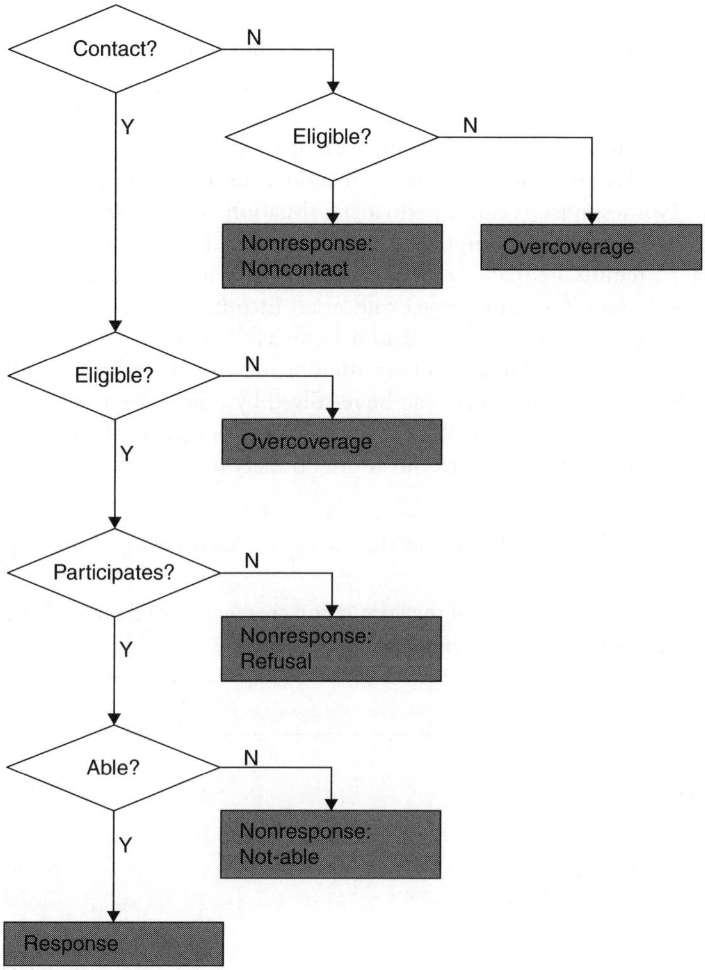

FIGURE 1.1 Possible survey outcomes.

part of the sample, this is nonresponse due to *noncontact*. If the selected contact does not belong to the target population (i.e., is not eligible), and should not be included in the sample, this is a case of overcoverage (see Section 1.3). That contact can therefore be excluded from the survey. Note that it is often not possible to determine in real-life situations whether a noncontact belongs to the target population, and this complicates the calculation of response rates.

Once there is contact with a selected person, we need to establish whether that contact belongs to the target population. If not, that case can be dismissed as an instance of overcoverage.

Once contact is established, that contact's cooperation is needed to get the required information. If the contacted person refuses to cooperate, this is a case of nonresponse due to *refusal*.

Once there is contact with someone who cooperates, there may still be an issue of that person not providing the required information. The reasons may range from illness to language problems. This is an instance of nonresponse due to *not-able*.

Last, say an eligible contact cooperates and is able to provide all the requested information; this is in fact a legitimate case of *response*.

Figure 1.1 shows the three causes of nonresponse: noncontact, refusal, and not-able. Nonresponse is not a permanent situation. In the case of noncontact, another contact may be attempted at some later time. In some surveys as many as six contact attempts are made before a case is closed as a noncontact. Also a refusal may be temporary if an interviewer calls at an inconvenient moment. It may be possible to suggest to a more convenient time for a follow-up. If someone is not able to participate because of illness, a later attempt may be made after the illness has passed. Also a language problem can be re-solved by translating the questionnaire or sending an interviewer capable of speaking the language of the respondent. Nevertheless, many refusals turn out to be nonnegotiable.

EXAMPLE 1.2 Results of the Survey on Well-Being of the Population

The Survey on Well-being of the Population that was carried out by Statistics Netherlands in 1998 had the following results:

Outcome	Frequency
Overcoverage	129
Response	24,008
Noncontact	2,093
Refusal	8,918
Not-able	1,151
Other nonresponse	3,132
Total	39,431

The category "not-able" includes nonresponse due to illness, physical handicap, or language problems. The additional nonresponse category, "other nonresponse," includes cases not followed up by interviewers because of workload. People who had moved and could not be located are also included in this category.

In some surveys the not-able category is split in two subcategories: not-able due to language problems and not-able due to other reasons. This is because these two types of nonresponse pertain to different groups of people and can have different effects on estimates.

The types of nonresponse given in Figure 1.1 do not exhaust the reasons for nonresponse. For example, it may happen that selected persons are not even contacted because of capacity problems of interviewers. This is sometimes called *administrative nonresponse*.

1.2.2 ERRORS IN SURVEYS

Nonresponse is just one thing that can go wrong in a survey. There are many more areas of data collection and data processing that can introduce errors and so affect the quality of the results.

There will always be some error in survey estimates of population characteristics. This error can have many explanations. Bethlehem (2009) gives some possible causes, as are presented in Figure 1.2. The taxonomy derives from a version given by Kish (1967).

The ultimate result of all errors is a discrepancy between the survey estimate and the population characteristic to be estimated. This discrepancy is called the *total survey error*. Two broad categories can be distinguished contributing to this total error: sampling errors and nonsampling errors.

Sampling errors are due to the sampling design. They are introduced when estimates are based on a sample and not on a complete enumeration of the population. Sampling errors could be avoided by investigating an entire population. However, only a part of the population is used for computing population characteristics. Because this is not a complete data set, estimates are only an approximation of the values of population characteristics, and some loss of precision results. The sampling error may be one of two types: a selection error or an estimation error.

The *estimation error* can occur when using a sample based on a random selection procedure. Every new selection of a sample will produce different respondents, and thus a different value of the estimator. The estimation error can be controlled through the sampling design. To reduce the error in an estimate,

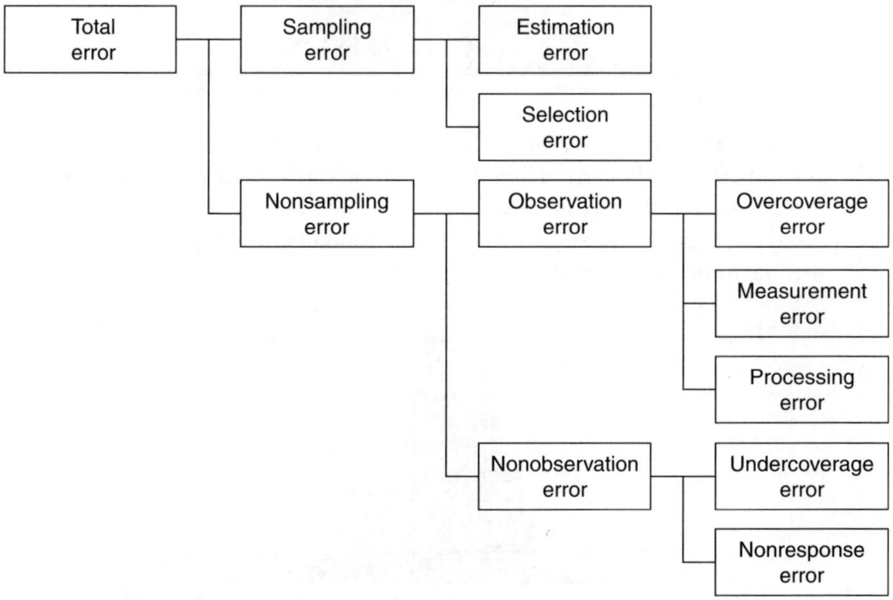

FIGURE 1.2 Types of survey errors.

the sample size could be increased, or selection probabilities could be taken proportional to the values of some well-chosen auxiliary variable.

A *selection error* can occur when incorrect selection probabilities are used in an estimation procedure. For example, true selection probabilities may differ from anticipated selection probabilities if elements have multiple occurrences in the sampling frame. Selection errors are hard to avoid without thorough investigation of the sampling frame.

Nonsampling errors may occur even if the whole population is investigated. They denote errors made during the process of obtaining answers to questions asked. Nonsampling errors can arise from both observation and nonobservation errors.

EXAMPLE 1.3 Effect of Sample Size

The effects of selection error can be illustrated by a simulation experiment. From the working population of the small country of Samplonia 1000 samples of size 20 are selected. For each sample the mean income is computed as an estimate of the mean income in the population. The distribution of these 1000 estimates is displayed below:

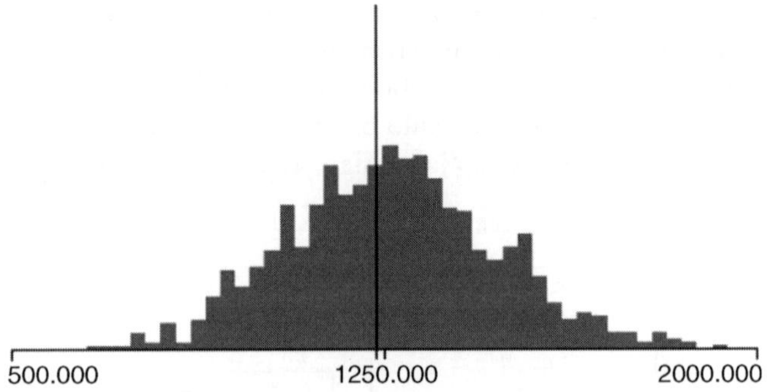

There is a lot of variation in the estimates around the population mean, as is indicated by the vertical line. This variation can be reduced be increasing the sample size. The figure below shows the distribution of 1000 estimates based on sample of the size 40. Notice that doubling the sample size reduces the magnitude of the error.

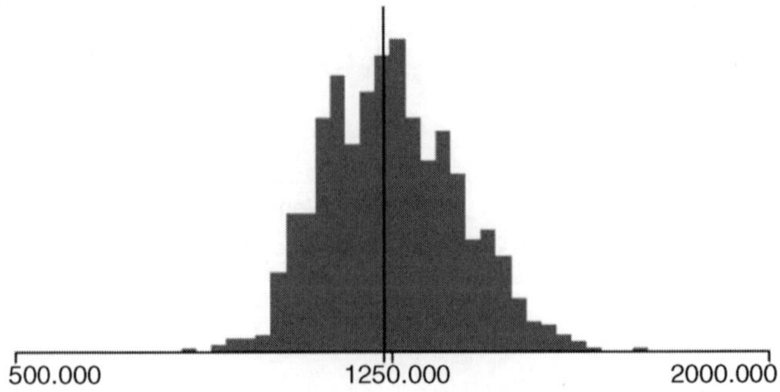

Observation errors are one form of nonsampling errors. They are errors that are made during the process of obtaining and recording answers. An *overcoverage error* occurs when elements are included in the survey do not belong to the target population. A *measurement error* occurs when a respondent does not understand a question, or does not want to give the true answer, or when the interviewer makes an error in recording the answer. Also interviewer effects, question wording effects, and memory effects belong to this group of errors. A measurement error results from a discrepancy between the true value and the value processed in the survey. A *processing* error is an error made during data processing, such as in data entry.

Nonobservation errors are errors made when the intended measurements are not obtained. *Undercoverage* occurs when elements of the target population do not have a corresponding entry in the sampling frame. These population members cannot ever be contacted. Another type of nonobservation error is *nonresponse*. This phenomenon occurs if the sampled person does not provide the required information.

Figure 1.2 clearly shows that many things can go wrong during the data collection process, and usually they do. Some errors can be avoided by taking preventive measures at the design stage. However, some errors will remain. Therefore it is important to check collected data for errors, and where possible, to correct these errors. This activity is called *data editing*. Data editing procedures do not correct for every type of survey error; they can detect and remove measurement errors, processing errors, and possibly overcoverage errors. Phenomena like selection errors, undercoverage, and nonresponse require the use of adjustment weights in estimation procedures, and not correction of individual values in recorded data.

There are two ways in which nonresponse errors can be minimized. First is by *nonresponse reduction*, whereby every effort is made to prevent nonresponse from occurring in the field. Ideally, if everyone sampled responds, there will be no nonresponse error. Nonresponse reduction measures can include better contact strategies, application of refusal conversion techniques, and deployment of interviewers speaking different languages. Unfortunately, nonresponse can never be eliminated completely. Nonresponse reduction is the topic of Chapter 3. The second way is by *nonresponse correction*, which recognizes that it is not possible to obtain 100% response, then a technique must be applied to reduce the bias of the estimators. An example of such a correction technique is adjustment weighting. This is the topic of Chapter 8.

1.2.3 NONRESPONSE AND UNDERCOVERAGE

From the survey errors shown in Figure 1.2, we see that some people may be missing from the sample for reason of nonresponse or undercoverage. It is important to make a distinction between these two types of missing information.

> *Nonresponse* denotes the situation where a member of the target population (and thus eligible for the survey) does not submit the required information.
>
> *Undercoverage* is created by the sampling frame from which the sample is selected. Undercoverage denotes the situation where the sampling frame does

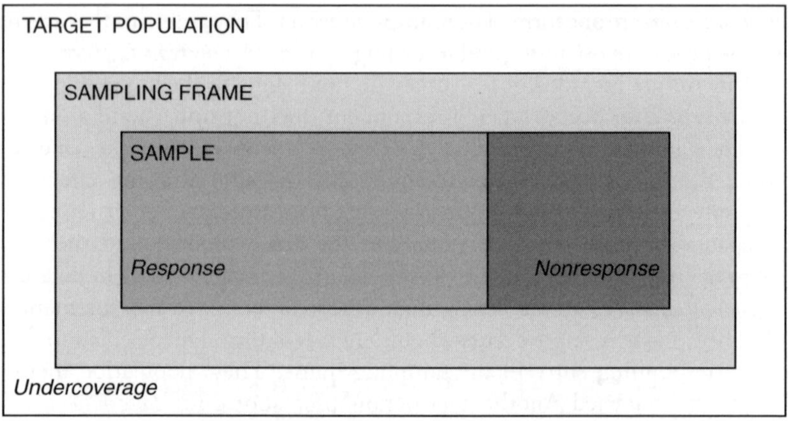

FIGURE 1.3 Nonresponse and undercoverage.

not cover completely the target population of the survey. There are persons in the population who do not appear in the sampling frame. They will, as a consequence, never be sampled.

The difference between nonresponse and undercoverage is shown in Figure 1.3. Undercoverage is a defect of the sampling frame that is often difficult to detect in practical survey situations. If people from some subpopulations do not appear in the sample, it may be simply that the sampling mechanism happened not to have selected them.

Undercoverage can also occur if the sample selected is from a different population than the one intended. The consequences for the outcomes of the survey is that the conclusion drawn from the survey does not apply to the original target population but to the population that was contacted through the sampling frame (sometimes called the *frame population*).

EXAMPLE 1.4 Nonresponse and Undercoverage

Suppose that a telephone survey is conducted. The target population consists of all adults in a certain country. The sampling frame is a telephone directory. There is undercoverage, because people without a listed number will never be selected in the sample. There will also be nonresponse because some calls to the selected persons will not be answered. And if calls are answered, persons may refuse to cooperate.

Suppose that a Web survey is conducted. The target population consists of all adults in a certain country. The sampling frame used is the population register. There is no undercoverage because the sample completely covers the population. The frame population is identical to the target population. There will be nonresponse. Not-able may be an important cause of nonresponse, since those people without Internet access will not be able to respond.

1.2.4 RESPONSE RATES

Because of the negative impact that nonresponse may have on the quality of survey results, the *response rate* is regarded as an important indicator of the quality of a survey. Response rates are frequently used to compare the quality of surveys and also to explore the quality of a repeated survey over time.

Presently there is no internationally accepted standard definition for the response rate. The definition we use here is based on one introduced by Lynn et al. (2002): the response rate is defined as the proportion of eligible contacts in the sample who completed the questionnaire. Referring to Figure 1.3, we therefore write the initial sample size n_I as

$$n_I = n_{NC} + n_{OC} + n_{RF} + n_{NA} + n_R, \tag{1.1}$$

where n_{NC} denotes the number of noncontacts, n_{OC} the number of noneligible contacts (i.e., cases of overcoverage), n_{RF} the number of refusers, n_{NA} the number of not-able respondents, and n_R the total number of respondents.

The *response rate* is defined as the total number of respondents divided by the number of n_E eligible contacts in the sample:

$$\textit{Response rate} = \frac{n_R}{n_E}. \tag{1.2}$$

There is a problem in computing the number of eligible elements. This problem arises because the noncontacts consist of eligible noncontacts and noneligible noncontacts. It is not known how many of these noncontacts are eligible. If it is assumed that all noncontacts are eligible, then $n_E = n_{NC} + n_{OC} + n_{RF} + n_{NA} + n_R$. So the response rate is equal to

$$\textit{Response rate} = \frac{n_R}{n_{NC} + n_{RF} + n_{NA} + n_R}. \tag{1.3}$$

This is usually not a realistic assumption. Another assumption is that the proportion of eligibles among the noncontacts is equal to the proportion of eligibles among the contacts. Then the response rate would be equal to

$$\textit{Response rate} = \frac{n_R}{n_{NC}\dfrac{n_{RF} + n_{NA} + n_R}{n_{OC} + n_{RF} + n_{NA} + n_R} + n_{RF} + n_{NA} + n_R}. \tag{1.4}$$

Response rate definitions like (1.3) or (1.4) can be used in a straightforward way for surveys where one person per household is selected. The situation becomes more complicated when the survey population consists of households for which several or all of its members have to provide information. Then there is risk of partial response: some eligible household members may respond, but for other eligible members it may be impossible to obtain response. How to define response at the household level? There are examples of surveys where an outcome is defined as response only if all eligible members respond. So, response rates for households may differ from response rates for persons.

Another complication arises from self-administered surveys. These are surveys where there are no interviewers. Examples of such surveys are mail surveys (pen-and-paper surveys) or Web surveys. For such surveys it is not possible to distinguish among the different sources of nonresponse. There are only two possible outcomes: response and nonresponse. The response rate simplifies to

$$\textit{Response rate} = \frac{n_R}{n_R + n_{NR}}. \tag{1.5}$$

Self-administered surveys also do not control for variation in the eligible population. The extreme example is the self-selection survey. Internet surveys are often self-selection surveys, as it is a convenient way for an organization to survey a large group of people. No proper sample is selected for such surveys. The survey questionnaire is simply put on the Internet. Respondents are those with Internet access who visit the survey website and decide to complete the questionnaire. The survey researcher has no control over the selection process. It is unclear whether respondents belong to the target population of the survey.

EXAMPLE 1.5 Computing the Response Rate

The Dutch Survey on Well-being of the Population had the following fieldwork results in 1998 (see also Example 1.4):

Outcome	Frequency
Overcoverage	129
Response	24,008
Noncontact	2,093
Refusal	8,918
Not-able	1,151
Other nonresponse	3,132
Total	39,431

The category "Not-able" included nonresponse due to illness, handicap, or language problems. The extra nonresponse category "Other nonresponse" consisted of cases not processed by interviewers due to workload (administrative nonresponse). Also people who had moved and could not be found any more are included in this category.

If it is assumed that the noncontacts are all eligible, the response rate of this survey is

$$100 \times \frac{24008}{24008 + 2093 + 8918 + 1151 + 3132} = 61.09\,\%.$$

If it is assumed that the proportion of eligibles among contacts and noncontacts is the same, the response rate is equal to

$$100 \times \frac{24008}{24008 + 2093 \times \dfrac{39431 - 2093 - 129}{39431 - 2093} + 8918 + 1151 + 3132} = 61.11\%.$$

The differences in both response rates are small. This is due to the small amount of overcoverage.

Yet another complication that can affect the definition of the response rate is the use of sampling designs with unequal selection probabilities. On the one hand, because the response rate is used as an indicator of the quality of survey outcomes, the sizes of the various outcome categories should reflect the structure of the population. Consequently observations should be weighted with inverse selection probabilities. This leads to so-called *weighted response rates*. On the other hand, because the response rate is used as an indicator of the quality of the fieldwork, and more specifically the performance of interviewers, *unweighted response rates* may be more appropriate.

Response rates have declined over time in many countries. Table 1.1 contains (unweighted) response rates for a number of surveys of Statistics Netherlands. The definition of response rates is more or less the same for each survey. It is not easy to explain differences in response rates between surveys. Response rates are determined by a large number of factors, such as the topic of the survey, the target population, the time period, the length of the questionnaire, the quality of the interviewers, and the organization of the fieldwork.

The response rates for different surveys cannot be readily compared. Different surveys may have different target populations. For this reason response rates for interviewers or interviewer regions are usually adjusted for the composition of the population in the interviewer area.

As Table 1.1 shows, nonresponse can be a big problem. Nonresponse has become more of a problem in recent years. It has raised the cost of conducting surveys since more effort has to be expended to obtain estimates with the precision specified in the survey design.

The Labor Force Survey (LFS) is the most important survey of Statistics Netherlands. It has been subjected to many re-designs, the most comprehensive re-design taking place in 1987. Before 1986, data collection was carried out by means of a paper questionnaire (PAPI, paper and pencil interviewing). In 1987 Statistics Netherlands changed to computer-assisted interviewing, which was facilitated by the Blaise System. With this development, the LFS introduced computer-assisted personal interviewing (CAPI). Also, before 1986, the fieldwork for the LFS was carried out by municipal employees who were not professional interviewers. From 1987, each month about 400 interviewers equiped with laptops visited 12,000 addresses.

In 1987 all changed. The questionnaire of the LFS was completely redesigned, and the fieldwork began to be done by professional interviewers. Notice in Table 1.1 the large drop in the response rate of the LFS in 1987. Because this

TABLE 1.1 Response rates of some surveys of Statistics Netherlands

Year	Labor Force Survey	Consumer Sentiments Survey	Survey on Well-being of the Population	Mobility Survey	Holiday Survey
1972		71			
1973	88	77			
1974		75	72		
1975	86	78			86
1976		72	77[a]		87
1977	88	69	70		81
1978		64		67	78
1979	81	63	65[b]	69	74
1980		61	61	68	74
1981	83	65		68	74
1982		60	64[a]	66	71
1983	81	63	58	66	74
1984		65[c]		64	69
1985	77	69		61	68
1986		71	59	59	66
1987	60[c]	71		59	
1988	59	68		55	
1989	61	68	44	58	
1990	61	68	47	55	
1991	60	69	46	57	
1992	58	69	45	57	
1993	58	72	46	56	
1994	59	70	52[c]	55	
1995	60	67	54	54	
1996	58	67	52	52	
1997	56	57	63	50	
1998	54	64	60		
1999	56	62	60		
2000	56	61	57		
2001	58	64	60		
2002	58	65			
2003	59	65	62		

[a] *Young only.*
[b] *Elderly only.*
[c] *Change in survey design.*

was when so many things were changed in the survey design and the survey fieldwork, no single cause could explain this drop.

Another important survey of Statistics Netherlands is the Survey of Well-being of the Population (SWP). It is a survey that samples every month a group of 3000 selected persons. The survey has a modular in structure; a base module contains questions for all sampled persons, and in addition there are a number of modules about specific themes (e.g., employment situation, health, and justice). The sampled persons are selected for one of the thematic modules; the base module is answered by everyone. The SWP was created in 1997; before that year all the modules were separate surveys.

The Consumer Sentiments Survey (denoted by CSS) measures consumer confidence (e.g., in changing economic circumstances). Since April 1986, it is performed monthly by means of CATI (computer-assisted telephone interviewing). Before 1984, the survey was conducted by telephone interviewers using pen and paper (PAPI) to record responses. Every month 1500 households are selected in a simple random sample. Telephone numbers are obtained for the selected addresses from listed numbers of landline telephones. This is only possible for about two-thirds of the addresses. The phone numbers are then passed to the CATI interviewers. Only one person in every household is interviewed. The response rates of these three major surveys are graphically presented in Figure 1.4.

Notice that from 1972 to 1983 the response percentages of the CSS and the SWP show similar, falling trends. After 1983 the response percentage for the CSS stabilized, whereas for the SWP it kept on falling. Both rates start to converge in 1993 and settle into a similar pattern over the last six years. The two breakpoints coincide with re-designs of these surveys (CSS in 1984 and SWP in 1997). The re-design of the CSS in 1984 caused a temporary increase in response rates. The same is true for the re-design of the SWP in 1997.

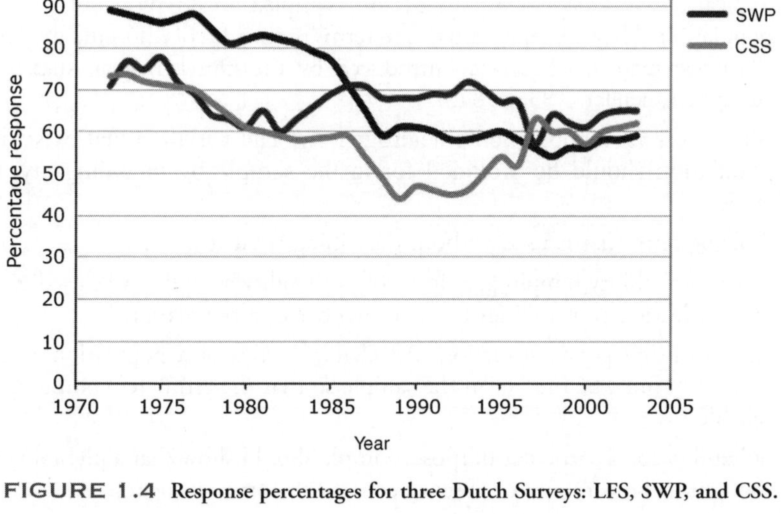

FIGURE 1.4 Response percentages for three Dutch Surveys: LFS, SWP, and CSS.

The response percentage of the LFS was initially higher than that of the other two surveys, but around 1983 and 1984 it decreased and reached the same level as the rates of the CSS and SWP. From 1987, responses took more or less stable paths. As mentioned before, 1987 was the year of a comprehensive re-design of the LFS.

1.2.5 REPRESENTATIVITY

Conducting a survey means two selection processes take place. First, a sampling design is chosen for selecting the sample and so is based on some form of probability sampling. Next, as nonresponse occurs in the fieldwork, only answers obtained from respondents (the survey response) can be used for analysis. The question is whether this data set allows for proper inference with respect to the population. Is it possible to draw reliable conclusions?

It is often said that the survey response must be *representative*, but what does it mean? Kruskal and Mosteller (1979a, b, c) present an extensive review of what representative is supposed to mean in nonscientific literature, in scientific literature excluding statistics, and in the current statistical literature. They compiled the following ways to consider if a sampling is truly "representative":

1. Validation for data. It means not much more than a general assurance, without evidence, that the data are supportable. This meaning of "representative" is typically used by the media, without explaining what it exactly means.
2. Absence of bias. No elements or groups of elements were favored in the selection process, either consciously or unconsciously.
3. Miniature of the population. The sample is a small-scale model of the population. The sample has the same characteristics as the population. The sample proportions are in all respects similar to population proportions.
4. Typical or ideal subject(s). The persons sampled are "typical" of a certain population. They are representative in terms of the idea of *l'* homme moyenne ("average man"), which was introduced by the Dutch/Belgian statistician Adolphe Quetelet (1835, 1846).
5. Allowances for a population's heterogeneity. The variation that exists in a population should be provided for in the sample by including atypical subjects.
6. A vague term that is used without describing what it means.
7. Equal probability sampling. A form of probability sampling was used giving equal selection probabilities to each member of a population.
8. As permitting good estimation. All characteristics of a population and its variability must be present in the sample so that the estimates computed are reliable.
9. Suitability for a particular purpose. Sample should show that a phenomenon thought to be very rare or absent occurs with some frequency.

Because the term "representative" can have many different interpretations, it is best not to use it in practice unless it is made clear what meaning is intended. In this book the term "representative" is used in two ways.

First, we say that the survey response is *representative with respect to* a variable if its relative distribution in the survey response is equal to its relative distribution in the population. For example, a sample is representative with respect to the variable "gender" if the percentages of males and females in the survey response are equal to the percentage of males and females in the population. Chapter 8 covers weighting adjustment techniques to correct for nonresponse problems. These techniques help make the survey responses representative of as many variables as possible. Of course the survey responses should be representative with respect to all other survey variables as well.

Second, we say that the response mechanism is representative if each member of a population in the sample would have the same probability of response. This implies that there are no biases active in the selection process. Chapter 7 covers indicators used for measuring representativity. These indicators estimate response probabilities and attempt to determine whether or not they are the same.

1.3 Application

Throughout this book theoretical concepts are applied to real survey data. The data are derived from a Dutch survey that was carried out by Statistics Netherlands. To avoid the disclosure of sensitive individual information, the data set has been anonymized. It is called the General Population Survey (GPS).

The fieldwork of the GPS covered a period of two months. In the first month, selected persons where approached by means of CAPI (computer-assisted personal interviewing). For persons that could not be contacted or refused and who had a listed phone number, a second attempt was made in the second month now using CATI (computer-assisted telephone interviewing). Table 1.2 shows the fieldwork

TABLE 1.2 Fieldwork results of the GPS

Result	Frequency	Percentage
Sample size	32,019	100.0%
Response	18,792	58.7%
First month	14,870	46.4%
Second month	3,922	12.2%
Nonresponse	13,227	41,3%
Unprocessed cases	2,456	7.7%
Noncontact	1,847	5.8%
Refusal	7,890	24.6%
Not able	1,034	3.2%

results. Note that there is a nonresponse category "Unprocessed." This denotes nonresponse due to unprocessed cases. Such cases were assigned to interviewers but were not undertaken by interviewers because of capacity problems or illness of the interviewer.

The selection of participants was by a stratified two-stage sampling process. In the first stage, municipalities were selected within regions with probabilities proportional to the number of inhabitants. In the second stage, an equal probability sample was drawn in each selected municipality. Sampling frames for the persons were the population registers of the municipalities. The sampling design was such that each person had the same probability of being selected (a so-called self-weighting sample). The sample of the GPS consisted of 32,019 persons. The number of respondents was 18,792.

Statistics Netherlands has an integrated system of social statistics which is called the *Social Statistics Database* (SSD). The SSD contains a wide range of information on each person who lives in The Netherlands. There are data on demography, geography, income, labor, education, health, and social protection. These data are obtained by combining data from registers and other administrative data sources. For more information about the SSD, see Everaers and Van Der Laan (2001).

SSD records can be linked to survey data records by way of personal identification numbers. This can be done for both respondents and nonrespondents, so demographic variables like gender, age, province of residence, and ethnicity are available for all sampled persons, and also socioeconomic variables like employment and various types of social security benefits.

The Netherlands is divided in approximately 420,000 postal code areas. A postal code area contains, on average, 17 addresses. These areas are homogeneous with respect to social and economic characteristics of its inhabitants. Using information from the population register, Statistics Netherlands has computed some demographic characteristics for these postal code areas. Since postal codes are included in the survey data file for both respondents and nonrespondents, these characteristics can be linked to the survey data file. Examples of such variables are degree of urbanization, town size, and percentage of people with a foreign background (nonnatives). From another source average house values can be included.

During their fieldwork the interviewers kept records of all contact attempts. For each attempt the contact result was recorded (successful, or not). If contact was established, the result of the cooperation request was recorded—response or nonresponse, and in case of nonresponse the reason of nonresponse. Other information was included, like the mode of contact used in the fieldwork attempt (CAPI or CATI). All the fieldwork information is available for analysis.

Table 1.3 gives an overview of all variables in the survey data file of the GPS. The values of the target variables are only available for the respondents. The auxiliary variables are available for both respondents and nonrespondents.

There were 150 cases in the initial sample that did not belong to the target population of the GPS. These cases are not contained in GPS survey data file. The amount of overcoverage is needed to be able to compute the response rate of

TABLE 1.3 Variables in the GPS survey data file

Variable	Description	Cats	Type	Source
PC	PC in household	2	Target variable	Survey
Move	Wants to move within a year	6	Target variable	Survey
Health	General health condition	5	Target variable	Survey
Newspap	Has newspaper subscription	2	Target variable	Survey
Club	Is active in a club	2	Target variable	Survey
Politics	Is interested politics	4	Target variable	Survey
Employed	Employment situation	3	Target variable	Survey
Joblevel	Job level	6	Target variable	Survey
Educat	Level of education	8	Target variable	Survey
Ownhouse	Owns house	2	Target variable	Survey
Religion	Religious denomination	5	Target variable	Survey
Gender	Gender	2	Auxiliary variable	SSD
Marstat	Marital status	4	Auxiliary variable	SSD
Married	Is married	2	Auxiliary variable	SSD
Age3	Age in 3 age groups	3	Auxiliary variable	SSD
Age13	Age in 13 age groups	13	Auxiliary variable	SSD
Nonnativ	Is nonnative	2	Auxiliary variable	SSD
Ethnic	Type of nonnative	5	Auxiliary variable	SSD
HHSize	Size of the household	5	Auxiliary variable	SSD
HHType	Type of household	5	Auxiliary variable	SSD
Children	Children in household	2	Auxiliary variable	SSD
Phone	Has listed phone number	2	Auxiliary variable	SSD
Hasjob	Has a job	2	Auxiliary variable	SSD
Socall	Has social allowance	2	Auxiliary variable	SSD
Disaball	Has disability allowance	2	Auxiliary variable	SSD
Unempall	Has unemployment allowance	2	Auxiliary variable	SSD
Allowan	Has an allowance	2	Auxiliary variable	SSD
Region	Region of the country	5	Auxiliary variable	SSD
Urban	Degree of urbanization	5	Auxiliary variable	SSD
Houseval	Average house value in neighborhood	12	Auxiliary variable	SSD
Pnonnat1	Percentage nonnatives in neighborhood	8	Auxiliary variable	SSD
Pnonnat2	Percentage nonwestern nonnatives in neighborhood	7	Auxiliary variable	SSD
Resp1	Response in first month of fieldwork	2	Fieldwork variable	SSD
Result	Final result of the fieldwork attempt	5	Fieldwork variable	SSD
Response	Response	2	Fieldwork variable	SSD

TABLE 1.4 Percentage of people with a job in GPS survey data file

	Job	No job	Total
Response	52.6%	47.4%	100.0%
Nonresponse	48.8%	51.2%	100.0%
Total	51.0%	49.0%	100.0%

the GPS. As was discussed in Section 1.2.4, the response rate can only be computed if the number of eligible cases among the noncontacts is known. If it is assumed that all noncontacts are eligible, the response rate is equal to

$$100 \times \frac{18792}{18792 + 2456 + 1847 + 7890 + 1034} = 100 \times \frac{18792}{32019} = 58.69\%.$$

If it is assumed that the proportion of eligibles among the contacted persons is the same as the proportion among the noncontacts, the response rate becomes

$$100 \times \frac{18792}{18792 + 2456 + 1847 \times \dfrac{32019 - 1847}{32019 - 1847 + 150} + 7890 + 1034} = 58.71\%.$$

The differences are minimal. This is because the amount of overcoverage is very small. So rounded to one decimal, the response rate is 58.7%.

The nonresponse of the GPS is selective. For example, Table 1.4 contains the distribution of the (register) variable *Hasjob*. This variable records whether a person has a job.

Among the respondents 52.6% of the people have a job, and among the nonrespondents only 48.8%. Apparently, those without jobs are less inclined to respond. Table 1.5 gives more detail. It shows the different causes of nonresponse. As is clear from this table, the nonresponse is mostly due to persons not able to respond. Among those persons only 17.2% have a job. This low percentage is not surprising as people unable to respond often are so disabled that they are unable to work.

TABLE 1.5 Types of job-related responses and nonresponse of the GPS survey date file

	Job	No job	Total
Response	52.6%	47.4%	100.0%
Unprocessed	53.5%	46.5%	100.0%
Noncontact	51.4%	47.4%	100.0%
Refusal	50.9%	49.1%	100.0%
Not-able	17.2%	82.8%	100.0%
Total	51.0%	49.0%	100.0%

1.4 Summary

A survey is an instrument to collect information about a specific populations. Such populations may consist of persons, households, companies, or other elements. Typically not all people are asked to participate in a survey, only a sample.

With a good survey design accurate estimates of population characteristics can be computed. Also the accuracy of estimates can be computed. A good design suggests also that the sample is selected by means of probability sampling.

Nonresponse is a phenomenon that may affect the quality of the survey outcomes. It occurs when the people who are selected as eligible for the sample do not provide the requested information, or when the provided information is not usable. Nonresponse can cause estimators of the population characteristics to be biased. This occurs when specific groups are over- or underrepresented, and these groups may behave differently with respect to the survey variables.

Nonresponse is mainly due to noncontact, refusal to answer, and not-able to answer. It is important to distinguish among these different causes because they may have different impacts on estimates.

Unfortunately, computation of the response rate is not straightforward. This is because the proportion of eligible elements among the noncontacts is not known. An estimate of the response rate can be obtained if assumptions are made about this component.

Response rates have been decreasing in The Netherlands over the last few decades. So nonresponse has become a serious problem. It not only affects the quality of the survey outcomes, but attempts to reduce the problem also increases survey costs.

1.5 KEY TERMS

Eligible To be eligible, the sample elements selected must belong to the target population of the survey.

Item nonresponse Some questions have been answered, but no answer is given for other, possibly sensitive, questions. So the questionnaire form has been only partially completed.

Noncontact A type of nonresponse where it is not possible to establish contact with a sampled population member.

Nonresponse The selected person for the sample is eligible for the survey but does not provide the requested information, or provides information that is not usable.

Nonresponse correction An attempt to compensate for the nonresponse problem by adjusting estimates using survey responses and other information about the population.

Nonresponse reduction An attempt to compensate for the nonresponse problem by reducing the amount of nonresponse in the field.

Nonsampling error The difference between the estimate and the true value caused by other phenomena than sampling. Such errors may also occur if the complete population is investigated. Nonresponse is one type of nonsampling error.

Not-able A cause of nonresponse where contact is established with a sampled person who is not able to cooperate for reason of illness or language problems.

Overcoverage Happens if the sampling frame includes persons who do not belong to the target population of the survey. These persons should be excluded from the survey.

Refusal A cause of nonresponse where contact is established with a sampled persons who refuses to cooperate.

Response rate The number of responding eligible persons in the sample divided by the total number of eligible persons in the sample. Their response rate can be weighted or unweighted.

Representative with respect to a variable The distribution of this variable in the survey response is equal to the distribution of the variable in the population.

Representative response All members of the population have the same probability of response.

Sampling error The difference between the estimate and the true value caused by estimates that are based on a sample and not on the complete enumeration of the population.

Selective nonresponse Nonresponse that is selective if, due to nonresponse, some groups in the population are over- or underrepresented in the sample, and these groups behave differently from the characteristics being sampled.

Undercoverage Happens if a sampling frame does not cover completely the target population of the survey. There are persons in the population who do not appear in the sampling frame. They will never be included in the sample.

Unit nonresponse A type of nonresponse that occurs when a selected person does not provide any information at all, meaning the questionnaire form remains empty.

1.6 REFERENCES

AAPOR (2000), *Standard Definitions: Final Dispositions of Case Codes and Outcome Rates for Surveys*. American Association for Public Opinion Research, Ann Arbor, MI.

Bethlehem, J.G. (2009), *Applied Survey Methods: A Statistical Perspective*. Wiley, Hoboken, NJ.

Bethlehem, J.G., and Kersten, H.M.P. (1985), On the Treatment of Non-response in Sample Surveys. *Journal of Official Statistics* 1, pp. 287–300.

Cochran, W.G. (1977), *Sampling Techniques*, 3rd ed. Wiley, New York.

Deming, W.E. (1950), *Some Theory of Sampling*. Wiley, New York.

Everaers, P., and Van Der Laan, P. (2001), *The Dutch Virtual Census*. E-Proceedings of the 53th Session. International Statistical Institute, Seoul, Korea.

Groves, R.M., and Couper, M.P. (1998), *Nonresponse in Household Interview Surveys*. Wiley, New York.

Hansen, M.H., Hurvitz, W.N., and Madow, W.G. (1953), *Survey Sampling Methods and Theory*. Wiley, New York.

Kish, L. (1967), *Survey Sampling*. Wiley, New York.

Kruskal, W., and Mosteller, F. (1979a), Representative Sampling I: Non-scientific Literature. *International Statistical Review* 47, pp. 13–24.

Kruskal, W., and Mosteller, F. (1979b), Representative Sampling II: Scientific Literature Excluding Statistics. *International Statistical Review* 47, pp. 111–123.

Kruskal, W., and Mosteller, F. (1979c), Representative Sampling III: Current Statistical Literature, *International Statistical Review* 47, pp. 245–265.

Lynn, P., Beerten, R., Laiho, J., and Martin, J. (2002), Towards Standardisation of Survey Outcome Categories and Response Rate Calculations. *Research in Official Statistics* 1, pp. 63–86.

Quetelet, L.A.J. (1835), *Sur l'*Homme et le Développement de ses Facultés, Essai de Physique Sociale. Paris, France.

Quetelet, L.A.J. (1846), *Lettre à S.A.R. le Duc Régant de Saxe Coburg et Gotha sur la Théorie des Probabilités, Appliquée aux Sciences Morales at Politiques*. Brussels, Belgium.

1.7 EXERCISES

Exercise 1.1 Which phenomenon makes it difficult, if not impossible, to computer the response rate of a survey?

(a) Overcoverage

(b) Undercoverage

(c) Noncontact

(d) Refusal

Exercise 1.2 A survey is usually carried out to measure the state of a target population at a specific point in time (the reference date). The survey outcomes are supposed to describe the population at this date. Ideally the fieldwork of the survey should take place at that date. This is not possible in practice. Interviewing usually takes place in a period of a number of days or weeks around the reference date.

Suppose that a sample of persons is selected from a sampling frame (a population register) two weeks before the reference date. Interviewing takes place in the period of four weeks: the two weeks between sample selection and the reference date and the two weeks after the reference date.

Explain for each of the situations described below what kind of problem has occurred. You can choose between nonresponse, undercoverage, overcoverage, and any other sampling frame errors.

- The contact attempt takes place between the sample selection date and the reference date. It turns out the person died before the sample selection date.
- The contact attempt takes place between the sample selection date and the reference date. It turns out the person died after the sample selection date.

- The contact attempt takes place after the reference date. It turns out the person died before the sample selection date.
- The contact attempt takes place after the reference date. It turns out the person died between the sample selection date and the reference date.
- The contact attempt takes place after the reference date. It turns out the person died after the reference date.

Exercise 1.3 A town council wants to do something about the traffic problems in the center of town. There is a plan to turn it into a pedestrian area. Because cars will not be permitted into to center any more, the town council wants to know what companies think of this plan. A simple random sample of 1000 companies is selected. A representative of each selected company is invited to participate in the survey. They are asked whether they are in favor of the plan. Also the location of the company is recorded (town center or suburb). The results of the survey are summarized in the table below:

	Suburbs	Town center
In favor	120	80
Not in favor	40	240

(a) Determine the response percentage.

(b) Determine the percentage respondents in favor of the plan.

(c) Compute a lower bound and an upper bound for the percentage in favor of the plan in the complete sample.

Exercise 1.4 The local authorities in a town want to know how satisfied citizens are with the public transport facilities. They conduct a survey. The target population is defined as all citizens that used public transport at least once in the last year. A sample is selected from the population register of the town. The results of the survey are summarized in the table below:

Result	Frequency
Overcoverage	320
Refusal	240
Noncontact	80
Not able	40
Response	440
Total	1120

Compute the response rate of the survey. Clearly show how the response rate was computed and the assumptions made.

Exercise 1.5 Suppose that a telephone directory is used as a sampling frame and a survey researcher is not aware of the fact that some people appear more than once in this directory. What type of error could this phenomenon cause?

(a) Overcoverage

(b) Estimation error

(c) Selection error

(d) Processing error

Exercise 1.6 Analyze the dataset of the GPS that is provided on the website of the book. Use Tables 1.4 and 1.5 in Section 1.3 to find other variables that show a large difference between respondents and nonrespondents.

CHAPTER TWO

Basic Theoretical Concepts

2.1 Introduction

There is ever-growing demand today for statistical information about the economic, social, political and cultural states of countries. Such information is required by policy makers and others who must make informed decisions for a better future. Sometimes such statistical information can be retrieved from existing administrative sources, but more often such sources are lacking. A survey is an instrument to collect new statistical information.

A *survey* investigates a well-defined population. Such populations consist of objects, like persons, households, farms, companies, and schools. Typically the survey information is collected by asking questions of persons representing population elements. A uniform and consistent way to do this is to develop a questionnaire.

A way to obtain information about an entire population is to collect data about all its members. Such a survey is called a *census* or *complete enumeration*. This form of survey has a number of disadvantages. In the first place, it is very expensive. To survey a very large number of people, many more interviewers and other resources must be used. In the second place, a census is very time-consuming. Collecting and processing a large amount of data takes time. This affects the timeliness of the results. Less timely information is less useful. Finally, large surveys raise the response burden for people. As many people are more frequently asked to participate in a survey, they will experience survey fatigue and view the procedure an unnecessary burden. So they will be less inclined to cooperate.

When a survey is designed to collect information on only a small part of the population, that small part is called the *sample*. In principle, the sample only

Handbook of Nonresponse in Household Surveys, By Jelke Bethlehem, Fannie Cobben, and Barry Schouten

provides information on the sampled members of the population. No information is obtained on the nonsampled members. Still, if the sample is selected in a "clever" way, it is possible to make inference about the population as a whole. In this context, "clever" means that the sample is selected using probability sampling methods. If it is clear how the selection mechanism works and it is possible to compute the probabilities of being selected in the sample, reliable and precise conclusions can be drawn about the population as a whole. The principles of probability sampling have been successfully applied in official and academic statistics since the 1940s, and to a lesser extent also in commercial market research e.g., see Bethlehem, 2009, (for an overview of the history of survey sampling).

Nonresponse is one of a number of phenomena that can affect the accuracy of survey results. To illustrate the possible consequences of nonresponses, a theoretical framework is described in this chapter. First, the basic concepts of survey sampling are introduced. Some descriptions of sampling designs and estimation procedures follow, and then it is shown how nonresponse can be included in the theory. Two models are presented: the *fixed response model* and the *random response model.* The application of this theory should make it clear that nonresponse prevents standard confidence intervals from being used to measure the accuracy of the estimates. Last, three mechanisms are introduced that demonstrate how missing values can occur. These mechanisms help elucidate how detrimental nonresponse can be, and how it is possible to correct survey outcomes for nonresponse.

2.2 Theory

2.2.1 BASIC CONCEPTS OF SAMPLING

The first step in setting up a survey is to define the *target population.* This is the population to be investigated, and about which conclusions are to be drawn. Such a population need not necessarily consist of people, but people are needed to provide the responses. Examples are all computer companies in a certain state, all cows on a certain farm, all cheese produced by a certain country in a certain year, or all cars that use a certain road in a specific period.

The definition of the target population must be unambiguous. It must always be possible to determine whether a certain element does or does not belong to the target population. Both failure to include relevant elements in the population, and failure to exclude irrelevant ones, may affect the survey results.

EXAMPLE 2.1 Target Population of a Labor Force Survey

Suppose that a labor force survey is to be carried in a specific region. It must be clear which people belong to the target population. Should people be included from the region who work abroad? And what about foreign workers that have temporary jobs in the region? Should illegal workers be included? And what about employees of foreign embassies?

The target population is denoted by U. It is assumed to be finite. The number of elements N is assumed to be known. However, this is not always the case. Examples are the number of foreign visitors in a country, or the size of the population of a developing country.

The elements of the target population must be *identifiable*. This means they can be uniquely assigned sequence numbers $1, 2, \ldots, N$. For each element encountered in practical situations, it must be possible to determine its sequence number. In some cases, this process is straightforward. An example is a population of persons that can be identified by means of their social security numbers. In other cases it is not so easy. Take a population of whales. The target population is denoted by

$$U = \{1, 2, \ldots, N\}. \tag{2.1}$$

In the survey design phase, the objectives of the survey must be translated into concrete operational procedures. This involves defining the *target variables* of the survey. These variables measure the various aspects of the phenomena to be investigated.

EXAMPLE 2.2 Target Variables

> Suppose that a research organization carries out a holiday survey. Then it may want to measure target variables like the length of the holiday, destination, means of transport, type of accommodation, or amount of money spend.

A target variable is denoted by the letter Y. The values of this variable for all elements in the population U are indicated by

$$Y_1, Y_2, \ldots, \; Y_N. \tag{2.2}$$

So Y_k is the value of Y for element k, where $k = 1, 2, \ldots, N$. Usually an additional number of variables are measured in the survey. These variables are called *auxiliary variables*. These variables assist in differentiating the survey results for various subpopulations. They can also be very useful in improving estimates of population characteristics. Examples of auxiliary variables are demographic characteristics like gender, age, and marital status. An auxiliary variable is denoted by the letter X. The values of this variable for all elements in the population U are indicated by

$$X_1, X_2, \ldots, \; X_N. \tag{2.3}$$

So X_k is the value of X for element k, where $k = 1, 2, \ldots, N$. Here Y_k and X_k indicate single values. Sometimes, however, this notation is also used to indicate vectors of values.

Target variables and auxiliary variables can be one of three types:

- *Quantitative variables.* These variables measure quantities, amounts, sizes, or values. It is possible to carry out meaningful computations on these values, like calculating totals and averages. Examples of such variables are income, ages of persons, and the number of cars they own.

- *Qualitative variables.* These variables divide the target population into subpopulations. The values denote labels. Elements with the same label belong to the same group. It is not meaningful to carry out computations on the values of a qualitative variable. Examples of qualitative variables are race, religion, marital status, and region of residence.
- *Indicator variables.* Such a variable measures whether an element has a certain property. It can only assume two values 0 and 1. The value is 1 if an element has the property, and the value is 0 if it does not. An example of an indicator variable is employment. If a person has a job, the value of the variable is 1, and otherwise, its value is 0.

The aim of a survey is to get information about the target population. This information is quantified in the form of *population parameters.* A population parameter is a function that only depends on the values in the population of one or more variables. These variables can be target variables as well as auxiliary variables.

One simple example of an often used population parameter for a quantitative variable is the *population total*:

$$Y_T = \sum_{k=1}^{N} Y_k = Y_1 + Y_2 + \ldots + Y_N. \tag{2.4}$$

Suppose that the target population consists of all farms in a country, and Y denotes the number of cows on a farm. Then the population total is the total number of cows in the country. Related to the population total is the *population mean*:

$$\bar{Y} = \frac{1}{N}\sum_{k=1}^{N} Y_k = \frac{Y_1 + Y_2 + \ldots + Y_N}{N} = \frac{Y_T}{N}. \tag{2.5}$$

The population mean is simply obtained by dividing the population total by the population size. If the target population consists of all inhabitants of a town, and Y denotes the age of a person, the population is the average age in this town. Another important population parameter is the *adjusted population variance*. It is defined by

$$S^2 = \frac{1}{N-1}\sum_{k=1}^{N} (Y_k - \bar{Y})^2. \tag{2.6}$$

This quantity gives an indication of the amount of variation of the target variable. If all values of Y are equal, the variance is 0. The more the values of Y are apart, the larger the variance will be. The adjusted population variance also appears in formulas for the variance of estimators.

For indicator variables, the population total counts the number of elements in the population having a certain property. The population mean is the fraction of elements with that property. The population percentage is defined by

$$P = 100\bar{Y} = \frac{100}{N}\sum_{k=1}^{N} Y_k = 100\frac{Y_1 + Y_2 + \ldots + Y_N}{N} = 100\frac{Y_T}{N}. \qquad (2.7)$$

Note that for indicator variables the adjusted population variance reduces to

$$S^2 = \frac{1}{N-1}\sum_{k=1}^{N}(Y_k - \bar{Y})^2 = \frac{P(100-P)}{N-1}. \qquad (2.8)$$

There are no specific population parameters for qualitative variables. Of course, totals, fractions, and percentages of elements can be estimated within categories. This actually comes down to replacing the qualitative variable by a set of indicator variables. The focus in this book is on estimating population means and population percentages.

To estimate population parameters, a random sample is selected from the population. The elements in this sample are obtained by means of a random selection procedure. This procedure assigns to every element in the target population a fixed, positive, and known probability of selection. The most straightforward way to select a random sample is giving each element the same probability of selection. Such a random sample is called a *simple random sample.*

Samples can be selected with replacement or without replacement. In *with replacement* selection procedures, a sample element is selected, and returned to the population after its characteristics have been recorded, before the next element is drawn. It is possible to select an element more than once. Since selecting an element more than once does not produce more information than selecting it once, selection *without replacement* is usually preferred.

It is assumed in this book that the sample is selected without replacement. It means that each element can appear at most once in the sample. Therefore the sample can be represented by a set of indicators

$$a = a_1, a_2, \ldots, a_N. \qquad (2.9)$$

The kth indicator a_k assumes the value 1 if element k is selected in the sample, and otherwise, it assumes the value 0. The expected value of a_k is denoted by

$$\pi_k = E(a_k). \qquad (2.10)$$

The quantity π_k is called the *first-order inclusion probability* of element k (for $k = 1, 2, \ldots, N$). For deriving variance formulas, also second-order inclusion probabilities are required. The *second-order inclusion probability* of elements k and l (with $k \neq l$) is equal to

$$\pi_{kl} = E(a_k a_l), \qquad (2.11)$$

and by definition $\pi_{kk} = \pi_k$. The *sample size*, meaning the number of selected elements, is denoted by n. Since the indicators a_k have the value 1 for all elements in the sample, and the value 0 for all other elements, the sample size can be written as the sum of the values of the indicators:

$$n = \sum_{k=1}^{N} a_k. \tag{2.12}$$

2.2.2 BASIC CONCEPTS OF ESTIMATION

The values of the target variable are observed for all sampled elements. These values must be used to estimate population characteristics. The recipe to compute such an estimate is called an *estimator*, and the result of this computation is called the *estimate*. To be useful, an estimator must have a number of properties.

- The estimator must be *unbiased*. This means that average value of the estimates over all possible samples must be equal to the value of the (unknown) population parameter to be estimated. On average, the estimator will result in the correct value. It will never underestimate or overestimate the population value in a systematic way. Consequently the expected value $E(z)$ of an estimator z must be equal to the value of the population parameter Z to be estimated: $E(z) = Z$.
- The estimator must be *precise*. This means that the variation in possible outcomes must be small. Consequently the variance $V(z)$ of an estimator z of a population parameter Z must be small over all possible samples.
- *For reasons of simplicity*, linear estimators are preferred. This means that an estimate is computed as a linear combination of the observed values of the target variable.

Imposing the conditions of unbiasedness and linearity leads to the estimator introduced by Horvitz and Thompson (1952). This estimator for the population mean is defined as

$$\bar{y}_{HT} = \frac{1}{N} \sum_{k=1}^{N} a_k \frac{Y_k}{\pi_k}. \tag{2.13}$$

The indicators a_k see to it that only the available sample values of the target variable are used in the computation of the estimate. Note that each value Y_k is weighted with the inverse selection probability π_k. Thus the estimator is corrected for the fact that elements with a large inclusion probability are overrepresented in the sample.

The Horvitz–Thompson estimator (2.13) is (in the case of complete response) an unbiased estimator of the population mean. The variance of this estimator is equal to

$$V(\bar{y}_{HT}) = \frac{1}{N^2} \sum_{k=1}^{N} \sum_{l=1}^{N} (\pi_{kl} - \pi_k \pi_l) \frac{Y_k}{\pi_k} \frac{Y_l}{\pi_l}. \tag{2.14}$$

For samples of fixed size n and selected without replacement, the variance can be rewritten in the form

$$V(\bar{y}_{HT}) = \frac{1}{2N^2} \sum_{k=1}^{N} \sum_{l=1}^{N} (\pi_k \pi_l - \pi_{kl}) \left(\frac{Y_k}{\pi_k} - \frac{Y_l}{\pi_l} \right)^2. \tag{2.15}$$

This expression shows that the variance can be reduced by taking the first-order inclusion probabilities proportional to the values of the target variable.

The *variance of the estimator* is just one way to quantify its precision. A small variance means high precision, and a large variance small precision. But what is small and what is large? The unit of measurement is a problem hampering easy interpretation of the magnitude of the variance. If, for example, the survey measures monthly income in euros, then the variance measures variation in squared euros. This problem can be solved by taking the square root of the variance. The result is called the *standard error of the estimator.* So the standard error of the Horvitz–Thompson estimator is defined by

$$S(\bar{y}_{HT}) = \sqrt{V(\bar{y}_{HT})}. \tag{2.16}$$

Still, the problem of interpretation remains. When is the standard error small, and when is it large? The solution is the confidence interval. The *confidence interval* is a range of possible values of the population mean. The interval encompasses the true value of the population mean with a high probability if an estimator is unbiased. This probability is called the *confidence level.* It is denoted by $(1 - \alpha)$, where α is a small probability. Often the value $\alpha = 0.05$ is used. The confidence level is then 95%.

The distribution of many estimators (including the sample mean) can, for large sample sizes, be approximated by a normal distribution. This makes it easier to compute confidence intervals. Only the standard error of the estimator is required. The 95% confidence interval of the Horvitz–Thompson estimator is equal to

$$\left(\bar{y}_{HT} - 1.96 \times S(\bar{y}_{HT}); \bar{y}_{HT} + 1.96 \times S(\bar{y}_{HT})\right). \tag{2.17}$$

For a 99% confidence interval, the value 1.96 must be replaced by 2.58. One problem is that usually the value of the standard error is not known. It can only be computed if all values of the target variable in the population are available. The way out is that an unbiased estimate of the standard error can be computed just using the sample data. This leads to an estimated confidence interval

$$((\bar{y}_{HT} - 1.96 \times s(\bar{y}_{HT}); \bar{y}_{HT} + 1.96 \times s(\bar{y}_{HT})). \tag{2.18}$$

where $s(\bar{y}_{HT})$ is the estimate for $S(\bar{y}_{HT})$.

The best known and probably most often used sampling design is a *simple random sample* without replacement. First-order inclusion probabilities of all elements are taken equal. It can be shown for sampling without replacement that all first-order inclusion probabilities sum up to n. Therefore $\pi_k = n/N$, for

$k = 1, 2, \ldots, N$. Furthermore all second-order inclusion probabilities sum up to $n(n-1)$. Therefore $\pi_{kl} = n(n\text{-}1)/N(N-1)$ for $k, l = 1, 2, \ldots, N$ and $k \neq l$.

Substitution of these values of the inclusion probabilities in expression (2.13) results in a simple estimator, the *sample mean*:

$$\bar{y} = \frac{1}{n}\sum_{k=1}^{N} a_k Y_k = \frac{1}{n}\sum_{i=1}^{n} y_i, \tag{2.19}$$

where $y_1, y_2, \ldots, y_n$ denote the n observations that have become available in the sample. This is an unbiased estimator with variance

$$V(\bar{y}) = \frac{1-f}{n} S^2, \tag{2.20}$$

where $f = n/N$ is the *sampling fraction* and S^2 the population variance. From expression (2.20) it is clear that an increased sample size produces more precise estimators. The standard error of the sample mean is equal to

$$S(\bar{y}) = \sqrt{V(\bar{y})} = \sqrt{\frac{1-f}{n} S^2}. \tag{2.21}$$

To compute an estimated 95% confidence interval, an unbiased estimator for S^2 is required. The sample variance

$$s^2 = \frac{1}{n-1}\sum_{i=1}^{n} (y_i - \bar{y})^2 \tag{2.22}$$

can be used for this.

EXAMPLE 2.3

The effect of the sample size on the precision of an estimator can be shown by means of a simulation experiment. From the working population of the small country of Samplonia 1000 simple random samples without replacement of size 20 are selected. For each sample the mean income is computed as an estimate of the mean income in the population. The distribution of these 1000 estimates is displayed in the graph on the left:

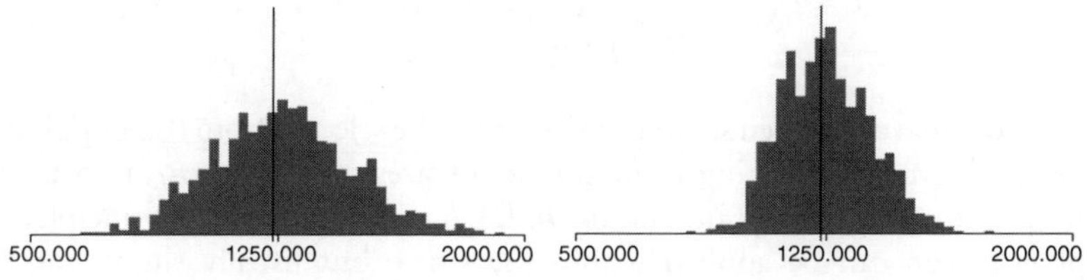

There is a lot of variation of the estimates around the population mean indicated by the vertical line. This variation can be reduced by increasing the sample size. The graph on the right shows the distribution of 1000 estimates based on samples of the size 40. Indeed, doubling the sample size reduces the variance.

Another well-known sampling design is *stratified sampling*. The target population U is divided into L subpopulations (strata) $U_1, U_2, \ldots, U_L$ of sizes $N_1, N_2, \ldots, N_L$, respectively. The strata are nonoverlapping and together comprise the whole population, so that

$$\sum_{h=1}^{L} N_h = N_1 + N_2 + \ldots + N_L = N. \tag{2.23}$$

The idea is to select a sample in each stratum. For each stratum an unbiased estimate of the stratum mean can then be computed. These stratum estimators can be combined into an unbiased estimator of the population mean.

There can be several reasons why one would want to apply stratified sampling. One reason is that only sampling frames for the subpopulations are available and not for the population as a whole. Another reason is that one is interested in estimates for subpopulations. And a third reason is that stratified sampling may lead to more precise estimators.

If the N_h values of the target variable Y in stratum h are denoted by

$$Y_1^{(h)}, Y_2^{(h)}, \ldots, Y_{N_h}^{(h)}, \tag{2.24}$$

the mean in stratum h can be written as

$$\bar{Y}^{(h)} = \frac{1}{N_h} \sum_{k=1}^{N_h} Y_k^{(h)}, \tag{2.25}$$

and the population mean becomes

$$\bar{Y} = \frac{1}{N} \sum_{h=1}^{L} N_h \bar{Y}^{(h)}. \tag{2.26}$$

So the population mean is a weighted average of the stratum means. The (adjusted) variance in stratum h is equal to

$$S_h^2 = \frac{1}{N_h - 1} \sum_{k=1}^{N_h} (Y_k^{(h)} - \bar{Y}^{(h)})^2. \tag{2.27}$$

Suppose that a stratified sample of size n must be selected from this population. This is a achieved by selecting L subsamples of sizes $n_1, n_2, \ldots, n_L$, respectively, where n_h is the sample size in stratum h, for $h = 1, 2, \ldots, L$. In principle, any sampling design can be applied within the strata, but usually simple random samples without replacement are used. If the n_h observations in stratum h are denoted by

$$y_1^{(h)}, y_2^{(h)}, \ldots, y_{n_h}^{(h)}, \tag{2.28}$$

the sample mean

$$\bar{y}^{(h)} = \frac{1}{n_h} \sum_{i=1}^{n_h} y_i^{(h)} \tag{2.29}$$

in stratum h is an unbiased estimator of the population mean in stratum h. Then the stratum estimators are combined into an estimator for the population mean of Y. Using expression (2.26) it can be shown that

$$\bar{y}_S = \frac{1}{N} \sum_{h=1}^{L} N_h \bar{y}^{(h)} \tag{2.30}$$

is an unbiased estimator of the population mean. The variance of the sample mean in stratum h is equal to

$$V(\bar{y}^{(h)}) = \frac{1-f_h}{n_h} S_h^2, \tag{2.31}$$

where $f_h = n_h/N_h$. Since the subsamples are selected independently, it can be shown that the variance of estimator (2.30) is equal to

$$V(\bar{y}_S) = \frac{1}{N^2} \sum_{h=1}^{L} N_h^2 \frac{1-f_h}{n_h} S_h^2. \tag{2.32}$$

This variance is small if the population variances S_h^2 are small. This is the case if there is little variation in the values of the target variable within strata, namely if the strata are homogeneous with respect to the target variable. The variance of the estimator is influenced by the sample sizes n_1, n_2, . . ., n_L in the strata, which is the sample *allocation*. The variance is minimal and therefore called a *Neyman allocation* (e.g., see Cochran, 1974) if the n_h are taken proportional to $N_h S_h$.

There are many other sampling designs. A number of them are mentioned here. The first is *sampling with unequal probabilities*. When the Horvitz–Thompson estimator was introduced, it became clear that its variance is small when first-order inclusion probabilities are more or less proportional to the values of target variable, namely when Y_k/π_k is approximately constant for all k. This is difficult to realize in practice because it requires knowledge of all values of the target variable in the population. If all these values were known, there was no reason to carry out the survey. However, sometimes the values in the population of an auxiliary variable X are known. So first-order inclusion probabilities could be taken proportional to the values of this variable. If there is a strong correlation between the target variable Y and the auxiliary variable X, then the result will be a precise estimator. An example is a shoplifting survey where shops are sampled according to their floor size, assuming there is more shoplifting in larger shops than in smaller shops.

Another sampling design is *cluster sampling*. This type of sampling can be applied if there is no sampling frame for the elements in the population, but there is one for clusters of elements. In this situation a sample of clusters can be selected, and all elements in each selected cluster can be observed. A typical example is an address sample. People are selected in the sample by first selecting addresses, and then including all people living at the selected addresses. Cluster sampling does not necessarily produce precise estimators. To the contrary, the more elements within clusters resemble each other, the less efficient the estimators will be. One other disadvantage of cluster sampling is that there is no control over the sample size. It simply depends on the numbers of elements in the selected clusters.

One way to get more control is to select a *two-stage sample*. First a sample of clusters is selected, and then a sample of elements is drawn from each selected cluster. The reasons for applying this sampling design are also more practical. Again, this procedure will not produce very accurate estimates, but it may be necessary to do this when there is lack of a proper sampling frame. Another reason to apply a two-stage sampling design may be to keep costs down. Travel costs less for face-to-face surveys if the addresses of selected persons are concentrated in clusters.

The precision of an estimator may be improved by using auxiliary information. A good example is sampling with inclusion probabilities proportional to the values of an auxiliary variable. If there is a strong correlation between target variable and auxiliary variable, the variance of the Horvitz–Thompson estimator will be small. And stratified sampling comes down to using qualitative auxiliary variables that divide the population into homogeneous groups.

In the examples above auxiliary information is used in the sampling design. Auxiliary information can also used in a different way, i.e. in the estimator itself. Some examples of improved estimation procedures are described here. It is assumed that simple random sampling without replacement is applied.

The *ratio estimator* assumes that an auxiliary variable X is available, whose values are more or less proportional to the values of the target variable:

$$Y_k \approx BX_k \tag{2.33}$$

for some constant B. The ratio estimator is defined by

$$\bar{y}_Q = \frac{\bar{y}}{\bar{x}}\bar{X}, \tag{2.34}$$

where $\bar{x}$ and $\bar{y}$ are the sample means of X and Y, and $\bar{X}$ is the population mean of X. The estimator is asymptotically unbiased, and its variance is approximately equal to

$$V(\bar{y}_Q) \approx \frac{1-f}{n}\frac{1}{N-1}\sum_{k=1}^{N}\left(Y_k - \frac{\bar{Y}}{\bar{X}}X_k\right)^2. \tag{2.35}$$

It can be shown that if this variance is smaller, the better condition (2.33) is satisfied. An even better estimator is the *regression estimator*. It assumes a linear relationship

$$Y_k \approx A + BX_k \tag{2.36}$$

between the values of the target variable and the auxiliary variable. A en B are constants that have to be estimated using the sample data. This can be the done with ordinary least squares. The regression estimator is defined by

$$\bar{y}_R = \bar{y} - b(\bar{x} - \bar{X}), \tag{2.37}$$

where

$$b = \frac{\sum_{i=1}^{n}(x_i - \bar{x})(y_i - \bar{y})}{\sum_{i=1}^{n}(x_i - \bar{x})^2} \tag{2.38}$$

and $x_1, x_2, \ldots, x_n$ denote the n observations for X that have become available in the sample. The estimator is asymptotically unbiased, and its variance is approximately equal to

$$V(\bar{y}_R) \approx \frac{1-f}{n} S^2(1 - R_{XY}^2), \tag{2.39}$$

where R_{XY} is the correlation between X and Y in the population. It is clear from expression (2.39) that the variance of the regression estimator is never larger than that of the simple sample mean. The stronger the correlation is, the smaller the variance will be.

EXAMPLE 2.4

The target population of a survey consists of 200 dairy farms in the rural part of Samplonia. The objective of the survey is to estimate the average daily milk production on each farm. To that end a simple random sample is selected. Two estimators are compared: the simple sample mean and the regression estimator. The regression estimator uses the number of cows per farm as the auxiliary variable. This seems not unreasonable as one may expect milk production per farm to be more or less proportional to the number of cows per farm.

Selection of a sample of size 40 and computation of the estimator has been repeated 500 times for both estimators. This gives 500 values of each estimator. The distribution of these values has been plotted in the graphs below. The histogram on the left shows the distribution of the sample mean. The distribution of the regression estimator is shown on the right.

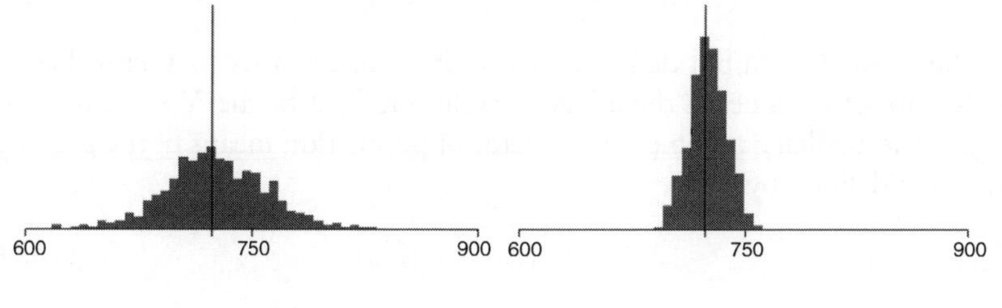

The regression estimator performs much better than the sample mean. The distribution of its values concentrates much more around the true value. The standard error of the sample mean is equal to 35.6, whereas it is 12.4 for the regression estimator. So a more precise estimator can be obtained with the same sample size if the proper auxiliary information is available.

The ratio estimator and the regression estimator both use a quantitative auxiliary variable. It is also possible to use a qualitative auxiliary variable. A well-known example of this is the *poststratification estimator.* Suppose that this auxiliary variable divides the target population U into L subpopulations (strata) U_1, U_2, . . ., U_L of sizes N_1, N_2, . . ., N_L, respectively. After a simple random sample has been selected, the sample mean can be computed in each stratum. Then these stratum estimates can be combined into an estimate

$$\bar{y}_{PS} = \frac{1}{N} \sum_{h=1}^{L} N_h \bar{y}^{(h)} \tag{2.40}$$

for the population mean. Note that this expression is identical to expression (2.30) for stratified sampling. However, estimator (2.40) has different statistical properties because the underlying selection mechanism is different. It can be shown that the poststratification estimator is approximately unbiased and that its variance is equal to

$$V(\bar{y}_{PS}) = \frac{1-f}{n} \sum_{h=1}^{L} W_h S_h^2 + \frac{1}{n^2} \sum_{h=1}^{L} (1-W_h) S_h^2, \tag{2.41}$$

where $W_h = N_h/N$, and S_h^2 is the population variance of the target variable in stratum h. If the strata are homogeneous with respect to the values of the target variable (i.e., there is little variation within strata), this variance will be small. Note that when n becomes larger, expression (2.41) will be closer to expression (2.32) as the second term in (2.41) decreases quickly in size.

A more general estimator can be defined of which the regression estimator and the poststratification estimator are special cases. This *generalized regression estimator* is introduced here because it is used in nonresponse correction techniques.

Suppose there are p auxiliary variables available. The p-vector of values of these variables for element k is denoted by

$$X_k = (X_{k1}, X_{k2}, \ldots, X_{kp})'. \tag{2.42}$$

The prime (′) symbol denotes transposition of a matrix or vector. Let Y be the N-vector of all values of the target variable, and let X be the $N \times p$-matrix of all values of the auxiliary variables. The vector of population means of the p auxiliary variables is defined by

$$\bar{X} = (\bar{X}_1, \bar{X}_2, \ldots, \bar{X}_p)'. \tag{2.43}$$

If the auxiliary variables are correlated with the target variable, then for a suitably chosen vector $B = (B_1, B_2, \ldots, B_p)'$ of regression coefficients for a best fit of Y on X, the residuals $E = (E_1, E_2, \ldots, E_N)'$, defined by

$$E = Y - XB, \tag{2.44}$$

vary less than the values of the target variable itself. Application of ordinary least squares results in

$$B = (X'X)^{-1}XY' = \left(\sum_{k=1}^{N} X_k X_k'\right)^{-1} \left(\sum_{k=1}^{N} X_k Y_k\right). \tag{2.45}$$

For any sampling design, the vector B can be estimated by

$$b = \left(\sum_{k=1}^{N} a_k \frac{X_k X_k'}{\pi_k}\right)^{-1} \left(\sum_{k=1}^{N} a_k \frac{X_k Y_k}{\pi_k}\right). \tag{2.46}$$

The estimator b is an asymptotically design unbiased (ADU) estimator of B. It means the bias vanishes for large samples. Using expression (2.46), we can define the generalized regression estimator by

$$\bar{y}_{GR} = \bar{y}_{HT} + (\bar{X} - \bar{x}_{HT})'b, \tag{2.47}$$

where $\bar{x}_{HT}$ and $\bar{y}_{HT}$ are the Horvitz–Thompson estimators for the population means of X and Y, respectively. The generalized regression estimator is an ADU estimator of the population mean of the target variable. If there exists a p-vector c of fixed numbers such that $Xc = J$, where J is a vector consisting of 1's, the generalized regression estimator can also be written as

$$\bar{y}_{GR} = \bar{X}'b. \tag{2.48}$$

This situation occurs if X contains an intercept term or a set of dummy variables corresponding to all categories of a qualitative variable. It can be shown that the variance of the generalized regression estimator can be approximated by

$$V(\bar{y}_{GR}) = \frac{1}{N^2} \sum_{k=1}^{N} \sum_{l=1}^{N} (\pi_{kl} - \pi_k \pi_l) \frac{E_k}{\pi_k} \frac{E_l}{\pi_l}. \tag{2.49}$$

This is the variance of the Horvitz–Thompson estimator, but with the values Y_k replaced by the residuals E_k. This variance will be small if the residual values E_k are small. Hence use of auxiliary variables which can explain the behavior of the target variable will result in a precise estimator.

Given simple random sampling without replacement and use of just one quantitative auxiliary variable, the generalized regression estimator reduces to the regression estimator defined in (2.37).

Suppose that a qualitative auxiliary variable is available with p categories. Then this variable can be replaced by p dummy variables. Associated with each element k

is a vector $X = (X_1, X_2, \ldots, X_p)'$ of dummy values. The hth dummy X_{kh} assumes the value 1 if element k belongs to stratum h, and it assumes the value 0 if it belongs to another stratum. In this case, B turns out to be equal to

$$B = (\bar{Y}^{(1)}, \bar{Y}^{(2)}, \ldots, \bar{Y}^{(L)})', \tag{2.50}$$

and this vector can be estimated unbiasedly by the vector

$$b = (\bar{y}_{HT}^{(1)}, \bar{y}_{HT}^{(2)}, \ldots, \bar{y}_{HT}^{(L)})' \tag{2.51}$$

of Horvitz–Thompson estimators of the stratum means. The vector of population means of the L auxiliary variables turns out to be equal to

$$\bar{X} = (W_1, W_2, \ldots, W_L)', \tag{2.52}$$

where $W_h = N_h/N$. If we substitute these quantities in expression (2.48), the result is

$$\bar{y}_{PS} = \frac{1}{N}\sum_{h=1}^{L} N_h \bar{y}_{HT}^{(h)}. \tag{2.53}$$

This is the poststratification estimator (2.40) but written down for arbitrary sampling designs. So the poststratification estimator is indeed a special case of the generalized regression estimator.

2.2.3 THE FIXED RESPONSE MODEL

To be able to investigate what possible impact nonresponse may have on estimators of population characteristics, the phenomenon of nonresponse must be incorporated in sampling theory. Two approaches are described to do this. One is the *random response model*, and the other is the *fixed response model*. Both approaches are discussed, for example, in Lindström et al. (1979), Kalsbeek (1980), Cassel et al. (1983), and Bethlehem and Kersten (1986). Both models give insight on conditions under which nonresponse causes estimators to be biased. The effect of biased estimators on the validity of confidence intervals is also explored.

The *fixed response model* assumes the population to consist of two mutually exclusive and exhaustive strata: the response stratum and the nonresponse stratum. Subjects from this population in the response stratum would participate in the survey with certainty if selected in the sample. Subjects in the nonresponse stratum would not participate with certainty if selected. Under this model a set of response indicators

$$R_1, R_2, \ldots, R_N \tag{2.54}$$

is introduced, where $R_k = 1$ if the corresponding element k is part of the response stratum, and $R_k = 0$ if element k is part of the nonresponse stratum. So $R_k = 1$ means response if selected, and $R_k = 0$ means nonresponse if selected. The size of the response stratum is denoted by

$$N_R = \sum_{k=1}^{N} R_k, \tag{2.55}$$

and the size of the nonresponse stratum by

$$N_{NR} = \sum_{k=1}^{N} (1-R_k), \tag{2.56}$$

where $N = N_R + N_{NR}$. The mean of the target variable Y in the response stratum is equal to

$$\bar{Y}_R = \frac{1}{N_R} \sum_{k=1}^{N} R_k Y_k. \tag{2.57}$$

Likewise the mean of the target variable in the nonresponse stratum can be written as

$$\bar{Y}_{NR} = \frac{1}{N_{NR}} \sum_{k=1}^{N} (1-R_k) Y_k. \tag{2.58}$$

Furthermore the *contrast* K is introduced as the difference between the means of the target variable in response stratum and the nonresponse stratum:

$$K = \bar{Y}_R - \bar{Y}_{NR}. \tag{2.59}$$

The contrast indicates how much respondents and nonrespondents differ on average for a specific variable.

Suppose that a simple random sample of size n is selected without replacement from this population. This sample is denoted by the set of indicators $a_1, a_2, \ldots, a_N$, where $a_k = 1$ means that element k is selected in the sample, and otherwise, $a_k = 0$. Because it is unknown beforehand to which of the two strata the elements belong, the result is

$$n_R = \sum_{k=1}^{N} a_k R_k \tag{2.60}$$

subjects from the response stratum, and

$$n_{NR} = \sum_{k=1}^{N} a_k (1-R_k) \tag{2.61}$$

subjects from the nonresponse stratum, where $n = n_R + n_{NR}$. Only the values of the n_R selected elements in the response stratum become available. The mean of these values is denoted by

$$\bar{y}_R = \frac{1}{n_R} \sum_{k=1}^{N} a_k R_k Y_k. \tag{2.62}$$

Theoretically it is possible for no observations at all to become available. This occurs when all who are sampled happen to fall in the nonresponse stratum. In practical situations, this event has a very small probability of happening if n is large. Therefore it will be ignored. For our purposes it can then be shown that the expected value of the response mean is equal to

$$E(\bar{y}_R) = \bar{Y}_R. \tag{2.63}$$

This is not surprising, since the responding elements can be seen as a simple random sample without replacement from the response stratum.

It is usually not the objective to estimate the mean of the response stratum, but the mean in the population. If both means are the same, there is no problem, but this is generally not the case. Therefore estimator (2.63) will be biased, and this bias is equal to

$$B(\bar{y}_R) = \bar{Y}_R - \bar{Y} = \frac{N_{NR}}{N}(\bar{Y}_R - \bar{Y}_{NR}) = QK, \tag{2.64}$$

where K is the contrast and $Q = N_{NR}/N$ is the relative size of the nonresponse stratum. From expression (2.64) it is clear that the bias is determined by two factors:

- The amount to which respondents and nonrespondents differ, on average, with respect to the target variable. The more they differ, the larger the bias will be.
- The relative size of the nonresponse stratum. The larger the group of nonrespondents is, the larger the bias will be.

EXAMPLE 2.5 The Fixed Response Model

The fixed response model is applied to data from the Dutch Housing Demand Survey. Statistics Netherlands carried out this survey in 1981. The sample size was 82,849. The number of respondents was 58,972, which comes down to a response rate of 71.2%.

The population characteristic to be estimated was the percentage of people with the intention to move within two years. To obtain more insight in the nonresponse, a follow-up survey was carried out among the nonrespondents. They were also asked whether they intended to move within two years. The results are summarized in the table below:

Do you intend to move within 2 years?	Response	Nonresponse	Total
Yes	17,515	3,056	20,571
No	41,457	20,821	62,278
Total	58,972	23,877	82,849

The percentage in the response stratum is estimated using the response data. The estimate is equal to $100 \times 17517/58972 = 29.7\%$.

The percentage in the nonresponse stratum is estimated using the nonresponse data. The estimate is equal to $100 \times 3056/23877 = 12.8\%$.

Hence the contrast K is equal to $29.7 - 12.8 = 16.9\%$. Apparently, the intention to move is much higher among respondents than among nonrespondents.

The relative size of the nonresponse stratum is estimated by $23877/82849 = 0.288$. Therefore, the bias of the estimator just based on the response data is equal to $16.9 \times 0.288 = 4.9\%$.

2.2.4 THE RANDOM RESPONSE MODEL

The *random response model* implies that every element k in the population has an (unknown) response probability ρ_k. If/when an element k is selected in the sample, a random mechanism is activated that results with probability ρ_k in response and with probability $1 - \rho_k$ in nonresponse. For this model a set

$$R_1, R_2, \ldots, R_N \tag{2.65}$$

of response indicators is introduced, where $R_k = 1$ if the corresponding element k responds, and where $R_k = 0$ otherwise. So $P(R_k = 1) = \rho_k$, and $P(R_k = 0) = 1 - \rho_k$.

Now suppose that a simple random sample without replacement of size n is selected from this population. This sample is again denoted by the set of indicators $a_1, a_2, \ldots, a_N$. The response only consists of those elements k for which $a_k = 1$ and $R_k = 1$. Hence the number of available cases is equal to

$$n_R = \sum_{k=1}^{N} a_k R_k. \tag{2.66}$$

Note that this realized sample size is a random variable, the same as for the fixed response model. The number of nonrespondents is equal to

$$n_{NR} = \sum_{k=1}^{N} a_k (1 - R_k), \tag{2.67}$$

where $n = n_R + n_{NR}$. Only the values of the n_R the responding elements become available. The mean of these values is denoted by

$$\bar{y}_R = \frac{1}{n_R} \sum_{k=1}^{N} a_k R_k Y_k. \tag{2.68}$$

Again, it is possible theoretically that no observations at all become available. This happens when all sample elements do not respond. In practical situations, this event has a very small probability of happening. Therefore it will be ignored. It can be shown (see Bethlehem, 1988) that the expected value of the response mean is approximately equal to

$$E(\bar{y}_R) \approx \tilde{Y}, \tag{2.69}$$

where

$$\tilde{Y} = \frac{1}{N}\sum_{k=1}^{N}\frac{\rho_k Y_k}{\bar{\rho}} \tag{2.70}$$

and

$$\bar{\rho} = \frac{1}{N}\sum_{k=1}^{N}\rho_k \tag{2.71}$$

is the mean of all response probabilities in the population. From expression (2.70) it is clear that the expected value of the response mean is unequal to the population mean to be estimated. Therefore this estimator is biased. This bias is approximately equal to

$$B(\bar{y}_R) = \tilde{Y} - \bar{Y} = \frac{R_{\rho Y}S_\rho S_Y}{\bar{\rho}}, \tag{2.72}$$

where $R_{\rho Y}$ is the correlation between the target variable and the response probability, S_Y is the standard deviation of the variable Y, and S_ρ is the standard deviation of the response probabilities. A number of conclusions can be drawn from expression (2.72):

- The bias disappears if there is no relationship between the target variable and response behavior. Then $R_{\rho Y} = 0$. The stronger the relationship between target variable and response behavior, the larger the bias will be.
- The bias disappears if all response probabilities are equal. Then $S_\rho = 0$ and $R_{\rho Y} = 0$. In this situation the nonresponse is not selective. It just leads to a reduced sample size.
- The magnitude of the bias increases as the mean of the response probabilities decreases (provided that the relationship between target variable and response behavior remains unchanged). Put in practical terms, this means that lower response rates will lead to larger biases.

EXAMPLE 2.6 The Random Response Model

The random response model can be illustrated by a simulation experiment. From the working population of the small country of Samplonia 1000 samples of size 40 are selected. For each sample the mean income is computed as an estimate of the mean income in the population. The distribution of these 1000 estimates is displayed in the graph on the left.

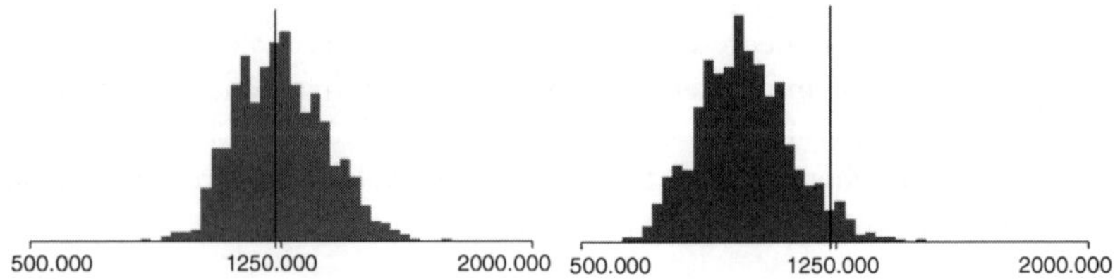

As the experiment is repeated, also nonresponse is generated. This is computed in the following way: the response probability of elements is linearly related to income. People with the lowest income have a response probability of 0.95, and people with the highest income have a response probability of 0.05. So, the higher the income, the lower the probability of response. Again 1000 samples of (initial) size 40 are generated. The resulting distribution of the estimates is displayed in the graph on the right.

Notice that the distribution has shifted to the left. This suggests that people with lower incomes are overrepresented, and people with high incomes are underrepresented. The vertical line representing the population mean is no longer in the center of the distribution. The average of the sample mean of all 1000 samples is equal to 970, whereas the population mean is equal to 1234. Clearly, the estimator has a substantial bias.

2.2.5 THE EFFECT OF NONRESPONSE ON THE CONFIDENCE INTERVAL

The precision of an estimator is usually quantified by computing the 95% confidence interval. First, the situation is considered in which all sampled elements respond. Then the sample mean is an unbiased estimator for the population mean. Since the sample mean has (approximately) a normal distribution, the 95% confidence interval for the population mean is equal to

$$I = (\bar{y} - 1.96 \times S(\bar{y}); \bar{y} + 1.96 \times S(\bar{y})), \tag{2.73}$$

where $S(\bar{y})$ is the standard error of the sample mean. The probability that this interval contains the true value, is by definition (approximately) equal to

$$P(\bar{Y} \in I) = 0.95. \tag{2.74}$$

In the case of nonresponse, only the response mean $\bar{y}_R$ can be used for computing the confidence interval. If this confidence interval is denoted by I_R, it can be shown that

$$P(\bar{Y} \in I_R) = \Phi\left(1.96 - \frac{B(\bar{y}_R)}{S(\bar{y}_R)}\right) - \Phi\left(-1.96 - \frac{B(\bar{y}_R)}{S(\bar{y}_R)}\right), \tag{2.75}$$

in which Φ is the standard normal distribution function. Table 2.1 gives the values of this probability as a function of the *relative bias* $B(\bar{y}_R)/S(\bar{y}_R)$.

As the table shows, the confidence level is much lower than expected if the relative bias is large. If the bias is equal to the standard error, meaning the relative bias is 1, the confidence level is only 0.83. As the relative bias differs more from zero, the situation becomes worse. The conclusion is that with increasing nonresponse correct interpretation of the confidence interval becomes more and more difficult.

TABLE 2.1 Confidence level of the 95% confidence interval as a function of the relative bias

$B(\bar{y}_R)/S(\bar{y}_R)$	$P(\bar{Y} \in I_R)$
0.0	0.95
0.2	0.95
0.4	0.93
0.6	0.91
0.8	0.87
1.0	0.83
1.2	0.78
1.4	0.71
1.6	0.64
1.8	0.56
2.0	0.48

EXAMPLE 2.7 The Effect of Nonresponse on the Confidence Interval

The effect of nonresponse on the confidence interval can be shown by means of a simulation experiment. From the working population of the small country of Samplonia, samples of size 40 are selected. Nonresponse is generated the same way as in Example 2.6. People with the lowest income have a response probability of 0.95. Nonresponse increases with income. People with the highest income have a response probability of 0.05.

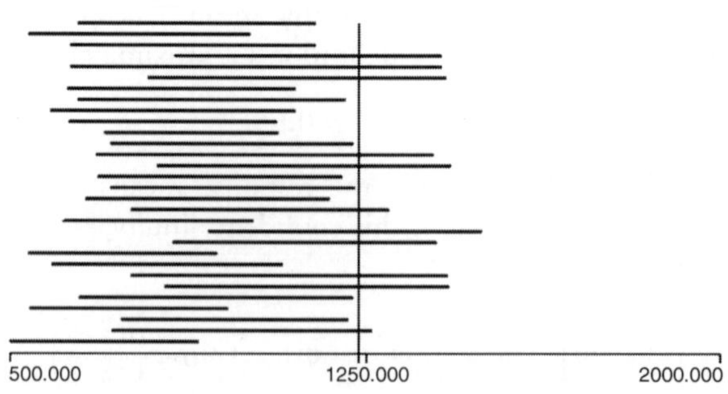

The confidence interval is computed for each sample. The graph above shows the result of the first 30 samples. Each confidence interval is indicated by a horizontal line. The vertical line denotes the true population mean to be estimated. Note that only 10 out of 30 confidence intervals contain the population mean. This suggests a confidence level of 33.3% instead of 95%. In fact the bias of the estimator is equal to 267 and the standard error of the estimate is 139. So the relative bias is 1.9, which corresponds to a confidence level of approximately 0.50.

Clearly, as nonresponses increase, the confidence interval can no longer be used to give an indication of the precision of the estimates.

The relative bias can be used to determine whether the bias is too large. Kersten and Bethlehem (1981) demonstrated this using the 1981 Labor Force Survey of Statistics Netherlands. This survey collected data by two-stage sampling. Municipalities were selected in the first stage. Because response rates differed substantially from municipality to municipality, a criterion was required to indicate when nonresponses were so high (and the quality of estimates was so low) that additional measures would be necessary. For the Labor Force Survey this meant going back to the municipality and doing additional data collection. Kersten and Bethlehem (1981) proposed the criterion

$$\left|\frac{B(\bar{y}_R)}{S(\bar{y}_R)}\right| \geq 1 \tag{2.76}$$

for callbacks. They concentrated on estimating the fractions for a fixed response model, using Y as the indicator variable. Because the variance of the response stratum can be approximated by the variance in the population, they took the variance of the response mean to be approximately equal to

$$V(\bar{y}_R) \approx \left(1-\frac{n}{N}\right)\frac{\bar{Y}_R(1-\bar{Y}_R)}{n(1-Q)}, \tag{2.77}$$

where $Q = N_{NR}/N$ is the relative size of the nonresponse stratum. This variance cannot be larger than

$$V(\bar{y}_R) \approx \left(1-\frac{n}{N}\right)\frac{1}{4n(1-Q)}, \tag{2.78}$$

so the callback was to take place if the contrast K satisfied the condition

$$|K| \geq \frac{1}{2Q}\sqrt{\frac{(N/n)-1}{N(1-Q)}}. \tag{2.79}$$

EXAMPLE 2.8 The Callback Criterion

The callback criterion as applied to data of the 1981 Labour Force Survey is illustrated in the graph below. Each curve corresponds to a specific value of the contrast (in %), and a point on the curve represents the situation in which equality in equation (2.79) exactly holds, which is when the response rate and population size match. For a point above the line the criterion is satisfied and callbacks must made.

Corresponding callback criteria for a number of large cities are also plotted in the graph. For example, if a contrast of 0.02 is considered possible

($K=2$), callback must be made in Amsterdam when the nonresponse exceeds 15%, and Rotterdam needs callback when the nonresponse is 20% or more. Higher nonresponse rates may be tolerated in smaller cities, given the same contrast. This is caused by higher variances due to smaller sample sizes.

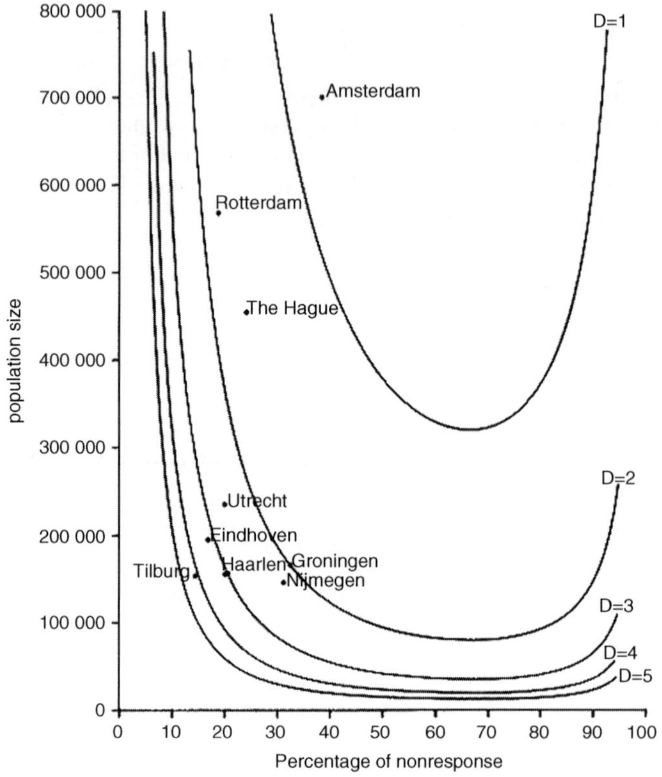

2.2.6 MISSING DATA MECHANISMS

Both unit and item nonresponses result in missing data. A strategy to use is to make assumptions about the mechanisms causing data to be missing and about relationships between the variables involved. These assumptions can be used to construct estimators that correct a nonresponse bias. The discussion in this subsection follows Little and Rubin (2002) and Schafer and Graham (2002).

A first step is to identify *patterns of missing data*. These patterns describe which values are observed and which ones are missing. The different patterns that missing data take require different treatments. Here we consider a number of ways to account for the *missing data*. Our chief concern will be to study the relationships between the missing values and the recorded values in the data sets.

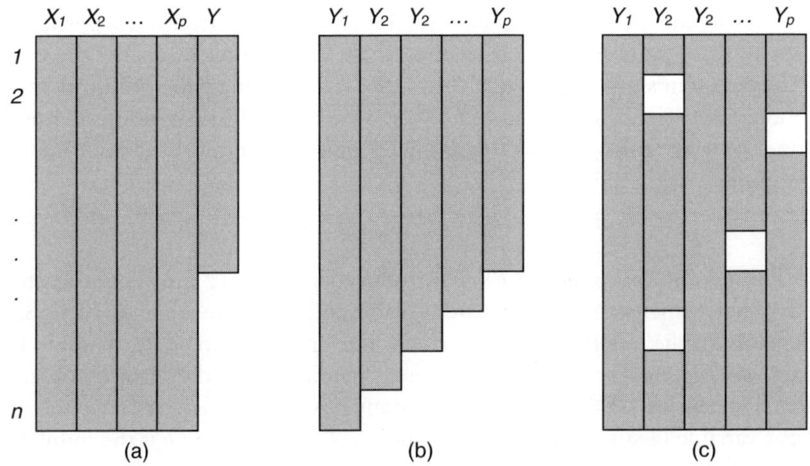

FIGURE 2.1 Nonresponse patterns.

Most survey data sets can be configured in rectangular or matrix form, so that the rows correspond to the observational elements and the columns correspond to the variables. Schafer and Graham (2002) identify three nonresponse patterns as shown in Figure 2.1.

Figure 2.1*a* shows the case of *univariate* missing data. The auxiliary variables X_1, X_2, X_P are observed for all sample elements. The column Y represents a group consisting of one or more target variables of the survey. The values of these variables are either completely observed for all variables or missing for all variables.

Figure 2.1*b* shows the case of a *monotone* missing data pattern. The target variables (or groups of target variables), $Y_1, \ldots, Y_p$ may be ordered in such a way that if Y_j is missing for an element, then $Y_{j+1}, \ldots, Y_p$ are missing as well.

Figure 2.1*c* shows the case of an *arbitrary* missing data pattern. There is no structure or ordering in the missingness of the values of the target variables.

EXAMPLE 2.9 Nonresponse Patterns

Unit nonresponse refers to the case where data are missing according to a *univariate* pattern. All values of variables corresponding to questions on the questionnaire form are therefore missing. Variables obtained from registers or sampling frames may be added to the survey data file, and for these variables, all values are available.

The *monotone* nonresponse pattern can arise in longitudinal studies with attrition, where subjects drop out prior to the end of the study and do not return.

Missing values due to item nonresponse typically have an *arbitrary* pattern; any set of variables may be missing for any unit.

Note that the patterns in Figure 2.1 are just basic patterns. It is possible for all three patterns to occur in one survey. An example is a panel survey. Univariate missing data (Figure 2.1*a*) may occur during panel recruitment, monotone missing data (Figure 2.1*b*) may be caused by wave nonresponse, and arbitrary missing data (Figure 2.1*c*) may be the result of item nonresponse within a wave.

The missing data indicator R was introduced in Section 2.2.3. This indicator is used to show which variable values are available and which are missing. The form of R depends on the missing data pattern. For univariate missing data, R would be a binary variable that for each sample element i indicates whether data are observed ($R_i = 1$) or missing ($R_i = 0$). For the monotone missing data pattern, R would be an integer variable (assuming one of the values $1, 2, \ldots, p$, where p is the number of [groups of] variables. Now $R_i = j$ means that values of $Y_1, Y_2, \ldots, Y_j$ are observed; the values of $Y_{j+1}, Y_{J+2}, \ldots, Y_p$ are not observed. For arbitrary missing data, R would be a matrix of binary indicators of the same dimension as the data matrix, with elements of R set to 1 or 0 according to whether the corresponding data values are observed or missing.

The *missing data mechanism* can be used to describe the causes of missing values in terms of the patterns of observed and missing values in the survey data set. In an analysis of data with missing values due to nonresponse, the properties of estimators and the success of bias correction techniques entirely depend on our learning the dependencies within these mechanisms based on assumptions about the missing data.

The mechanism for missing data can therefore be formalized as follows: The dataset Y is divided in an observed part Y_{obs} and a missing part Y_{mis}. So $Y = (Y_{obs}, Y_{mis})$. The distribution of the missingness is characterized by the conditional distribution of R given Y:

$$P(R|Y) = P(R|Y_{obs}, Y_{mis}) \tag{2.80}$$

Three situations can be distinguished:

- The conditional distribution of R given Y does not depend on the data at all. This situation is called *missing completely at random* (MCAR). This implies that $P(R|Y) = P(R)$.
- The conditional distribution of R given Y depends on the observed data, but not on the missing data. This is called *missing at random* (MAR). It implies that $P(R|Y) = P(R|Y_{obs})$. So $P(R|Y_{obs}, Y_{mis})$ cannot be simplified any further. The causes for missing data are identified by the observed data,
- The conditional distribution of R given Y depends on both the observed and missing data. This situation is called *not missing at random* (NMAR). This implies that $P(R|Y_{obs}, Y_{mis})$ cannot be simplified any further. The causes for the missing data are not fully identified by the observed data.

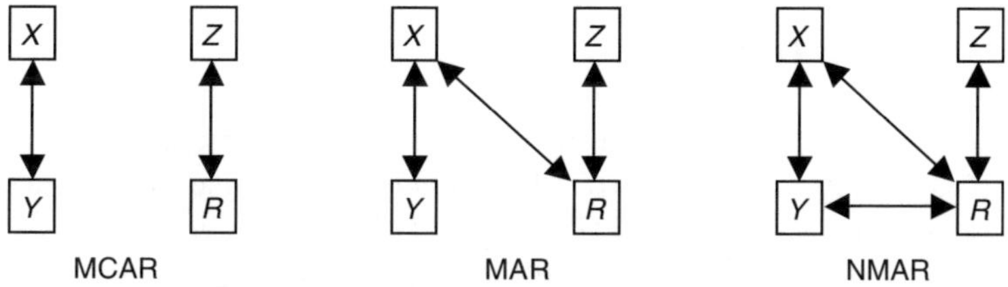

FIGURE 2.2 Missing data mechanisms.

The schema for the three missing data mechanisms is shown in Figure 2.2. In the figure we let X represent a set of auxiliary variables that are completely observed. Their values could have been obtained from another source such as a sampling frame or a register. We let Y be a target variable that is partly missing, with Z representing causes of the missing data completely unrelated to X and Y, and R representing the missing data state.

As the figure shows, the case of MCAR is brought on by a phenomenon Z that is completely unrelated to X and Y. Nonresponse has a minor effect; it only reduces the sample size. Estimates for parameters Y will not be biased

The case of MAR is caused partly by an independent phenomenon Z and partly by an auxiliary variable X. So there is an indirect relationship between Y and R, and as a result estimates for Y are biased. Fortunately, it is possible to correct for such a bias by using a technique that takes advantage of the availability of all values of X, both for respondents and nonrespondents. One has to assume the relation between X and Y can be identified just using the observed data.

The case of NMAR suggests a relationship between Z and R and between X and R. But there is also a direct relationship between Y and R that cannot be accounted for by X. This situation leads to biased estimates for Y. Unfortunately, correction techniques using X will not be able to remove the bias.

EXAMPLE 2.10 Missing Data Mechanisms

For the case of unit nonresponse in a sample survey, the missing data pattern is univariate. The auxiliary variables $X_1, X_2, \ldots, X_p$ have been measured for every person in the sample, but the target variable Y is missing for nonrespondents. The explanation may be one of the following:

The probability of response for an element does not depend on the values of X and Y, which means MCAR is the missing data mechanism.

The probability of response does depend on X but not on Y, which means that MAR is the mechanism.

The probability of response depends on the values of Y as well, which means that NMAR is the mechanism. For instance, a person does not have time to participate in the labor force survey because he/she has to go to work.

Remember, all three mechanisms can operate in the same survey if the survey contains several target variables Y.

2.3 Application

Some of the concepts introduced in this chapter are illustrated using an actual survey data set. This data set is derived from a Dutch survey that was carried out by Statistics Netherlands, and is called the General Population Survey (GPS).

The persons contacted were selected by way of a stratified two-stage sampling procedure. In the first stage, municipalities were selected within regions with probabilities proportional to the number of inhabitants. In the second stage, an equal probability sample was drawn in each selected municipality. The sampling frames for selecting survey participants were population registers of the municipalities. The sampling design was such that each person had the same probability of being selected (called a self-weighting sample). The sample of the GPS consisted of 32,019 persons. The number of respondents was 18,792.

2.3.1 THE FIXED RESPONSE MODEL

The fixed response model assumes that the population consists of two mutually exclusive and exhaustive strata: the response stratum and the nonresponse stratum. Persons in the response stratum would participate in the survey with certainty, if selected in the sample. Persons in the nonresponse stratum would not participate with certainty, if selected. The strata sizes are not known, but they can be estimated unbiasedly. The response fraction n_r/n is an unbiased estimator of the population fraction N_r/N. The population size is equal to 12,170,001. Therefore the estimate of the size of the response stratum is equal to $(18{,}792/32{,}019) \times 12{,}170{,}001 = 7{,}142{,}592$. Consequently the size of the nonresponse is estimated by $12{,}170{,}001 - 7{,}142{,}592 = 5{,}027{,}409$.

Statistics Netherlands has an integrated system of social statistics. This system is called the *Social Statistics Database* (SSD). The SSD contains a wide range of characteristics on each individual in The Netherlands. There are data on demography, geography, income, labor, education, health, and social protection. These data are obtained by combining data from registers and other administrative data sources. Linking the sample file to the SSD allows the values of a large set of auxiliary variable to become available for both respondents and nonrespondents. Table 2.2 contains the estimates for three of these variables.

The percentage of people with a social allowance in the nonresponse stratum is twice as large as the percentage in the response stratum. Substantial differences can

TABLE 2.2 Fixed response model applied to the GPS

	Response stratum	Nonresponse stratum	Population
Receives a social allowance	2.9%	6.0%	4.2%
Has a job	52.6%	48.8%	51.0%
Is a nonnative	12.1%	18.4%	14.7%

also be observed for people with a job and for nonnatives. This immediately suggests that the composition of the nonresponse stratum differs from that of the response stratum. Therefore one can expect other estimates also to be biased.

The contrast for, for example, the percentage nonnatives is equal to $K = 12.1\% - 18.4\% = -6.3\%$. The relative size Q of the nonresponse stratum = 5,027,409/12,170,001 = 0.413. Hence the bias is equal to $Q \times K = 0.413 \times -6.3\% = -2.6\%$. This value is also obtained by subtracting the population estimate (14.7%) in Table 2.2 from the response stratum estimate (12.1%).

Suppose that a 95% confidence interval is computed for the percentage of nonnatives based on just the response data. With a response size of 18,792 and 2272 nonnatives among them, this interval will be equal to (11.6%; 12.6%). The true value will be very close to 14.7%. It is clear that the confidence interval does not contain the true value. Due to nonresponse the confidence interval is not a reliable indicator of the accuracy of an estimator.

2.3.2 THE RANDOM RESPONSE MODEL

The random response model assumes that every element k in the population has an (unknown) response probability ρ_k. If element k is selected in the sample, a random mechanism is activated that results with probability ρ_k in response and with probability $1 - \rho_k$ in nonresponse. The response probabilities ρ_k are unknown theoretical quantities. It is possible, however, to estimate them using the available auxiliary information. Rosenbaum and Rubin (1983) introduced the technique of propensity scores to do this.

The *propensity score* $\rho(X)$ is the conditional probability that an individual with observed characteristics X responds in a survey when invited to do so ($R = 1$):

$$\rho(X) = P(R = 1|X). \tag{2.81}$$

This approach assumes that within subpopulations defined by values of the observed characteristics X, all individuals have the same response probability. This is the "missing at random" (MAR) assumption introduced in Section 2.2.6. Often the propensity score is modeled by the logit model

$$\log\left(\frac{\rho(X_k)}{1-\rho(X_k)}\right) = \alpha + \beta' X_k. \tag{2.82}$$

The quantity α and the vector β are the model coefficients that have to be estimated. X_k is the vector of values of the auxiliary variables for element k. The logit transform on the left-hand side of the equation guarantees that predicted values of the propensities $\rho(X_k)$ are within the interval [0,1]. Other models can be used too. Examples are the probit model or even the linear model. Dehjia and Wahba (1999), however, conclude that different models often produce similar results. The logit model can be fitted with maximum likelihood estimation.

The data in the GPS survey data file are now used to fit a model for the response probabilities. To find such a model, auxiliary variables are required. These variables must have been measured for both respondents and nonrespondents. Since the

TABLE 2.3 Auxiliary variables in logit model

Variable	Description	Categories
Gender	Gender	2
Married	Is married	2
Age13	Age in 13 age groups	13
Ethnic	Type of nonnative	5
HHSize	Size of the household	5
HHType	Type of household	5
Phone	Has listed phone number	2
Hasjob	Has a job	2
Region	Region of the country	5
Urban	Degree of urbanization	5

GPS survey data file has been linked to the Social Statistics Database (SSD) of Statistics Netherlands, many such variables are available. Table 2.3 contains the subset of variables that turned out to have a significant contribution in the model.

Notice that all variables in this table are of a categorical nature. To include them in model (2.82), they have to be replaced by sets of dummy variables, where there is a dummy variable for each category of each variable. Furthermore, to be able to estimate the model parameters, we must impose extra restrictions. This is usually accomplished by setting one of the parameters for each set of dummies to 0.

A logit model has been fitted with the variables in Table 2.3 as explanatory variables. Just main effects are included in the model; no interaction effects. The estimated model is used to estimate the response propensity for each sample person. This distribution of these probabilities is displayed in Figure 2.3.

As the figure shows, there is a wide variation in these probabilities. The values of the response propensities fall roughly between 0.2 and 0.8. The average value is 0.587. This is equal to the response rate. Table 2.4 lists the characteristics of persons with the highest and the lowest response propensities.

The person with the lowest response propensity is a unmarried middle-aged nonnative male. He lives in a big city and has no job. There are two people in the household, but the type of household is unclear. He does not have a listed phone number.

The person with the highest response probability is a native young girl. She is one of the children in a larger household (5 persons or more) living in a rural area. She has a job and a listed phone number.

More insight can be obtained from studying the relationship between the estimated response propensities and other variables. Two examples are given here. Figure 2.4 shows box plots of the response propensities for each level of urbanization. To avoid clutter in the graph, outlying values are not displayed. The graph shows a clear relationship between response propensities and urbanization level. Response propensities are substantially lower in higher urbanized areas.

TABLE 2.4 Lowest and highest response propensity

Variable	Value for lowest propensity	Value for highest propensity
Gender	Male	Female
Is married	No	No
Age in 13 age groups	45–49	18–19
Type of nonnative	First generation non-Western	Native
Size of the household	2	5 or more
Type of household	Other	Couple with children
Has listed phone number	No	Yes
Has a job	No	Yes
Region of the country	Metropolis	Woodlands
Extent of urbanization	Very strong	Not
Response propensity	0.109	0.798

Figure 2.5 shows box plots of the response propensities for the percentage of nonnatives in the neighborhood. There is a clear pattern. The response propensities drop as the number of nonnatives living in the neighborhood increases.

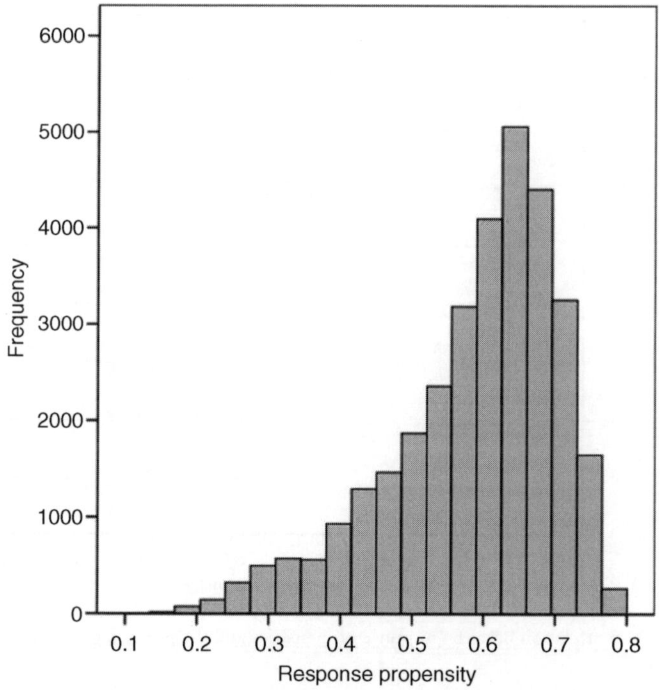

FIGURE 2.3 Distribution of the estimated response propensities.

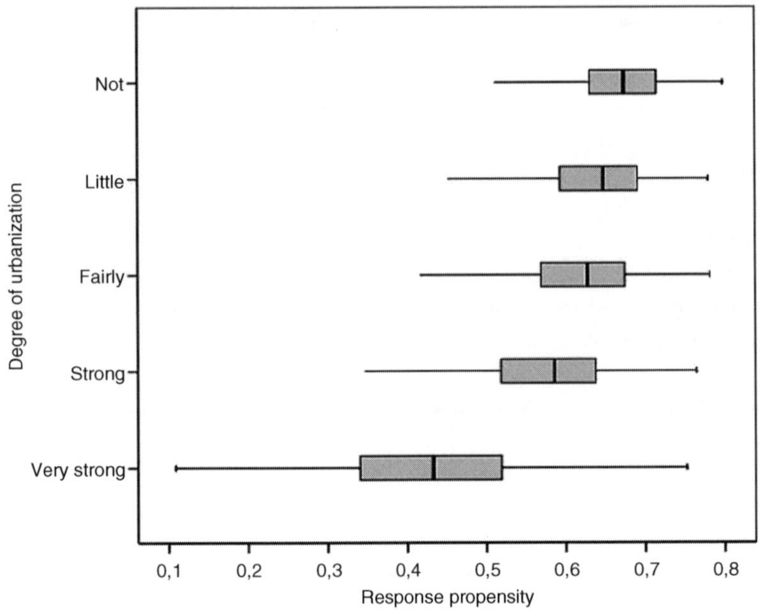

FIGURE 2.4 Relation between response propensities and extent of urbanization.

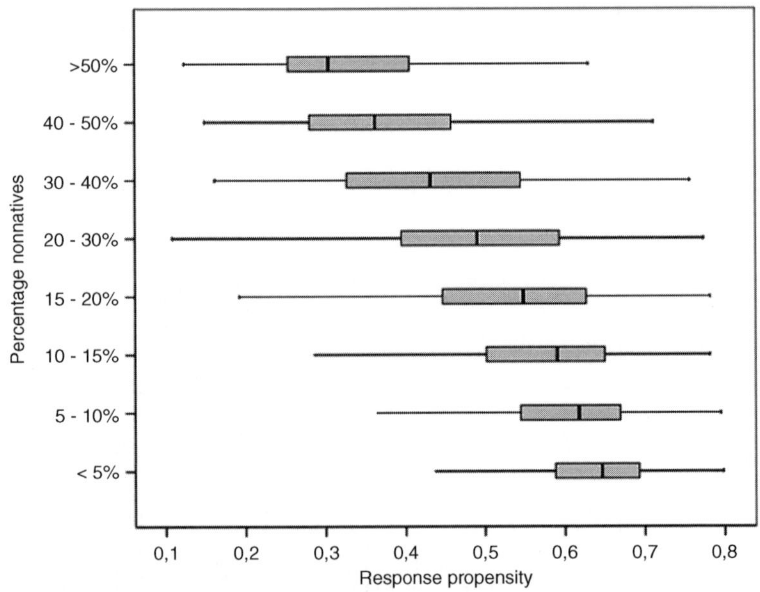

FIGURE 2.5 Relation between response propensities and the percentage of nonnatives in the neighborhood.

As these examples show, response probabilities are not equal for every person in the population. Moreover, response probabilities are correlated with other variables. So there is a serious risk that estimators are biased.

2.4 Summary

A survey is in effect an instrument used to collect information about a specific populations. Typically not all members of the selected population are contacted in a survey, just a sample. This sample is selected by means of probability sampling. There are various ways to do this. The usual sampling methods are simple random sampling, stratified sampling, sampling with unequal probabilities, cluster sampling, and two-stage sampling.

If every person in a population has a known and positive probability of selection, it is possible to define a unbiased estimator. This is known as a Horvitz–Thompson estimator. This estimator can be improved by taking advantage of available auxiliary variables. Examples of such estimators are the ratio estimator, the regression estimator, and the poststratification estimator.

The phenomenon of nonresponse must be included in sampling theory. There are two ways to analyze its effects. The first one is by way of the fixed response model. It assumes the population to consist of two mutually exclusive and exhaustive strata: the response stratum and the nonresponse stratum. Persons in the response stratum always respond and persons in the nonresponse stratum never respond. The second one is the random response model. It assumes every member of the population to have an (unknown) response probability. The bias of an estimator is in both models determined by two factors: the response rate and the relationship between response behavior and the target variables of the survey.

The standard error of an estimator decreases as the sample size increases. The expressions for the nonresponse bias show that its value does not depend on the sample size. This implies that the nonresponse error tends to dominate the sampling error in large surveys. Consequently confidence intervals become meaningless.

The possible effect of nonresponse is determined by the missing data mechanism. Three mechanisms are distinguished. Missing completely at random (MCAR) means that a nonresponse is independent of all survey variables; estimators will not be biased. Missing at random (MAR) means that a nonresponse depends on auxiliary variables only; estimators will be biased but a correction is possible if some technique is used that takes advantage of this auxiliary information. Not missing at random (NMAR) means that nonresponse depends directly on the target variables of the survey; estimators may be biased and correction techniques may not be successful.

2.5 KEY TERMS

Arbitrary missing data A missing data pattern where data are missing completely random.

Cluster sampling A sampling design for which the population is divided into a number of groups. A sample of the groups is drawn and all persons in the selected groups are included in the sample.

Fixed response model A model for nonresponse for which the population is assumed to consist of two mutually exclusive and exhaustive strata: the response stratum and the nonresponse stratum. Persons in the response stratum always respond and persons in the nonresponse stratum never respond.

Horvitz–Thompson estimator An unbiased estimator that can be computed if the selection probabilities of all elements are known and positive.

Missing at random (MAR) Nonresponse that depends on auxiliary variables only. Estimators will be biased, but a correction is possible if some technique is used that takes advantage of this auxiliary information.

Missing completely at random (MCAR) Nonresponse that happens completely independent of all survey variables. Estimators will not be biased.

Monotone missing data A missing data pattern where more values are missing for each subsequent variable in a data matrix. The set of missing values of a variable is a subset of the set of missing variables of the next variable.

Not missing at random (NMAR) Nonresponse that depends directly on the target variables of the survey. Estimators will be biased and correction techniques will not be successful.

Poststratification estimator An estimator that can be computed if all strata sizes in a stratified population are available. It is a precise estimator if the strata are homogeneous.

Random response model A model for nonresponse that assumes every element in the population to have an (unknown) response probability.

Ratio estimator An estimator that can be computed if the population mean of an auxiliary variable is available. It is a precise estimator if the target variable and auxiliary variable are (approximately) proportional.

Regression estimator An estimator that can be computed if the population mean of an auxiliary variable is available. It is a precise estimator if there is a (approximate) linear relationship between target variable and auxiliary variable.

Simple random sampling A sampling design where persons are selected with equal probabilities.

Stratified sampling A sampling design where the population is divided into a number of strata and where a random sample is drawn from each stratum.

Two-stage sampling A sampling design where the population is divided into a number of clusters. A sample of clusters is drawn. From each selected cluster a sample of elements is drawn.

Unequal probability sampling A sampling design whereby persons are selected with probabilities that are proportional to the values of some auxiliary variable.

Univariate missing data Data that are missing in the same rows of the data matrix for a set of variables.

2.6 REFERENCES

Bethlehem, J.G. (1988), Reduction of Nonresponse Bias through Regression Estimation. *Journal of Official Statistics 4*, pp 251–260.

Bethlehem, J.G. (2009), *Applied Survey Methods: A Statistical Perspective*. Wiley, Hoboken NJ.

Bethlehem, J.G., and Kersten, H.M.P. (1986), Werken met Non-respons. *Statistische Onderzoekingen* M 30. Statistics Netherlands, Voorburg.

Cassel, C.M., Särndal, C.E., and Wretman, J.H. (1983), Some Uses of Statistical Models in Connection with the Nonresponse Problem. In: Madow, W.G. , and Olkin, I. (eds.): *Incomplete Data in Sample Surveys*, Vol. 3 Proceedings of the Symposium. Academic Press, New York.

Cochran, W.G. (1974), *Sampling Techniques*, 3rd ed. Wiley, New York.

Dehjia, R., and Wahba, S. (1999), Causal Effects in Non-experimental Studies: Re-evaluating the Evaluation of Training Programs. *Journal of the American Statistical Association* 94, pp. 1053–1062.

Horvitz, D.G., and Thompson, D.J. (1952), A Generalization of Sampling without Replacement from a Finite Universe. *Journal of the American Statistical Association* 47, pp. 663–685.

Kalsbeek, W.D. (1980), A Conceptual Review of Survey Error due to Nonresponse. American Statistical Association, *Proceedings of the Section on Survey Research Methods*, pp. 131–136.

Lindström, H., Wretman, J., Forsman, G., and Cassel, C. (1979), *Standard Methods for Non-response Treatment in Statistical Estimation*. National Central Bureau of Statistics, Stockholm.

Little, R.J.A., and Rubin, D.B. (2002), *Statistical Analysis with Missing Data*, 2nd ed. Wiley Interscience, New York.

Rosenbaum, P.R., and Rubin, D.B. (1983), The Central Role of the Propensity Score in Observational Studies for Causal Effects. *Biometrika* 70, pp. 41–50.

Schafer, J.L., and Graham, J.W. (2002), Missing Data: Our View of the State of the Art. *Psychological Methods* 7, pp. 147–177.

2.7 EXERCISES

Exercise 2.1 Suppose that a simple random sample of size 100 is selected from a population of size 1000. What is the value of the second-order inclusion probability for every pair of elements?

(a) 0.1

(b) 0.09

(c) 0.01

(d) 0.0099

Exercise 2.2 Suppose that a simple random sample of size 1000 is selected from a population of size 16,000,000. What would happen to the variance of the sample mean if the sample size is doubled to 2000?

(a) It would be twice as small.

(b) It would be approximately twice as small.

(c) It would be twice as large.

(d) It would be approximately twice as large.

Exercise 2.3 Under which condition will stratified sampling lead to more precise estimates than the simple sample mean?

(a) The values of the target variable vary little within strata.

(b) The stratum means vary little.

(c) The strata are all of approximately the same size.

(d) The subsamples in all strata are of the same size.

Exercise 2.4 Which of the statements below does not apply to cluster sampling?

(a) It is not clear beforehand how large the sample size will be.

(b) It can reduce travel costs of interviewers.

(c) No sampling frame at all is necessary.

(d) Generally, it will lead to more precise estimators than simple random sampling.

Exercise 2.5 Under which condition is the variance of the regression estimator smaller than the variance of the simple sample mean?

(a) Always.

(b) Only if the correlation between target variable and auxiliary variable is greater than 0.

(c) Only if the correlation between target variable and auxiliary variable is equal to 1.

(d) Only if the squared correlation between target variable and auxiliary variable is greater than 0.

Exercise 2.6 What can be said about the value of the nonresponse bias?

(a) It increases as the sample size increases.

(b) It increases as the sample size decreases.

(c) It is independent of the sample size.

(d) It is proportional to the square root of the sample size.

Exercise 2.7 What happens to the value of the relative bias as the sample size increases?

(a) It decreases.

(b) It increases.

(c) It remains constant.

(d) It approaches 0.

Exercise 2.8 What is the effect of an increasing sample size on the value of the confidence level of the confidence interval?

(a) It remains 0.95.

(b) It increases in size and approaches 1.00.

(c) It decreases in size and approaches 0.00.

(d) It remains 0.05.

Exercise 2.9 A survey investigates Internet use. Suppose that response behavior only depends on age (the probability of response decreases with age). Suppose also that Internet use only depends on age. What kind of missing data mechanism does apply in this survey?

(a) Not Missing at Random (NMAR).

(b) Missing at Random (MAR).

(c) Missing Completely at Random (MCAR).

(d) A mix of all three patterns.

Exercise 2.10 There are two target variables in a survey about shoplifting. The first variable records whether or not someone ever has committed some form of shoplifting. Not every respondent wants to answer this question. So there is item nonresponse. The second variable records what kind of product was stolen the last time. Only a subset of respondents who admitted shoplifting want to answer the second question. What kind of missing data pattern does apply?

(a) Univariate missing data.

(b) Monotone missing data.

(c) Arbitrary missing data.

(d) A mix of at least two mechanisms.

CHAPTER THREE

Reducing Nonresponse

3.1 Introduction

There are two ways to do something about the nonresponse problem. One way is to prevent nonresponse from happening during the fieldwork of the survey. If there is no nonresponse, the risk of biased estimates is avoided. However, it is almost impossible to obtain a response rate of 100%. Notwithstanding all the efforts, always some nonresponse will remain. The second way of doing something about nonresponse is to correct estimates for a possible bias.

This chapter is about reducing nonresponse in the field and focuses on the psychological aspects of nonresponse reduction research (e.g., see Biemer and Lyberg, 2003; Groves and Couper, 1998). Nonresponse can be anticipated from the behavior of potential respondents, so we start by asking the questions What causes persons not to participate in a survey? And how can this be prevented? The survey organization can translate this knowledge into fieldwork strategies that focus on obtaining higher, representative response.

Although nonresponse adjustment techniques can be applied to reduce bias after the data have been collected, nonresponse reduction can take place before and during the data collection process. This chapter looks at the psychological component of obtaining participation from persons for a survey and the practical implications that result in measures used to reduce nonresponse. In Section 3.2 a number of theories from the literature about survey participation are presented. In Section 3.3 it is shown how this theory can be implemented to reduce nonresponse.

Handbook of Nonresponse in Household Surveys, By Jelke Bethlehem, Fannie Cobben, and Barry Schouten

3.2 Theory

3.2.1 INTRODUCTION

Recall from Chapter 2 the main sources of nonresponse: noncontact, refusal, not-able (due to language problems), and not-able (due to long time illness). Excluding persons whose inability to cooperate is due to chronic illness, measures can be used to reduce the nonresponse of persons in the other categories. This discussion will therefore consider only the cases of nonresponse due to noncontact, refusal, or inability to cooperate due to language problems. Inability to cooperate because of *language problems* obviously mostly occurs among nonnative people.

EXAMPLE 3.1 Nonresponse Due to Language Problems

In The Netherlands, first- and second-generation nonnative persons are distinguished. First-generation nonnatives are born outside The Netherlands. Second generation nonnatives are born in The Netherlands; they are children of first generation nonnatives.

Table 3.1 shows the percentages of nonresponse arising from language problems for the 2004 Integrated Survey on Household Living Conditions.

TABLE 3.1 Nonresponse due to language problems

Population group	Size of the group	Percentage of nonresponse
Native	26,671	0.1 %
Nonnative, second generation, Western	1,597	0.3 %
Nonnative, second generation, non-Western	708	1.4 %
Nonnative, first generation, Western	1,144	8.4 %
Nonnative, first generation, non-Western	1,838	14.7 %
Total	31,958	1.3 %

The percentage of nonresponse due to language problems is highest for first-generation non-Western nonnatives. This is also the largest group of nonnatives.

Noncontact can be caused by various reasons. Groves and Couper (1998) present a model that relates contact to three causes:

1. The at-home pattern of the household.
2. The presence of physical impediments.
3. The call pattern of the interviewers.

The literature is unambiguous with regard to the process of making contact. The probability of making contact is strongly related to the probability that some member of the household is at home when the interviewer calls. Physical impediments make life harder for the interviewer. Physical impediments can result in lower contact rates in both face-to-face surveys (e.g., a locked central entrance to an apartment building, or an intercom) and telephone surveys (e.g., answering machines or number identification). In a face-to-face survey the impediments can be observed prior to the first contact. Households that have some sort of impediment have a lower probability of contact on the first attempt, and more households with impediments remain noncontacted at the end of the fieldwork period.

The call pattern concerns the timing and the frequency of calls made by the interviewer. Research on at-home patterns of households shows that during the evening, the probability of finding someone at home is much larger than during daytime (Groves and Couper, 1998). For various reasons, however, the percentage of calls that is made during the evening is smaller than during daytime. Of course, evening time is a shorter period than daytime. In a telephone survey where all calls are distributed to the interviewers by a central call management system, the number of evening calls is smaller than the number of daytime calls. In a face-to-face survey the interviewer makes the decision when to visit a household. A possible reason for the smaller proportion of visits made during the evening might be that interviewers feel less safe during evenings and therefore prefer daytime visits.

Refusals are different from noncontacts. Being in a position to refuse means that contact has already been made. When a person refuses participation, this is a conscious decision. Theories that can be used to explain survey participation are found in both the sociological and the psychological literature. Goyder (1987) describes two psychological views on the phenomenon of nonresponse in survey research: behavioral and voluntary. Survey researchers with a behavioral view of nonresponse believe that survey participation can be influenced by the survey design, and depends on sociodemographic characteristics, or social location, of the respondents. In the voluntary view, the respondents themselves decide whether or not they want to participate. This decision is hardly influenced by the survey researcher. In the extreme, the voluntary theory resolves into a view that persons act for reasons best known to themselves. The behavioral extreme would be that there is no conscious decision-making left within the social context of a survey request.

Blau (1964) describes a theory of social exchange that both Dillman (1978) and Goyder (1987) translate to the survey participation context. The social exchange theory (Blau, 1964) "considers the perceived value of the equity of long-term associations between persons, or between a person and societal institutions." Social exchange theory assumes that relationships between humans are based on a subjective cost–benefit analysis, and the comparison of alternatives.

Goyder (1987) finds a synthesis in the social exchange theory for both the behavioral and the voluntary viewpoint. He describes it as a number of obligations and expectations over a longer period of time between an individual and various institutions of society. Individuals balance cost and reward for survey participation, in which their behavior is based on rational decisions. Dillman (1978) takes a more narrow view on the social exchange by describing the social exchange between a survey organization and an individual.

Groves et al. (2000) have developed a theory of interaction between survey design features and socioeconomic characteristics of the sample person. This theory stipulates that individual persons assign different values to survey design features. These values are determined by the socioeconomic characteristics of the sample person. The interviewer tries to make some of the survey design features more salient to the sample person. The final response decision is determined by the value that the sample person attributes to the survey design features, multiplied by the saliency that the features have for the sample person. Groves et al. (2000) refer to this theory as the "leverage-salience" theory.

EXAMPLE 3.2 The Leverage-Salience Theory

Roose et al. (2007) analyze the effects of reminders and replacement questionnaires in a two-step data collection survey. They find that the positive leverage of reminders and replacement questionnaires decreases as the interest in the survey topic increases.

Groves, Cialdini, and Couper (1992) identify several components that influence survey participation: sociodemographic and survey design features, the interaction between the interviewer and the sample person, tailoring and maintaining interaction. These components will be discussed separately in the following sections.

3.2.2 INFLUENCES OF SOCIODEMOGRAPHIC AND SURVEY DESIGN FEATURES

Influences of sociodemographic features comprise sociodemographic factors and characteristics of the sample person. Besides attributes of the survey design features of the survey design include attributes of the interviewer and the respondent–interviewer interaction:

- *Sociodemographic factors* relate to the context within which the survey request takes place, and can influence both the interviewer and the respondent. They determine the expectations of both interviewer and respondent, but also the persuasion strategy (interviewer) and the decision-making process (sample person). Examples of these factors are the number of surveys conducted in society (survey climate) and the perceived legitimacy of surveys.
- *Attributes of the survey design.* The mode of data collection plays a major role in survey participation. An important distinction between modes involves the possible assistance of an interviewer; see also *attributes of the interviewer* and respondent–interviewer interaction (Section 3.2.3). The choice for a particular

mode determines the number of channels that can be exploited for communication. In a face-to-face survey, both visual and audio communication can be used, whereas in a telephone survey, there is only audio communication. In a paper or mail survey, the only communication is the design of the questionnaire and/or accompanying material (letter, brochure). Other attributes are the length of the interview as an indication of the burden of participation (the shorter the better); the length of the fieldwork period, which determines how flexible the interviewer can be in timing the interview; the topic of the survey to determine the respondent's interest in the survey; and fieldwork strategies such as the number and timing of contact attempts, reminders, and the use of incentives (see Section 3.3.4.2).

- *Attributes of the interviewer.* Interviewers with low expectations, such as those who believe that the survey is uninteresting or contains sensitive questions, obtain a lower response rate. Expectations of the interviewer are determined by their sociodemographic attributes (e.g., race, age, and gender) and the match to those of the sample person. Expectations are also determined by their personal reaction to the survey topic and their prior experience. The more experienced interviewer is believed to dispose of a wider range of behaviors that have proved to be successful in approaching different persons.

3.2.3 RESPONDENT–INTERVIEWER INTERACTION

In the interaction between the interviewer and the sample person, the factors described above come together. Both the interviewer and the sample person bring behaviors, attitudes, and expectations to the interaction. The strategy that the interviewer employs to persuade the sample person is determined not only by the interviewer's own ability and expectations but also by features of the survey design and by characteristics of the immediate environment and broader society. Similarly the response that the sample person makes to the request is affected by a variety of factors.

The understanding of the interaction between the interviewer and the sample person, as well as their behavior, attitudes, and expectations are elements of a theory of survey participation. Groves et al. (1992) discuss several psychological concepts that are relevant to survey participation. They focus on the social psychology literature that describes the compliance with requests. Cialdini (1988) argues that in addition to the consideration of the attractiveness of a requested activity, there are social and psychological factors that influence the decision to agree or disagree to performing an activity. He provides six heuristic rules for compliance within society:

- *Reciprocation.* People tend to respond with similar behavior as being approached with. A sample person may be more willing to participate in a survey when the compliance is seen as repayment of a gift or favor. This principle underlies the positive effect of incentives on survey participation. Advance letters and brochures may also enhance compliance.

- *Consistency.* People have a strong tendency to be consistent within their attitudes, beliefs, words, and deeds. Appealing to this consistency may enhance survey participation. If the interviewer can make a connection between the attitude of the respondent and survey participation, for instance, the sharing of knowledge, this may enhance survey participation.
- *Social validation.* People like to validate their behavior by comparing it to the behavior of similar other persons. With respect to survey participation; persons will be more inclined to participate when they believe that similar persons do so too.
- *Authority.* A request that comes from (a representative of) a legitimate authority will be more likely to obtain participation.
- *Scarcity.* Scarce opportunities are perceived as more valuable. Thus, suggesting that participating in a survey is an opportunity that is limited ("only one in every 300,000 persons is contacted" or "we only have a couple of days left to do the interview") may encourage people to comply with the request.
- *Liking.* People favor persons they like. Obviously they will be more inclined to comply with the request of a person that they like. In case of survey participation, this may also be the organization the interviewer represents.

Besides these six heuristic rules for compliance with a request, compliance theory offers two social psychological research principles that provide useful other ways of understanding survey participation: *helping tendency* and *opinion change.*

Helping tendency with respect to a survey refers to a nonemergency situation in which low-level aid is explicitly requested. Most compliance principles can be applied here as well, but in addition there is a so-called "norm of social responsibility" or a helping norm. This norm may differ per culture. Appealing to this norm may successfully enhance participation, for instance, by ending the request with "it would really help us out." Also three emotional states of the respondent are found to be correlated with the helping decision: anger, happiness, and sadness. In general, it is expected that anger reduces compliance, happiness enhances it, and the effect of sadness depends on the perceived response cost to benefit ratio (e.g., a long survey with a small expected value for the respondent has a high ratio and will reduce compliance).

Research on *opinion change* shows that it is useful to distinguish between systematic and heuristic decision processes. When the topic is of great personal interest, the decision to change opinion is made based on a systematic review of intrinsic features (i.e., logic and evidence). However, when the topic is of small personal importance, the decision will be made based on a quick review of extrinsic features. The last consideration resembles the compliance principle, with factors such as liking the interviewer and the authority of the survey organization. The act of participation in a survey is rarely of such great personal importance that it makes sample persons want to systematically review all the available information for their participation decision. Most of the time the sample person processes the information heuristically. The decision to participate is then based on a small number of extrinsic features, such as the length of the survey or the charm of the interviewer.

3.2.4 TAILORING AND MAINTAINING INTERACTION

The discussion in the previous sections was based on social and psychological theory. However, to a large extent obtaining participation in a survey has a more practical basis. The interviewers may be experts in the field, but they also contribute to the theory behind survey participation by adding the last component: *tailoring* and *maintaining interaction*.

Tailoring essentially means adapting the approach to the sample person or unit. It involves continuously searching for attributes that may be related to one of the psychological heuristics noted above. For instance, the compliance principle that similarity engenders liking encourages interviewers to dress in a manner consistent with the neighborhood. Experienced interviewers usually have a large repertoire of phrases and cues they can employ to request for survey participation. What phrase or cue to use depends on observations of the housing unit and the neighborhood and also the reaction upon first contact. All features of communication are relevant, not just words but also intonation, pacing, volume, and physical movements.

Maintaining interaction aims at maximizing the probability of obtaining participation from each sample person or unit. Another goal might be to maximize the number of acceptances per time unit. The second objective is common to quota sampling, whereas the first goal is the implicit aim of probability sampling. With probability sampling, the interviewer can spend a lot of time on each sample person/unit and thus apply the tailoring-technique several times. Maintaining interaction also relies on a commitment principle: as the length of the conversation grows, it becomes more difficult to end it without violating the principle of reciprocation.

But there is another reason why maintaining interaction is so important: the longer a conversation holds, the more the behavior of the sample person can be observed. And this information can be used by the interviewer to adapt his or her approach, namely to apply the tailoring technique. So tailoring and maintaining interaction are often used in combination.

3.3 Application

3.3.1 INTRODUCTION

In this section the measures used to reduce nonresponse are discussed. In the previous section three possible reasons why a person could be a nonrespondent were explored: language problems, noncontact, or refusal. For the methods to reduce nonresponse, these same three causes of nonresponse are followed. Luiten (2009) gives a detailed overview of measures to reduce nonresponse. She distinguishes between noncontact and refusals, as well as between interviewer-assisted and self-administered data collection modes.

3.3.2 LANGUAGE PROBLEMS

Nonresponse due to language problems can be reduced by employing multilingual interviewers and/or translated questionnaires. But there are practical limitations to these solutions. Feskens (2005) conducted an international comparison of methods used to reduce nonresponse among nonnatives. Information was collected from six National Statistical Institutes (The Netherlands, Belgium, Germany, France, United Kingdom, and Sweden) and several semigovernment agencies in Germany and Belgium (because the NSIs of Belgium and Germany could not provide the requested data). The nonresponse due to language problems is mostly seen among nonnatives.

It turned out that none of these institutes use multilingual interviewers or translated questionnaires because the reduction in nonresponse is considered too costly for the number of additional responses. In addition these measures raise the risk of measurement errors. A cheaper way of dealing with language problems is having a family member act as interpreter, as they do in the United Kingdom and Sweden.

3.3.3 NONCONTACT

As was mentioned in Section 3.2.1, noncontact is caused by characteristics of the fieldwork and the at-home patterns of sample persons. To reduce the noncontact rate, and be successful, the measures must tailor the fieldwork to the contactability of the selected persons.

Groves and Couper (1998) analyze "contact" from these two perspectives: characteristics of the household that are correlated with contact, and the number and timing of calls/visits. They also look at the role of the interviewer but conclude that there are no significant differences among interviewers with respect to establishing contact.

In accord with these two perspectives of noncontact, there are several measures that can be taken to reduce the number of noncontacts. The most obvious is to *increase the number of contact attempts*. There is a strong correlation between number of contact attempts and percentage of contacted sample elements. Increasing the number of contact attempts increases the probability of finding someone at home by providing the interviewer with more opportunities to discover the at-home pattern of households. Fieldwork organizations usually prescribe a minimum number of contact attempts that have to be made before the sample element can be considered as a noncontact. In general, this number is lower for face-to-face surveys than for telephone surveys, due to the higher costs associated with house visits. Then again, after a certain number of attempts, the increased costs of contact attempts no longer outweigh the increase in response.

With respect to the *timing of calls/visits*, it has been shown that evening calls are much more profitable than calls during the daytime. For various reasons, however, the percentage of calls that is made during the evening is smaller than during daytime. Of course, evenings are shorter than days. In a telephone survey where all calls are distributed to the interviewers by a central calling system, it is logical that the number of evening calls is smaller than the number of daytime calls. In a face-to-face survey, the interviewer makes the decision when to visit a household. A possible reason for fewer evening calls may be that interviewers feel less safe during evenings and therefore prefer daytime visits.

Extending the length of the fieldwork period also increases the probability of finding someone at home. Obviously the longer the fieldwork period, the higher is the number of contact attempts that can be made. Then again, extending the fieldwork period can also reduce noncontact related to the at-home pattern of the household, for instance, during a long holiday.

Another consideration may be to *lower the workload of the interviewers*. This affects the number of calls and the ability to choose times of day, as well as the attention the interviewer can give to hard-to-contact persons. A workload that is too high will evolve in a higher non-contact rate, as well as a higher refusal rate because there is less time to make an appointment in case the sample element has no time when first contacted.

Interviewers could even be allowed to *make contact in a different mode* than the original survey in order to increase the number of contacts. Especially with CAPI surveys, it is cost efficient to make contact attempts in a less expensive mode because this can save travel costs. Contact attempts by telephone have the advantage that they can be made frequently at low cost, and also more contact attempts can be made in the evening hours when the at-home probability is highest. The disadvantage of contact attempts by telephone is that persons can more easily refuse participation. Other combinations of modes may also be profitable, for instance, a telephone reminder for a Web survey or a request for a telephone number sent by post; see also Chapter 4.

Sending a reminder is crucial for self-administered surveys, i.e., postal and Web surveys. Dillman (2000) shows that not sending a reminder resulted in a 20% to 40% decrease in response. Reminders serve the purpose of conveying the message that the survey organization values the responder's participation in the survey. This includes leaving a voicemail in case of noncontact in a telephone survey, or leaving a card behind in case of no contact in a personal survey. The social exchange theory hypothesizes that this will increase response because the survey organization has made an obvious effort that should be repaid by participating in the survey.

Background information from the sample elements can be used to make informed contact attempts that maximize the probability of contact. Sociodemographic characteristics such as the number of persons in the household, the presence of children and the composition of the household (single, couple, with/without children) are an indication of the at-home pattern. Households with children tend to have someone at home to take care of young children, elderly people are at home more often because of retirement, and a person that lives alone is at home less often. The employment status of adults is also a good indicator of the at-home pattern. Interviewers can use this type of information to plan their visits. This will mostly result in a more efficient contact strategy. Groves and Couper (1998) distinguish the following relevant information:

- The number of adults in the household (the more adults, the higher the probability of contact).
- Persons aged 70 + present (increased probability of contact).
- Children under 6 years present (increased probability of contact).
- Single household (decreased probability of contact).
- All members of the household under age 30 (decreased probability of contact).

EXAMPLE 3.3 Effect of a New Contact Strategy

Statistics Netherlands changed its fieldwork strategy in March 2004. Before this date the interviewers had to make *at least* 3 contact attempts. This resulted in a variable number of attempts for every household, which could be as much as 18, and a large number of households with 3 contact attempts. Starting in March 2004, the strategy has been to make at most 6 contact attempts. Table 3.2 shows the nonresponse rates in the categories noncontact, refusal, and language problems for the 2004 Integrated Survey on Household Living Conditions.

TABLE 3.2 Nonresponse due to language problems

Period	Noncontact	Refusal	Language problems
January–February 2004	7 %	23 %	1 %
March–December 2004	5 %	22 %	1 %
Total 2004	5 %	23 %	1 %

As is clear, increasing the number of contact attempts has a positive effect on the noncontact rate. The other nonresponse categories show no increase. So there is no counteractive effect on the response rate.

Physical impediments are a characteristic of the household. Examples are a locked central entrance to an apartment building or an intercom system in a face-to-face survey, and answering machines or caller ID in a telephone survey. In a face-to-face survey the impediments can be observed prior to the first contact. Households that have some sort of impediment have a lower probability of contact on the first attempt, and more households with impediments remain uncontacted at the end of the fieldwork period. This could be circumscribed by the fieldwork strategy of making a telephone call instead of a house visit to establish the first contact.

The measures described above mostly apply to interviewer-assisted surveys, either by telephone or face-to-face. In self-administered surveys, like Web or pen-and-paper surveys, most of the foregoing measures can be applied too, but in addition some other factors are important. In Web surveys, the risk of dropout is a threat to the response rate. In addition it is less easy to control that the selected persons will actually log in to the Web survey.

Heerwegh (2005) describes a number of experiments to improve both the login ratio and the dropout rate in Web surveys. These are to personalize the letter of invitation; mentioning the duration of the survey in the letter; tailor the letter of invitation to the respondent; and send a reminder.

Personalizing the letter of invitation relates to the reciprocation principle. This measure has been shown to increase the login ratio but not the completion ratio. *Mentioning the duration of the survey*, as opposed to making some vague statement about the duration, has resulted in a higher login ratio but also in a higher dropout rate. Hence the final result was not changed. *Tailoring the letter of invitation to the respondent* has been implemented in the following ways:

- Including a section with Frequently Asked Questions (FAQ). Only a few respondents turned out to use this section. This measure hence did not result in a higher response.
- Including the statement that no knowledge of the subject was required to participate in the survey. This statement was included in a panel survey to be completed by panel respondents with a low level of education. It turned out that including the statement had no effect on the login ratio nor the dropout rate.
- Manipulating the login procedure by varying the number of access codes. With a semiautomatic login procedure the login ratio was higher, but the completion ratio was higher where the respondents had to give more than one access code. In addition there was an indication that sensitive questions were better answered with multiple access codes.

3.3.4 REFUSALS

Section 3.2.1 introduced some factors that can interest persons to participate in a survey. Here measures aimed at obtaining participation, and reducing the refusal rate, are explored more fully. These are: advance letters, incentives, interviewer training, mixed mode data collection, and proxy interviewing.

3.3.4.1 Advance Letters Sending an *advance letter* to alert sampled persons to on upcoming survey is a polite thing to do. When they are finally approached to participate in the survey, they are already aware of the nature of the request, unlike unannounced calls from a salesperson. Interviewers find that comforting because they do not have to make a lot of effort to explain who they are and what they want.

Advance letters that present information about the survey, can be a powerful tool in reducing nonresponse. Several compliance principles underlie the choice of wording and content of the letter. First of all, a letter is perfect for communicating the authority of the request. For instance, official stationary from the organization (this only works if the organization is indeed legitimized in asking for information) could be used, with the letter signed by a person with authority inside the organization. Further, a personalized letter, with the exact name and address of the sampled person, suggests that the organization has gone to some trouble to send a personal letter, especially if signed by the director, and hence reciprocity would be in order. A brochure with additional information could be added to avoid making the letter so long that the recipient won't read it all. The additional information may also be perceived as a gift that people may want to repay by participating (again: reciprocation).

An explanation that the survey is based on a sample of a limited number of persons may stress that it is a scarce opportunity to participate and that it is important that the sampled person participate. By the scarcity principle, there is value added to the opportunity to participate in the survey. It is further important to point out that the survey results are strictly confidential so to dispel the fear of circulation of private information. Last, information about the approximate duration of the interview and expected use of the survey could be offered to educate the sampled person about the request.

3.3.4.2 Incentives An *incentive*, or a small gift, is most frequently presented to prospective respondents, but incentives could also be used as a stimulus for interviewers. The incentive could be monetary, but also postage stamps, pencils, or even statistical information, are appropriate to accompany a request for participation.

The incentive could be used at different points in the fieldwork. A *prepaid incentive* is a gift included with the survey request, often with an advance letter. This sort of incentive is unconditional. A *promised incentive* is conditional on survey participation. Sampled persons are promised a reward for participating in the survey. Finally, incentives can be distributed to specific groups. For example, in a refusal conversion attempt where persons who initially refused participation are re-approached with an incentive. Groves and Couper (1998) present several experiments where incentives have successfully reduced nonresponse.

There are a number of hypotheses on how incentives work to enhance response. The first is based on the principle of reciprocation. The incentive is seen as a gift that should be repaid, and one way of repaying it is by complying with the survey request. Another is related to effort–cost consideration. For participating in the survey certain costs are made (time, effort), and the incentive can be seen as a payment. The implication then is that lower income households would be more willing to participate when an incentive is offered, as opposed to high income households that valuate their time at a higher price. This brings us to another important view of incentives. The incentive should, in general be unrelated to the topic of the survey. Analysis has shown that nonresponse is reduced in subgroups that do not respond because of an interest in the topic of the survey, or social obligations. In other words, to obtain a positive effect, the incentive should appeal to a group of people that otherwise would not have participated in the survey, thereby increasing the representativity of the response subset. However, use of incentives can result in a less balanced composition of the response.

Singer (2002) conducted a meta-analysis of incentive effects. She found that incentives lead to a higher response rate, and higher incentives lead to a higher response as well, but that relationship is not linear. It becomes weaker with the height of the incentive and can even become negative if the respondents feel that the incentive is too high compared to the response burden.

EXAMPLE 3.4 Some Effects of Incentives

Groves et al. (2000) studied the different effects of incentives on survey response, depending on the degree of involvement in the community. Highly involved persons in the community are more inclined to participate in a survey. Incentives also increase the probability of participation. However, the leverage-salience theory stipulates that when these correlates are combined, the positive effect of the incentive on survey participation will be smaller for persons with a high community involvement.

So paying an incentive may alter the composition of the response. In some cases this will lead to a more balanced composition of the response, but it may also lead to a less balanced response. Another conclusion of the analysis is that prepaid incentives have a higher effect on survey participation than promised incentives.

More recently Barón et al. (2009) consider the effect of a promised incentive in a telephone survey based on the socioeconomic status of the respondents. They find that a promised incentive improves response rates primarily among those sample elements with a long history of income support. The persons who participate because of the promised incentive would not have participated otherwise, and hence the composition of the response is improved by offering an incentive. They conclude that the incentive both increases the response and reduces nonresponse bias by equalizing response rates across different socioeconomic groups.

3.3.4.3 Interviewer Training Interviewer training is an important activity. Interviewers need to be made aware of their interactions with persons being interviewed. Training should focus on minimizing refusal. For instance, when a person seems reluctant to cooperate, the interviewer should not push toward a decision but instead retreat and return another time.

Tailoring and maintaining interaction, as discussed in Section 3.2.1.3, are two such important activities interviewers employ to increase the probability of obtaining participation. Interviewer training should provide the interviewer with a set of phrases and cues, as well as ways to observe the respondents' behaviors so that they can apply the appropriate phrases and cues to obtain participation, and hence minimize refusal.

Besides fieldwork, management should continually monitor the achievements of the interviewers to see whether more training is necessary. At Statistics Netherlands, it is customary for a Response Analysis Survey (RAS) to be performed. This survey is used on a random sample of respondents to inform about the course of an interview and study the interaction between the interviewer and the respondent. The information is then used to monitor the interviewers and evaluate the fieldwork.

Good interviewer training should cover the following topics (Luiten, 2009):

- *Administrative issues.* This involves writing down the number of hours worked, receiving and sending interview material, and communication with the fieldwork supervisors.
- *Identifying sample elements.* Interviewers should be instructed on how to identify the people that they are assigned to interview. Also they should be instructed how to correct errors in the information about the sample and how to report the number and result of each contact attempt. A good instruction should lead to all interviewers recording this information the same way, thereby increasing the quality of information on the data collection process.
- *Interviewing techniques—general.* This involves instructions on how to read the questions, note the answers, and probe for more exact answers without introducing interviewer effects.
- *Interviewing techniques—specific.* Interviewers should be instructed about the key items in the questionnaire, the reason why some questions are asked, the requirements for the answers, how the questionnaire is constructed, and how to act in difficult situations.
- *Obtaining participation.* Both general and survey-specific instructions on tailoring and maintaining interaction should be given. Survey-specific instructions include the purpose of the survey and the expected answers to certain questions. The privacy issue should be discussed as well. Sometimes a role game is used to teach interviewers how to handle reluctant respondents.

3.3.4.4 Mixed Mode Data Collection Each individual data collection mode has its shortcomings and its benefits. Mixing data collection modes provides an opportunity to compensate for the weaknesses of each mode. Mixed mode data collection is discussed in more detail in Chapter 4.

Mixing modes can reduce survey costs and at the same time increase response. It may even be possible to reduce the selectivity of the response beforehand. For this purpose sampled persons or households can be allocated to a specific mode based on known background characteristics. If a specific group in the population does not participate in one mode, these persons may be willing to participate in another mode. This may reduce the selectivity of the response.

The usual survey research modes are CAPI, CATI, mail, and Web surveys. *CAPI* stands for computer-assisted personal interviewing. The interviewer visits the respondents at their homes and administers the computerized questionnaire from a laptop. CAPI surveys allow for a long, complex routing structure of the questionnaire because the routes are controlled by the computer. This method allows for data to be checked and corrected during the interview. Besides that, more complex surveys can be conducted with the assistance of the interviewer, who can explain the questions and probe for accurate answers. Furthermore the interviewer can apply strategies that gain cooperation from the respondents, for example, by employing the leverage-salience theory described in Section 3.2.1. Visual aids can be used when needed, for example, show cards.

In general, CAPI surveys have a high response rate because of the face-to-face interaction between the interviewer and the respondent. The interviewer can build rapport and confidence while conversing with the respondent. CAPI surveys have a high coverage of the population. Hence, CAPI surveys have manifold advantages and therefore, in general, the highest data quality. However, there are also disadvantages associated with CAPI. These disadvantages are caused mainly by the assistance of the interviewers. First, there are high costs associated with interviewers travelling to respondents. Furthermore some types of questions are sensitive to a social desirability bias caused by the interviewer's presence. Additionally, the interviewer behavior may lead to interviewer effects, often referred to as interviewer variance. Interviewers are usually assigned a cluster of sample elements. Because interviewers have their own way of acting and reacting, a cluster effect, or interviewer variance, may result. To prevent interviewer variance, interviewer procedures must be standardized.

CATI stands for computer-assisted telephone interviewing. The same as in CAPI, the questionnaire is computerized, but in CATI it is administered over the telephone. The fieldwork is conducted from a central call center, where a call management system is used to distribute telephone numbers to the interviewers present. This allows for an optimal exploitation of the sample, in contrast to CAPI where there is usually a higher number of unprocessed cases due to interviewer illness or vacation time. In CAPI it is more cumbersome to re-distribute addresses to other interviewers in the event of interviewer illness, whereas with a centralized CATI system the call management system keeps track of all sample elements and can therefore easily re-distribute sample elements. Technologies that facilitate decentralized CATI surveys allow for interviewers to perform CATI surveys from home, as is the case in Sweden since the mid-1980s.

Because of the interviewer assistance and the computerized questionnaire, CATI is very similar to CAPI. However, there are some important differences between the two modes. First of all, CATI is less prone to interviewer variance and social desirability bias because of the less personal communication channel. Some research seems to indicate that respondents are more inclined to answer sensitive questions in CATI (than in CAPI). Also CATI surveys are easier to set up and hence can be performed more quickly than CAPI surveys. They are less expensive because no travel costs are involved and less time is needed for an interview.

Compared to CAPI, CATI is less flexible in that the questions cannot be very complicated, no visual aids can be used, and the optimal duration of an interview is shorter than in CAPI surveys (approximately 30 minutes). The interviewers in CATI cannot easily build rapport with the respondent. So less probing for accurate answers is possible, and less persuading power can be initiated to gain participation. Understandably, then, the response rates in CATI surveys are usually lower. Lastly, the coverage of CATI surveys is lower than CAPI surveys. Persons without telephones are not covered at all. Also, in case listed phone numbers are used by a linkage to the sample frame, unlisted number are not covered. Random digit dialing (RDD) is a solution to this form of undercoverage, but RDD lacks the possibility to link registration and frame data.

In a *mail survey*, a paper questionnaire is sent to a sample person or household. The questions are answered without interviewer assistance and sent back by mail. So mail surveys are self-administered. The main difference from CAPI and CATI is the absence of an interviewer. Also mail questionnaires are not computerized. In a mail survey the adequacy of responses depends heavily on the questionnaire's design. Then again, compared to CAPI and CATI surveys, mail surveys are inexpensive. Furthermore there is a reduced risk of social desirability bias. Compared to CATI surveys, yet some other advantages of mail surveys are the possibility to use visual aids, the respondents can answer the questions whenever they like, and they have more time to think about the questions and give more thoughtful answers.

Like CAPI surveys, mail surveys have a high coverage of the population. However, the response rate to a mail survey is lower than for interviewer-assisted surveys. The questionnaire cannot be too complicated and there is a larger risk of item nonresponse. Additionally mail surveys require a longer fieldwork period and longer processing time because of the data entry and editing upon their return. There is no control over the response process, and this complicates the calculation of response rates and the adjustments for nonresponse. The type of response is unknown, it is only observed whether or not the completed questionnaires are returned.

In a *Web survey* the respondent is invited to a Web page where the questionnaire is found. There is no interviewer present. Web surveys are self-administered surveys like mail surveys, thereby sharing the advantages of a lower risk of social desirability bias and lower costs. Likewise among the disadvantages of self-administered surveys there is the problem of no interviewer assistance to gain cooperation and to probe for more precise answers. Compared to mail surveys, Web surveys have more questionnaire design choices. Also the data collection process is entirely computerized, so routing and data editing during fieldwork are reduced as is the risk of measurement errors. Therefore Web surveys allow for more complex questionnaires to be used. Still, compared to mail surveys, Web surveys have a lower coverage. The part of the population that has no access to the Internet cannot be reached. Another issue in Web surveys concerns the confidentiality and security of sending information over the Internet. Finally, the risk of dropouts is high because there is no interviewer present to ensure participation.

CAPI, CATI, and Web surveys offer some of the same advantages of a computerized questionnaire. CATI and CAPI surveys have the advantage of interviewer assistance. Mail and Web surveys are cheaper, and do not have the disadvantages of interviewer variance and social desirability bias caused by the interviewers. In self-administered data collection modes like mail and web surveys, the survey topic has an influence on response so the response tends to be more selective. De Leeuw (1992) found the largest mode difference to be due to differences in administration: interviewer-assisted versus self-administered.

Switching modes can be an effective way to increase the probability of contact, as well as the probability of participation. For persons who are not at home very often, but do have regular access to email, switching from a personal to a Web survey can increase the number of contacts. Likewise, for persons who do not like to invite strangers into their homes, switching from a personal to a telephone survey can decrease the number of refusals.

Nonresponse is often selective with respect to demographic characteristics like age and household composition, and socioeconomic characteristics like income, educational level, and employment situation. Different modes lead to different response levels and thus to different compositions of the respondent pools within modes. So nonresponse bias could be reduced by choosing an optimal mix of data collection modes. Roberts (2007) finds that compared to CAPI respondents, CATI respondents are higher educated, have a higher income, are younger, and less frequently nonnative. Kwak and Radler (2002) conclude that Web/mail respondents are usually better educated and more literate than CATI respondents. Web respondents more often are male and, in general, younger than mail respondents. Correspondingly, some years earlier, De Leeuw (1992) found that in mail surveys, there are more male and married respondents. In face-to-face surveys, there are more women and divorced or widowed respondents.

3.3.4.5 Proxy Respondents When the person to be interviewed is not available, some other member of the household could be allowed to answer questions on behalf of the selected person. This is called a *proxy respondent*, or *proxy interviewing*. The proxy respondent has to be a member of the core of the household.

Proxy interviewing decreases nonresponse and saves time and money. However, proxy interviewing can also increase measurement errors because the questions are answered by a different person than the intended respondent.

Machielsen (2009) performed a literature research into proxy response behavior regarding work-related questions. From his research it is clear that the quality of the proxy response depends on the type of survey. Highton (2005) found that in a survey on voting behavior the social desirability bias is smaller in proxy interviewing compared to self-interviewing. Proxy respondents have less reason to give socially desirable answers. Then again, respondents themselves know better whether they will vote or have voted. Schwartz and Wellens (1997) note that the measurement error in proxy responses is larger for questions about recent events than for events that happened in the long past.

Besides the type of questions, the quality of the proxy response depends on the relationship between the proxy respondent and the intended respondent, as noted by Becker et al. (2004). Often the mother is a good proxy respondent.

Thomsen and Villund (2008) conclude that proxy interviewing leads to a more balanced pool of respondents because some persons that are hard to reach or to persuade to participate in the survey may be included in the survey by a proxy interview. The result is a smaller nonresponse bias that, despite any measurement error, raises the quality of the data.

3.4 Summary

In this chapter, a number of explanations are offered for the occurrence of three different types of nonresponses, these being noncontact, non-language-speaker problems, and outright refusals. Some ways to reduce these types of nonresponse in the field are discussed.

Inability to cooperate due to language problems is a category of nonresponse that mostly pertains to nonnative people. When it comes to actual contact, the survey literature agrees that the probability of making contact is strongly related to the probability that some member of the household is at home when the interviewer calls. Hence, noncontact is caused by the combination of fieldwork effort and the at-home pattern of the sampled persons. Physical impediments can make it hard to get an interview going.

Refusals are different from noncontacts and language problems. People have the freedom to refuse to cooperate after contact has been made and there are no language problems. When a person refuses to participate in the survey, this is a conscious decision. Theories on why this happens are found in the sociological and the psychological literature. Groves, Cialdini, and Couper (1992) consider several factors that can influence survey participation that relate to sociodemographic influences and survey design features as well as to the communicative style of the interviewer while interacting with the sample person, namely tailoring the conversation to maintain interaction.

Multilingual interviewers could be used where language is an issue, and/or questionnaires could be translated into the relevant language. Measures to reduce the noncontact rate could include increasing the number of contact attempts, lowering the workload of the interviewer or allowing the interviewer to make the contact in a different communications mode. Some measures that have proved effective to reduce refusal are sending an advance letter, giving a small gift as an incentive, training interviewers in refusal conversion, and allowing for proxy interviewing.

3.5 KEY TERMS

Advance letter A letter informing sampled persons about aspects of the upcoming survey.

Cognitive burden The burden of the processing of information related to the decision making while answering questions.

Incentive A small gift that is offered as a reward to sampled persons in order to gain their participation in a survey.

Maintaining interaction Keeping a level of conversation going with prospective respondents to maximize the probability of obtaining participation.

Mixed mode survey A survey conducted by using more than one data collection mode.

Prepaid incentive A small gift that is presented at the time of survey request in order to encourage the person's participation in the survey.

Promised incentive A small gift that is promised to the sampled persons after they complete the survey.

Proxy respondent A member of the household who answers questions on behalf of the selected person. This type of interviewing is also referred to as proxy interviewing.

Social desirability bias A bias that occurs if respondents change their "internal" answer to an answer that they perceive to be more socially desirable.

Tailoring Adapting the approach for the survey request to the sample person.

3.6 REFERENCES

Barón, J.D., Breunig, R.V., Cobb-Clark, D., Gørgens, T., and Sartbayeva, A. (2009), Does the Effect of Incentive Payments on Survey Response Rates Differ by Income Support History? *Journal of Official Statistics* 25, pp. 483–507.

Becker, C., Rodkin, S., O'Connor, K., and Moorman, J. (2004), *How Accurate are Proxy Reports: Results of a Verification Study.* AAPOR, Section on Survey Research Methods

Biemer, P.P., and Lyberg, L.E. (2003), *Introduction to Survey Quality.* Wiley, New York.

Blau, P. (1964), *Exchange and Power in Social Life.* Wiley, New York.

Cialdini, R.B. (1988), *Influence: Science and Practice.*, Scott, Foresman, Glenview, IL.

Cobben, F., and Bethlehem, J.G. (2005), *Adjusting Undercoverage and Nonresponse Bias in Telephone Surveys.* Discussion paper 05006. Statistics Netherlands, Voorburg.

De Leeuw, E.D. (1992). Data Quality in Mail, Telephone, and Face to Face Surveys. PhD thesis. T.T.-publikaties, Amsterdam.

Dillman, D.A. (1978), *Mail and Telephone Surveys: The Total Design Method.* Wiley, New York.

Dillman, D.A. (2000), *Mail and Internet Surveys: The Total Design Method.* Wiley, New York.

Feskens, R., Hox, J., Lensvelt-Mulders, G., and Schmeets, H. (2006), Collecting Data among Ethnic Minorities in an International Perspective. *Field Methods* 18, pp. 284–304.

Goyder, J. (1987), *The Silent Minority: Nonrespondents in Sample Surveys.* Polity Press, Cambridge, UK.

Groves, R.M. (1989), *Survey Errors and Survey Costs.* Wiley, New York.

Groves, R.M., Cialdini, R.B., and Couper, M.P. (1992), Understanding the Decision to Participate in a Survey. *Public Opinion Quarterly* 56, pp. 475–495.

Groves, R.M., and Couper, M.P. (1998), *Nonresponse in Household Interview Surveys.* Wiley, New York.

Groves, R.M., Singer, E., and Corning, A. (2000), A Leverage-Salience Theory of Survey Participation. *Public Opinion Quarterly* 64, pp. 299–308.

Heerwegh, D. (2005), Effects of Personal Salutations in E-mail Invitations to Participate in a Web Survey. *Public Opinion Quarterly* 69, pp. 588–598.

Highton, B. (2005), Self-reported versus Proxy-Reported Voter Turnout in the Current Population Survey. *Public Opinion Quarterly* 69, pp. 113–123.

Kwak, N.& Radler, B. (2002), A Comparison between Mail and Web Surveys: Response Pattern, Respondent Profile, and Data Quality. *Journal of Official Statistics* 18, pp. 257–273.

Machielsen, R. (2009), *Literatuurstudie proxy interviewen.* Research paper BPA DMV-2009-09-24-RMCN. Statistics Netherlands, The Hague.

Luiten, A. (2009), Benaderingsstrategien: Responsverhogende Maatregelen bij Sociale Statistieken. *Methodenreeks,* Statistics Netherlands, The Hague/Heerlen.

Roberts, C. (2007), *Mixing Modes of Data Collection in Surveys: A Methodological Review,* NCRM Methods review paper 008. ESRC National Centre for Research Methods, Swindon, UK.

Roose, H., Lievens, J., and Waege, H. (2007), The Joint Effect of Topic Interest and Follow-up Procedures on the Response in a Mail Questionnaire: An Empirical Test of the Leverage-Salience Theory in Audience Research. *Sociological Methods and Research*. 35, pp. 410–428.

Thomsen, I., and Villund, O. (2008), *Using Administrative Registers to Evaluate the Effects of Proxy Interviewing in the Norwegian Labor Force Survey.* Working paper. Statistics Norway, Olso.

Singer, E. (2002), The Use of Incentives to Reduce Non Response in Household Surveys. In: Groves, R.M., Dillman, D.A., Eltinge, J., and Little, R. (eds.), *Survey Nonresponse*, Wiley, New York, pp. 163–177.

Schwartz, N., and Wellens, T. (1997), Cognitive Dynamics of Proxy Responding: The Diverging Perspectives of Actors and Observers. *Journal of Official Statistics* 13, pp. 159–179.

3.7 EXERCISES

Exercise 3.1 Which statement about incentives is not correct?

(a) It is sometimes accepted as a gift that should by repaid by participating in the survey.

(b) It is sometimes accepted as payment for time and money spend on completing the questionnaire.

(c) It often increases the representativity of the response.

(d) Promised incentives work better than prepaid incentives.

Exercise 3.2 Which information should not be included in an advance letter?

(a) Signature of a person with authority inside the survey organization.

(b) Name and address of the sampled person.

(c) Name and phone number of the interviewer.

(d) A statement about the confidentiality of the survey results.

Exercise 3.3 Which is the worst time for making calls or visits?

(a) Evening

(b) Weekend

(c) Workdays between 9 am and 4 pm

(d) Saturday morning

Exercise 3.4 When is participation in a survey less likely?

(a) If compliance is seen as a gift of favor.

(b) If one believes similar persons will do so too.

(c) If the request comes from a nonlegitimate authority.

(d) If one likes the interviewer or the organization behind the interviewer.

Exercise 3.5 Which factor improves survey participation?

(a) Increased length of the interview.

(b) A shorter fieldwork period.

(c) Interviewers with more experience.

(d) As few channels of communication as possible.

Exercise 3.6 Which statement about proxy interviewing is not correct?

(a) It decreases nonresponse and saves time and money.

(b) It increases measurement errors.

(c) The quality of the proxy response depends on the type of survey.

(d) The mother is not a good proxy respondent.

Chapter Four

Nonresponse and the Mode of Data Collection

4.1 Introduction

4.1.1 THE EARLY HISTORY

The way surveys are conducted has changed over time. Changes were caused by developments in society. For example, survey sampling emerged in a period of industrialization, followed by urbanization. Emerging social movements and more centralized government created new demands for statistical information. More recent changes in survey taking are caused by the rapid developments in computer technology. This section gives a historical overview.

All these developments have led to various forms of data collection: face-to-face surveys, telephone surveys, mail surveys, CAPI, CATI, CASI, Web surveys, and mixed mode surveys. They all have a specific impact on the size and structure of the nonresponse problem. This chapter describes the several modes of data collection and the effect they have on nonresponse. A clear conclusion is that quality comes at a price: higher response rates require more expensive forms of data collection.

The idea of collecting data for compiling statistical overviews is very old. As far back as biblical Babylon censuses of agriculture were taken. Ancient China counted its people to determine the revenues and the military strength of its provinces. There is also evidence of statistical overviews compiled by Egyptian rulers long

Handbook of Nonresponse in Household Surveys, By Jelke Bethlehem, Fannie Cobben, and Barry Schouten

before the birth of Christ. All these overviews were based on complete enumeration of the population (censuses). The idea of sampling had not yet emerged.

The first recorded census took place in the fifth century B.C. in Rome. Under the rule of Servius Tullius, the sixth king of Rome, the population was counted and divided into classes according to wealth. During the reign of Caesar Augustus, the census was taken every five years. Families were to travel to the city of the husband's heritage for the purposes of taxation. This was the census that called Joseph and Mary to Bethlehem. The data were used to establish the political status of citizens and to assess their military and tax obligations to the state.

The first recorded reference to a census in the United Kingdom is that of the seventh century "Senchus fer n'Alba" in Gaelic Scotland, so translated as "tradition/census of the men of Alba." The original document is lost, but it was transcribed in the tenth century, partly as a genealogical record, partly as an inventory of the territories of the descendents of the Scottish king Eochaid Muinremor. It consists of a list of the numbers of men that the various families of Dalriada (the western part of modern Scotland) could provide for the navy.

Censuses were rare in the Middle Ages. The most famous was the census of England taken by order of William the Conqueror, king of England. The compilation of this Domesday Book started in the year 1086. The book records a wealth of information about every manor and village in the country.

Another interesting example can be found in the Inca empire that existed between 1000 and 1500 in South America. Each Inca tribe had its own statistician, called the Quipucamayoc. This man kept records of, for example, the number of people, the number of houses, the number of llamas, the number of marriages, and the number of young men that could be recruited for the army. All these facts were recorded on a quipu, a system of knots in colored ropes.

4.1.2 THE RISE OF SAMPLING

The idea of using sampling instead of a complete enumeration came up around the year 1895. In that year, Anders Kiaer (1895, 1997), the founder and first director of Statistics Norway, published his *Representative Method.* He proposed questioning only a (large) sample of persons who together formed a "miniature" of the population. Anders Kiaer stressed the importance of representativity. His argument was that if a sample is representative with respect to variables for which the population distribution is known, it is also representative with respect to other survey variables.

A basic problem of the representative method was that there was no way of establishing the accuracy of estimates. The method lacked a formal theory of inference. It was Bowley (1906, 1926), who took the first steps in this direction. He showed that for large samples, selected at random from the population with equal probabilities, estimators had an approximately normal distribution.

From this moment on, there were two methods of sample selection. The first one was Kiaer's representative method, based on purposive selection where representativity played a crucial role but no measure of the accuracy of estimates could be obtained. The second was Bowley's approach, based on simple random sampling for which an indication of the accuracy of estimates could be computed. Both methods existed side by side for a number of years. This situation lasted until 1934, in which year the Polish mathematician and statistician Jerzy Neyman published his now famous paper; see Neyman (1934). Neyman introduced a theory based on the concept of the confidence interval. By using random selection instead of purposive selection, there was no need any more to make prior assumptions about the population. Neyman also showed that the representative method based on purposive sampling failed to provide satisfactory estimates of population characteristics. As a result the method of purposive sampling fell into disrepute in official statistics.

The principles of probability sampling developed by people like Kiaer, Bowley, and Neyman form the basis for modern survey taking. They are vital for making valid inferences about the population being investigated. Where samples are not based on probability sampling, it is not possible to compute unbiased estimates, and it is also not possible to quantify margins of error. These principles have been successfully applied in official and academic statistics since the 1940s.

Application of probability sampling in more commercial market research has been much less widespread. Some forms of nonprobability sampling were and are still popular. One such form is quota sampling.

Quota sampling is a sampling technique where elements are selected in such a way that the distribution of some auxiliary variables matches the population distribution of these variables. For example, a quota sample could be selected such that the sample distribution by gender, age, and socioeconomic class reflects the population distribution of these three variables. Consequently the sample distribution is equal to the population distribution with respect to gender, age, and social class.

Quota sampling starts by choosing the total sample size. Then the sample size is distributed proportionally over the strata obtained by cross-classifying the auxiliary variables. This results in the quota. It is the task of the interviewers to locate persons and establish to which strata they belong. The fieldwork continues until the right quota in all strata are obtained.

A quota sample is much simpler to realize than a probability sample. It does not need a sampling frame and any complex sample selection procedure. Interviewers can just go out and interview people they happen to come across until all quota are filled.

The problem of quota sampling is that it is not a form of proper probability sampling. Selection probabilities are unknown, and they may be equal to 0 for some members of the population. Therefore probability theory cannot be applied. It is not possible to construct unbiased estimators and consequently the accuracy of estimators is also unclear.

Quota sampling produces a sample that is representative with respect to the auxiliary variables used to determine the quota. There is no guarantee at all that the sample is also representative with respect to the other variables measured in the survey.

EXAMPLE 4.1 A Commuter Survey

For a commuter survey a quota sample is to be selected. The quota are based on gender and age. The table below contains the (relative) population distribution of these two variables:

Age class	Male	Female
15–30 years	13.6%	13.2%
30–50 years	22.2%	22.1%
50–65 years	14,3%	14.6%

A quota sample of size 500 must contain $0.136 \times 500 = 68$ young males, $0.222 \times 500 = 111$ middle aged males, and so forth.

Interviewers could go to a railway station and collect their quota there. The result will be a sample that is representative with respect to gender and age, but certainly not with respect to commuting behavior. Car drivers will be seriously underrepresented in the sample.

Quota sample estimates can be substantially biased. The bias can come from two sources: (1) search behavior of interviewers, and (2) a high (generally unrecorded) number of nonresponses. It is normally left to the interviewers to follow their own procedures for finding respondents who meet the quota requirements. There is a risk of interviewers avoiding certain types of people or locations. Personal preferences of interviewers may also lead to systematic differences between them.

EXAMPLE 4.2 George Gallup and Quota Sampling

George Gallup was one of the first to apply quota sampling. With his opinion poll he predicted the winner of the 1936 presidential election to be Franklin Roosevelt. Hundreds of interviewers all over the United States were given quotas for different types of respondents: so many middle-class urban women, so many lower-class rural men, and so on. With a sample of only 3,000 respondents he could make a better prediction than the *Literary Digest* poll. This magazine had a sample of 10 million respondents. They were recruited from vehicle registrations and telephone directories. The sample was not representative because at that time cars and telephones were more often owned by middle and upper classes, and they preferred the Republican Landon. Poorer people favored Roosevelt. As a result *Literary Digest* predicted Landon to be the winner. Gallup turned out to be right: Roosevelt was re elected.

Quota sampling failed to predict the winner in the 1948 elections. George Gallup incorrectly predicted the victory of New York Governor Thomas Dewey over President Harry Truman. Quota sampling turned out to be ineffective. As a result use of quota sampling was no longer considered a valid method for opinion polls. Starting in 1956, it was replaced by probability sampling.

It is often overlooked that there can be a considerable amount of nonresponse in surveys based on quota sampling. On the surface there may not seem to be a problem because all quota strata have been filled. But this disregards the fact that all cases of nonresponse are simply substituted by responding cases. This means that people who cannot be contacted or refuse to cooperate are simply excluded from the surveys. If the respondents differ from nonrespondents, this will result in biased estimators.

The (hidden) nonresponse rate can be high in quota surveys. Typically no pre-notification letters are sent to announce the visit of the interviewers. Also no callback attempts are made to get the nonrespondents in the sample.

Therefore it is not possible to compute a proper response rate. For this, an initial sample size and a realized sample size are required. In quota sampling there is no initial sample size, only a final sample size. Of course, attempts of interviewers may fail because people refuse or are not at home. This is, however, often not recorded as nonresponse. Consequently the impression is created that there was no nonresponse.

4.1.3 THE IMPACT OF COMPUTER TECHNOLOGY

The instruments used for data collection have changed over the years. Until the 1970s paper questionnaires were used in face-to-face surveys, telephone surveys, and mail surveys. The rapid developments in computer technology changed all that.

Collecting high-quality survey data can be a complex, costly, and time-consuming process. One of the problems of traditional data collection with paper questionnaires was that the answers could contain errors. These errors must be detected and corrected. Cleaning questionnaire forms requires substantial resources. Rapid developments of information technology since the 1970s have made it possible to use microcomputers for data collection. Thus *computer-assisted interviewing* (CAI) was born. The paper questionnaire was replaced by a computer program containing the questions to be asked. The computer took control of the interviewing process, and it also checked answers to questions on the spot.

Like traditional interviewing, computer-assisted interviewing has different modes of data collection. Historically the first mode of data collection was *computer-assisted telephone interviewing* (CATI). Couper and Nicholls (1998) describe how the technology was developed in the United States in the early 1970s. The first nationwide telephone facility for surveys was conducted already in

1966. The idea was to automate and simplify case management, and not to design a new mode of data collection. These early systems evolved in subsequent years into full-featured CATI systems. Particularly in the United States, there was a rapid growth of the use of these systems. CATI systems were little used in Europe until the early 1980s.

The interview software controls the work of interviewers in a CATI survey. This software decides when and who to contact. If contact is established by telephone, and the person is prepared to participate in the survey, the interviewer starts the interviewing program. The first question appears on the screen. If this is answered, and no error is detected, the software proceeds to the next question on the route through the questionnaire.

Many CATI systems have facilities for call management. The main function of call management is to offer the right telephone number at the right moment to the right interviewer. This is particularly important for cases where the interviewer has made an appointment with a respondent for a specific time and date. Such a call management system also has facilities to deal with special situations like busy numbers (try again after a short while) or no answers (try again later). It should also help to increase the response as much as possible.

More about the use of CATI in the United States can be found in Nicholls and Groves (1986). De Bie et al. (1989) give an overview of the available software in the early stages of development.

The emergence of small portable computers in the 1980s made *computer-assisted personal interviewing* (CAPI) possible. It is face-to-face interviewing where interviewers take their laptop computer to the homes of the respondents. There they start the interview program and enter answers to the questions that appear on the screen.

Statistics Netherlands was one of the first in official statistics starting experiments with this mode of data collection. Laptop computers were tested in a survey on prices in 1984. This is a survey where interviewers visit shops and record prices of products. The results are used to compute the consumer price index. It turned out the interviewers were able to handle the hardware and software. Moreover respondents (shopkeepers) did not object to this kind of data collection.

The consumer price index experiment gave insight as to the conditions laptop computers have to satisfy in order to be useful for data collection. First, they cannot be too heavy. A weight of 3 kilograms was considered the maximum (often female) interviewers could handle. Second, the readability of the screen should always be sufficient, even in bad conditions, like a sunny room. Third, battery capacity should be sufficient to allow for a day of interviewing without recharging. And if this was not possible, interviewers should have spare batteries that can be replaced easily and quickly. Finally, the interviewers preferred a full-size keyboard. Small keys were considered too cumbersome and too error-prone.

In a second experiment, the laptops were tested in interviews held in the homes of respondents. The idea behind this experiment was to test whether respondents accepted this type of data collection. It turned out that use of laptops did not have any effect on response rates. Respondents simply accepted it as a form of progress in survey taking. At that time there was some concern in society about "big brother" effects. Use of computers might cause anxiety among respondents that they might have become part of a large government operation to collect large amounts of data about people, and that therefore their privacy was at stake. However, no such "big brother" effects could be observed. Another conclusion was that interviewers very rapidly became accustomed to using the new technology for their work.

After the success of these experiments, Statistics Netherlands introduced CAPI in its regular surveys. The Dutch Labor Force Survey became a CAPI survey in 1987. Interviewers were equipped with laptop computers. After a day of interviewing, they returned home, and connected their computer to the power supply to recharge the batteries. They also connected their laptop to a telephone and modem. At night, the computers automatically called Statistics Netherlands and uploaded the collected data. New address data were downloaded in the same session. In the morning, the computer was ready for a new day of interviewing. Respondents accepted computer-assisted interviewing as a mode of survey data collection. There was no increase in nonresponse rates.

More on the early years of CAPI at Statistics Netherlands can be found in CBS (1987) and Bethlehem and Hofman (2006). More information on CAPI in general can be found in Couper et al. (1998).

The computer-assisted mode of mail interviewing also emerged in the 1980s. It is called *computer-assisted self-interviewing* (CASI), or sometimes also *computer-assisted self-administered questionnaires* (CASAQ). The electronic questionnaire is send to the respondents. They run the software on their own computer, answer the questions, and send them back to the survey agency. Early CASI applications used diskettes or a telephone and modem to send the questionnaire, but nowadays it is common practice to download it from the Internet. The answers are returned electronically in the same fashion.

An early application in social surveys was the *Telepanel* (see Saris, 1998). The Telepanel was founded in 1986. It was a panel of 2000 households that agreed to regularly complete questionnaires with the computer equipment provided to them by the survey organization. A home computer was installed in each household. It was connected to the telephone with a modem. It was also connected to the television in the household so that it could be used as a monitor. After a diskette was inserted into the home computer, it automatically established a connection with the survey agency to exchange information (downloading a new questionnaire or uploading answers of the current questionnaires). Panel members had agreed to complete a questionnaire each weekend.

The rapid development of the Internet in 1990s has led to a new mode of data collection. Some call it *computer-assisted Web interviewing* (CAWI). The questionnaire is offered to the respondents through the Internet. Therefore such a survey is sometimes also called a *Web survey* or *online survey.* In fact such an online survey is a special type of CASI survey. At first sight, Web surveys have a number of attractive properties. Now that so many people are connected to the Internet, it is a simple matter to get access to a large group of potential respondents. Furthermore questionnaires can be distributed at very low costs. No interviewers are needed, and there are no mailing and printing costs. Finally, surveys can be launched very quickly. Little time is lost between the moment the questionnaire is ready and the start of the fieldwork. Consequently it is a cheap and fast way to access a large group of people.

However, online surveys also have some serious drawbacks. One is undercoverage (not everyone has access to Internet). Another is often that no proper sampling frames are available. Particularly market research companies therefore often rely on self-selection of respondents.

Application of computer-assisted interviewing for data collection has three major advantages. The first advantage is that it simplifies the work of interviewers. They no longer have to pay attention to choosing the correct route through the questionnaire. They can concentrate instead on asking questions, and assisting respondents in getting the answers.

The second advantage is that it improves the quality of the collected data, since answers can be checked and corrected during the interview. This is more effective than having to do the checking of answers afterwards in the survey agency.

The third advantage is that data are entered in the computer during the interview resulting in a clean record, so no more subsequent data entry and data editing are necessary. The time needed to process the survey data is greatly reduced, and so the timeliness of the survey results is improved.

EXAMPLE 4.3 The Blaise System

There are many software packages for computer-assisted interviewing. One of these packages is frequently used for data collection by national statistical institutes. The name of this package is Blaise. The first version of the Blaise System was developed in 1986 by Statistics Netherlands. The desire was to find an innovative alternative for the disadvantages of traditional data collection with paper questionnaires. Bethlehem (1997) gives more background information.

Questionnaires are defined in Blaise by a script language. This definition contains all possible questions, route instructions, checks to be carried out on the answers, and computations that may be needed in the course of the interview. The questionnaire definition is used by the Data Entry Program. This program can be tuned for various forms of data collection (CAPI, CATI, CASI, or CAWI) and also for data entry of paper forms. The figure below shows and example of a CAPI screen. The top part of the screen shows the current question to be answered. The bottom part gives an overview of part of the questionnaire. Some questions have already been answered, and others still have to be answered.

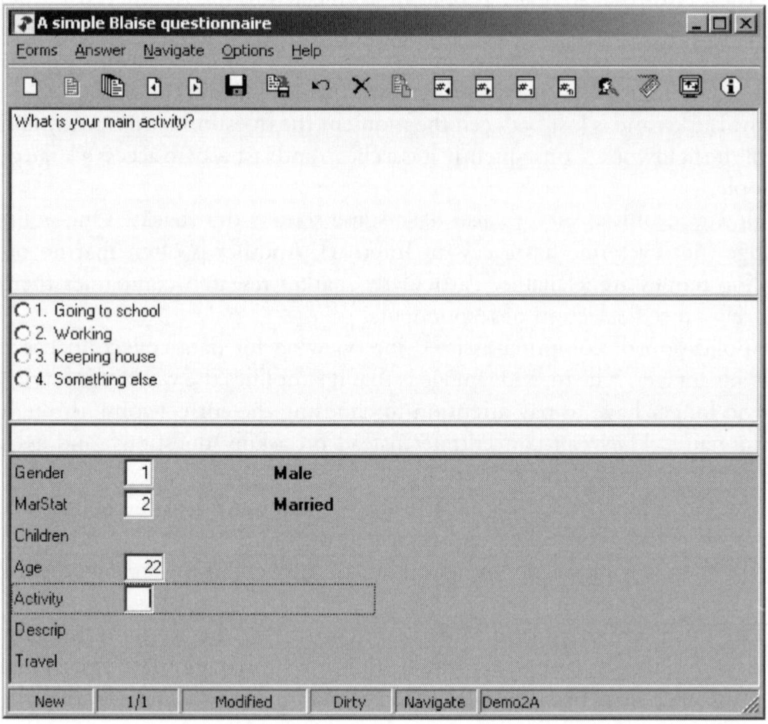

A strong point of systems like Blaise is that respondents are forced to follow a specified route through the questionnaire. It is not possible to inadvertently skip questions. This helps to reduce item nonresponse.

Another strong point is that checks on the answers can be built in. The figure below shows an example of a detected error. Such an error must be corrected before the next question will appear on the screen. At least one of the questions involved must be answered differently.

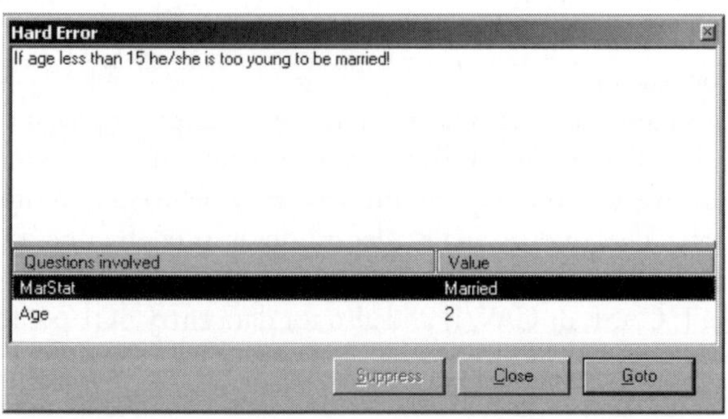

TABLE 4.1 Modes of data collection

Paper	Computer assisted
Face-to-face	CAPI
Telephone	CATI
Mail	CASI, CAWI

Table 4.1 gives an overview of the different modes of data collection introduced in this section. Each mode has its advantages and disadvantages. Each mode also has a different effect on size and composition of the nonresponse. Section 4.3 describes a project where various modes of data collection were compared in terms of response rates and costs.

4.2 Theory

4.2.1 FACE-TO-FACE SURVEYS

A face-to-face survey is a mode of data collection where interviewers visit the homes of the respondents, or another location convenient for the respondent. Together, the interviewer and the respondent complete the questionnaire.

An important aspect of face-to-face interviewing is that the interviewer and the selected person are together at the same location. This makes it easier for the interviewer to convince a person to participate. This is much more difficult to achieve by telephone or by sending written text. As a consequence response rates are higher for face-to-face surveys than for other types of surveys.

Goyder (1987) was among the first to compare response rates of different modes of data collection. He obtained response rates of 112 face-to-face surveys, 53 telephone surveys, and 385 mail surveys. These surveys were conducted in the United States and Canada in the period 1930 to 1980. They were mostly government surveys and academic surveys. The average response rate of face-to-face surveys was around 67%. The rate was lower for telephone surveys, 60%. The response was lowest for mail surveys, 58%. Goyder (1987) also observed that response rates for face-to-face surveys and telephone surveys decreased over time, while rates remained stable for mail surveys.

Similar conclusions were drawn by Hox and De Leeuw (1994). They compared response rates of 35 surveys in Europe, the United States, and Canada. They found that face-to-face surveys had the highest response rates (on average, 70%), followed by telephone surveys (67%), and mail surveys (61%). They also observed a trend of falling response rates for face-to-face and telephone surveys, while rates for mail surveys remained stable over time.

The main difference between face-to-face and telephone surveys, on the one hand, and mail surveys, on the other, is the presence or absence of interviewers. The presence of interviewers can have advantages. Interviewers can provide additional information about the surveys and the questions. Thus they can assist the respondents in answering the questions. This may help respondents to correctly interpret questions. So risk of item nonresponse is reduced, which has a positive impact on the quality of the collected data.

In a face-to-face survey the interviewer and the respondent are in a position where it is possible to use visual and graphical aids such as cards. Cards can be used to show a list of all possible answers to a question. This way respondents can read through the list and select the answer that reflects their situation or opinion best.

Groves and Couper (1998) demonstrate the importance of using additional information to make contact and get cooperation. For example, a sample selected from a sampling frame (e.g., a population register) can provide some additional information like composition of the household and ages of household members. Then the interviewer can choose an optimal approach strategy.

Another advantage of face-to-face interviewing is that the visit of the interviewer can be longer than a telephone call. It is not uncommon for respondents to allow interviewers to be in their homes for an hour or more. It is unlikely that a telephone call can take more than 30 minutes. If more time is available for an interview, more questions can be asked, and also the questionnaire can be made more complex.

The disadvantage of face-to-face surveys is that they cost much more than telephone surveys or self-interview surveys. Interviewers have to be employed all over the country, and they have to travel a lot. This costs time and money. Moreover, because the interviewers spend much of their time on travel, they can do fewer interviews a day compared to a telephone survey. Face-to-face interviewing is also more complicated from an organizational point of view. Supervising and monitoring interviewers calls for more logistics.

Sending an advance letter has proved to increase response rates of face-to-face surveys. Such letters announce the visit of the interviewer. They also contain background information about the survey and explain why it is important to participate. They take away the surprise of an unexpected visit and provide legitimacy for the survey. Biemer and Lyberg (2003) note that an advance letter can have a negative effect, however, as they give selected persons more time to think of an excuse not to participate. Still the literature shows that the prevailing effect is positive, the response rates are higher.

Physical impediments can lead to nonresponse. Doormen, gatekeepers, or locked gates can make it difficult, or impossible, for an interviewer to make contact with a person. Locked communal entrances or intercom systems prevent making face-to-face contact. As a result the interviewer cannot show an identification card, or a copy of the advance letter. It may even happen that the person is no longer at an address, and the interviewing then has no way of knowing this.

In the early days of CAPI, there was some concern about the "big brother" effect. Basically the fear was that respondents might become part of a government operation to collect large amounts of data on people, and therefore their privacy was at stake. An experiment by Statistics Netherlands (CBS, 1987) showed, however, that there was no effect on response rates. Respondents simply accepted this form of data collection as a form of routine survey taking.

Interviewer-assisted interviews nevertheless result in less accurate answers to questions about sensitive topics. The presence of an interviewer may prevent a respondent from giving an honest answer to a question about a potentially embarrassing topic. Self-administered interviews (mail, Web) perform better. It has been shown that a sensitive question in CAPI can lead not only to item nonresponse but also to socially desirable answer; see De Leeuw (2008) for more information.

4.2.2 TELEPHONE SURVEYS

A telephone survey is a mode of data collection where interviewers call selected persons by telephone. If contact is made with that person, and the person wants to cooperate, the interview is started and conducted over the telephone.

Telephone interviewing has a characteristic in common with face-to-face interviewing in that interviewers are used for data collection. Similarly a number of advantages and some disadvantages should be considered.

A big advantage is that interviewers can use their skills to persuade persons to participate. Another advantage is that interviewers can give additional information about the survey and its questions. They can assist respondents in finding the proper answer to a question. The risk of item nonresponse is reduced. Direct person-to-person communication generally also has a positive effect on the quality of the collected data. And, of course, higher response rates are possible.

As was already pointed out, all kinds of additional information can be used to get cooperation. Information like composition of the household or the ages of its members may help in choosing an optimal approach strategy. The amount of available information depends partly on the way in which the sample was selected. For example, Statistics Netherlands selects its samples for telephone surveys from the population register. Telephone numbers are added to the selected names/addresses by the phone company. All register information on gender, age, marital status, and residence location is available for everyone in the sample.

If the sampling frame contains address information, it is possible to send an advance letter. Such a letter would announce that an interviewer will call, and therefore take away the surprise of an unexpected call. The letter also should provide information about the survey and explain why it is important to participate. To be effective, advance letters should also mention the research agency, and they must have an official letterhead. Such advance letters can help increase response rates.

Advance letters can also have a negative effect on the response rate. It is an early warning of the survey. People may prepare themselves and decide not to participate. However, a meta-analysis by De Leeuw et al. (2007) showed that advance letters work well to increase the response rate in a telephone survey. The response rate was increased by, on average, 8%. The refusal rate was reduced by 11%.

It should be noted that only listed telephone numbers can be linked to addresses and other sampling frame information. Unfortunately, not every telephone number is listed. For example, only about two-thirds of the people in

The Netherlands have a listed number. So not everyone can be reached through listed telephone numbers. This presents a serious problem of undercoverage. Moreover the coverage of listed telephone numbers is steadily decreasing over time.

The fast rise of mobile telephones has not made telephone interviewing easier. Landline telephones are increasingly replaced by mobile telephones. Landline telephones are typically a means to contact households, whereas mobile telephones make contact with individual persons. Therefore the probability of contacting a member of the household is higher for landline telephones. Moreover, if persons can only be contacted through their mobile telephones, it is often in situations not fit for interviewing. An additional problem is that sampling frames (telephone directories) in many countries do not contain mobile telephone numbers. A final complication to be mentioned here is that in countries such as The Netherlands people often switch from one telephone company to another. Usually this means that they get a different telephone number. It is possible, but sometimes not easy, to keep the original number. For more information about the problems and possibilities of mobile telephones for interviewing, see Kuusela et al. (2006).

If a sampling frame of listed telephone numbers is not available, another approach can be used called random digit dialing (RDD). Taking into account the structure of telephone numbers, a computer algorithm generates random, but valid, telephone numbers. Such an algorithm will produce both listed and unlisted numbers. This guarantees complete coverage. Random digital dialing has drawbacks. In some countries it is not clear what an unanswered number means. It can mean that the number is not in use, which is a case of overcoverage. No follow-up is needed. It can also mean that someone simply does not answer the phone, which is a case of nonresponse and so has to be followed up.

Another drawback of RDD is that there is no information at all about the selected people. This does not help the interviewers in their preparation of calls. It becomes more difficult to persuade reluctant persons. Lack of information from a sampling frame also makes nonresponse correction (e.g., adjustment weighting) more difficult.

Physical impediments may prevent interviewers from making contact with respondents. One example is an answering machine. People may be at home but still have the answering machine switched on. It is not clear whether it is good idea for the interviewers to leave a message. It may or may not help get a contact at the next attempt. Groves and Couper (1998) note that sometimes answering machines give relevant information about the people living at the address. This information may be useful for the next contact attempt.

When no contact is established with a selected person, and also when contact results in a refusal, one or more subsequent attempts may be made to obtain participation. Fortunately, repeated call attempts are not so expensive (as compared to face-to-face surveys). So it is relatively easy to do this in practice. It is not uncommon for survey agencies to make six call attempts or more before a case is closed as nonresponse due to no contact.

Some CATI systems (e.g., Blaise) distinguish call attempts from contact attempts. A contact attempt consists of a series of call attempts within a short time interval. Several contact attempts, each with several call attempts, are made before a case is closed as nonresponse. For example, Statistics Netherlands makes at most three or four contact attempts each consisting of at most three call attempts. The time interval for contact attempts is one hour in the case of noanswer and five minutes in the case of a busy number.

Many telephone companies have a calling number identification (CNID) service. It transmits the survey organization's number to the telephone of the selected person during the ringing signal. This information then becomes visible on the telephone of that person. If persons do not recognize the number, or if the number is shown as an "unknown number," they may decide not to pick up to phone. Thus CNID can lead to nonresponse.

Response rates of telephone surveys suffer from telemarketing activities. Typically people are called around dinner time (when the contact probability is high) in attempts to sell products or services. This spoils the survey climate, and it has led to a hostile attitude toward interviewers. So it is therefore important that interviewers make clear at the very start of the interview that they are not selling anything. In some countries (e.g., the United States and The Netherlands) there are do-not-call registers. When people register, they will not be called by telemarketing companies. Such registers can help improve the survey climate.

Incentives can help increase response rates. Research has shown that incentives are most effective when they are given before the interview and not after it. To be able to give incentives, the addresses of the respondents must be available. This is typically not the case for RDD surveys. So the possibilities for offering incentives are limited.

CATI is the computerized form of telephone interviewing. It was one of the modes of data collection to be computerized early on. The first CATI survey was carried in 1966. From the point of view of the respondent, there is no difference between a traditional telephone survey and a CATI survey. They just answer questions by telephone without being aware of what is on the other end of the line. The interviewers can have a paper questionnaire form or a computer. Therefore the response rates will not differ.

An important component of a CATI survey is the call management system. The objective of such a system is to manage and schedule call attempts. It typically involves the following tasks:

- Dialing the right telephone number at the right moment, taking into account the quota that interviewers can handle.
- Handling busy signals, since often the number is offered again after a few minutes.
- Handling no answers or answering machine calls, since often the number is offered again at a different day and/or a different time of the day. For example, if there is no answer in the afternoon, the next call could be made in the evening.
- Managing appointments, since the interviewer makes an appointment with the respondent to call back at a specific date and time. The system must then offer the call to this interviewer at the correct data and time.
- Producing progress reports.

An effective call management system can reduce nonresponse due to no contact. Models can be developed to predict the optimal time to call. Of course, this requires auxiliary information about the selected persons.

Call management systems are able to display results of earlier attempts and other information on the computer the screen of the interviewers. This information can help in persuading reluctant respondents, so the refusal rate may go down. See Wagner (2008) for details on tuning CATI call management systems to reduce nonresponse rates.

4.2.3 MAIL SURVEYS

The mail survey is completed without any interviewers asking the questions and recording the answers. Respondents, if they so choose, read the questions and write down the answers. No interviewer is present to persuade a reluctant letter recipient to fill in the form. No interviewer is present to explain unclear aspects and to assist in answering the questions. As a result response rates are generally lower in mail surveys than in interviewer-assisted surveys like the CAPI and CATI surveys.

A special type of mail survey is one that uses questionnaire drop-off and pick-up. Personal delivery of a questionnaire by only slightly trained survey takers may increase the response. This approach also provides some information about the nonrespondents.

Nonresponse occurs in a mail survey if no completed form is returned to the survey agency. There is usually no information at all about the reason for the nonresponse. Here are some reasons for nonresponse:

- The letter did not arrive at the indicated address (no-contact).
- The people were not at home during the survey period (no-contact).
- The letters was received but ignored (refusal).
- The people at the address did not understand the language the letter was written in (not-able).

Recall from Chapter 3 that it is important to distinguish among the different types of nonresponse. The different types of nonresponse can have different impacts on survey estimates, and therefore each type requires different treatment. Unfortunately, in mail surveys it is difficult to correct for the effects of nonresponse.

To obtain a reasonable amount of response, special effort is required in contacting and persuading people to participate in the survey. The design of the questionnaire is very important, and also all other procedures like advance letters, cover letters, reminders, and incentives. Dillman et al. (2008) discusses mail survey in detail. Only the main points are summarized here.

A lot of attention has to be paid to the design of the questionnaire form. It must look attractive to the respondents. The more personalized it is, the better it works, and the more likely it is that respondents will complete it. The clearer it is, the less likely it is they will get confused and abandon it.

A questionnaire must be spacious. The pages should not be crammed with a lot of questions and other text. It is counterproductive to attempt to get more questions on a single page by using less space and a smaller font.

The questionnaire must be structured so that questions are separated from one another by space. There should be more space between the questions than between the components of the questions. Modest and functional use of color can help make the design clear.

Navigation instructions (routing instructions, skip patterns) are a potential source of error. If respondents fail to follow the correct route through the questionnaire, wrong questions are answered and right questions are left unanswered, which comes down to item nonresponse. Navigation instruction must be unambiguous. Dillman (2007) suggests use of arrows or other graphics to indicate jumps to other questions with similar meaning.

When the questionnaire is mailed, it should be accompanied by a compelling cover letter to the sample persons. This letter should at least explain why participation is important and for what purpose the data will be used. The letter should preferably come from a high official in the organization. The letter should not look like it has been photocopied, but resemble an original letter (including colored letterheads and a signature). It will also help increasing the response rate if the letter contains a clear statement that all collected data will be treated as confidential.

It should be as easy as possible for the respondents to return the completed questionnaire. Therefore enclose a postage-paid return-envelop in the letter to the sampled persons.

Reminders are important for increasing the response rate. If no response is obtained after two to four weeks, a letter can be sent. It reminds persons that their questionnaire has not yet been received and urges them to respond. This reminder letter should contain a replacement questionnaire.

Consider enclosing a token of appreciation in the first letter to the selected persons. Such incentives generally increase response rates. Incentives work better when sent in advance. Use of advance incentives is based on social exchange theory. According to this theory, people engage in social exchange with others when the perceived reward is at least as large as the expected costs. So they will respond when the reward (the value of the incentive) is sufficient to cover the costs (the effort of filling in the questionnaire form). Incentives can take various forms: cash payment, lottery tickets, postage stamps, pens, or a donation to a charity organization in the respondent's name. Some research seems to suggest that donations seem not to work as well as real monetary incentives. The larger the incentive, the higher the response rate, but the relationship is not linear. If the value of the incentive is doubled, the increase in response is not doubled. According to Armstrong (1975) and Fox et al. (1988), the concept of diminishing returns applies.

Survey research literature suggests that long questionnaires should be avoided. There is empirical evidence that long questionnaires reduce the response rate. For example, Dillman et al. (1993) show that shorter questionnaire forms increase response in census mail surveys.

A final practical word of advice is to avoid time periods when there is other heavy mail traffic. Examples are just before tax forms have to be submitted and before the Christmas holiday.

4.2.4 WEB SURVEYS

The rapid development of the Internet over the last decades has led to new survey data collection methods. Such surveys are usually denoted by Web surveys. Sometimes the terms Internet survey, online survey, or computer-assisted Web interviewing (CAWI) are used. The basic feature of a Web survey is that the questionnaire is designed as a website, which is accessed by respondents. Web surveys are almost always self-administered: respondents visit the website, and complete the questionnaire by answering the questions. Not surprisingly, many survey organizations now use, or are considering, Web surveys. Web surveys have become very popular in a short time. This is not surprising as Web surveys have some attractive advantages:

- Today many people are connected to the Internet, so a Web survey is a simple way to get access to a large group of potential respondents.
- Questionnaires can be distributed at very low costs. No interviewers are needed, and there are no mailing and printing costs.
- Surveys can be launched very quickly. Little time is lost between the moment the questionnaire is ready and the start of the fieldwork.
- Web surveys offer use of attractive multimedia, such as sound, pictures, animation, and movies.

Web surveys can be a fast, inexpensive way of collecting large amounts of data. There are nevertheless methodological challenges. Some of these have an impact on the response rate of the survey.

A big problem is undercoverage, in that not all of the target population may be represented in the sampling frame. Therefore it is difficult to select the sample, since the target population is usually wider than the Internet population. Table 4.2 shows the Internet coverage in the member states of the European Union (excluding Belgium).

There are large variations in Internet access among countries. The countries with the highest percentages of Internet access are The Netherlands (86%), Sweden (84%), and Denmark (82%). Internet access is lowest in Bulgaria (25%), Romania (30%), and Greece (31%). For more information, see Eurostat (2008). Broadband Internet access is even lower. As a result Web surveys with advanced features (requiring broadband) can only be conducted for a limited part of the population.

Unfortunately, Internet access is unevenly distributed over all populations. A typical pattern found in many countries is that elderly, low-educated, and ethnic minorities are severely underrepresented among those having access to Internet. Bethlehem (2009) shows that the bias of the response mean as an estimator of the population mean of a variable Y is equal to

TABLE 4.2 Percentage of Internet access and broadband Internet access in the countries of the European Union in 2008

Country	Access	Broadband
Bulgaria	25	21
Romania	30	13
Greece	31	22
Italy	42	31
Cyprus	43	33
Portugal	46	39
Hungary	48	42
Poland	48	38
Lithuania	51	43
Spain	51	45
Latvia	53	40
Slovakia	58	35
Estonia	58	54
Slovenia	59	50
Malta	59	55
France	62	57
Ireland	63	43
Austria	69	54
United Kingdom	71	62
Finland	72	66
Germany	75	55
Luxemburg	80	61
Denmark	82	74
Sweden	84	71
Netherlands	86	74
EU 27	60	48

Source: Eurostat (2008).

$$B(\bar{y}) = E(\bar{y}) - \bar{Y} = \bar{Y}_I - \bar{Y} = \frac{N_{NI}}{N}(\bar{Y}_I - \bar{Y}_{NI}), \tag{4.1}$$

where the subscript I denotes the subpopulation of people with Internet access and NI the subpopulation of people without Internet access. The size of this bias is determined by two factors:

- The relative size N_{NI}/N of the subpopulation without Internet. The bias decreases as Internet coverage increases.
- The contrast $\bar{Y}_I - \bar{Y}_{NI}$ between the means of the target variable in the Internet subpopulation and the non-Internet subpopulation. The more the mean of the target variable differs for these two subpopulations, the larger the bias will be.

Since Internet coverage is steadily increasing, the factor N_{NI}/N is decreasing. While this reduces the bias, it is not clear whether the contrast between persons with and without Internet also decreases. It is likely that the (small) group of people without Internet will be more and more different from the rest of the population. As a result substantial bias may still remain.

Not only Web surveys have coverage problems, but telephone surveys also have coverage problems. For example, only between 60% and 70% of the households in The Netherlands have a listed landline telephone. So one out of three households is missing if a sample is selected from a telephone directory. This situation will not improve over time. The problems for Web surveys may become less severe over time. From the point of view of coverage, a Web survey may be better than a telephone survey, at least in The Netherlands. For a telephone survey it is known in advance which people do not have a listed telephone number, so it may be better to do a mixed mode survey; see Section 4.2.5 for more information about mixed mode surveys.

When undercoverage in a Web survey is a problem, a possible solution may be to provide Internet access to those without Internet. An example of this approach is the LISS panel, described by Scherpenzeel (2008). This Web panel was constructed by selecting a random sample of households from the population register of The Netherlands. Selected households were recruited for this panel by means of CAPI or CATI. Cooperative households without Internet access were provided with equipment giving them access to Internet.

Another problem of Web surveys is self-selection. Many Web surveys, particularly those conducted by commercial market research organizations, are self-selection surveys. The survey is simply put on the Web. Participation requires that respondents are aware of the existence of the survey. They have to accidentally visit the website, or they have to follow up a banner, email message, or a call in another commercial. Then they have to make the decision to fill in the questionnaire on the Internet.

Self-selection means that no probability sample is selected. Survey participation is the result of a spontaneous decision of a respondent. This implies each element k in the target population has unknown probability ρ_k of participating in the survey. In fact self-selection also implies that the sample is the complete population. Bethlehem (2009) shows that the expected value of the response mean is equal to

$$E(\bar{y}) \approx \bar{Y}^* = \frac{1}{N}\sum_{k=1}^{N}\frac{\rho_k}{\bar{\rho}}Y_k, \tag{4.2}$$

where $\bar{\rho}$ is the mean of all response propensities. The bias of this estimator is equal to

$$B(\bar{y}) = E(\bar{y}) - \bar{Y} \approx \bar{Y}^* - \bar{Y} = \frac{R_{\rho Y} S_\rho S_Y}{\bar{\rho}}, \tag{4.3}$$

where $R_{\rho Y}$ is the correlation coefficient of the target variable and the response probabilities, S_ρ is the standard deviation of the response probabilities, and S_Y is the standard deviation of the target variable. It can be shown that in the worst case (S_ρ assumes it maximum value and the correlation $R_{\rho Y}$ is equal to either $+1$ or -1)

the absolute value of the bias is equal to

$$|B_{\max}(\bar{y})| = S_Y \sqrt{\frac{1}{\bar{\rho}} - 1}. \tag{4.4}$$

Bethlehem (1988) shows that (4.3) also applies in the situation where a probability sample has been drawn, and subsequently nonresponse occurs during the fieldwork. Consequently expression (4.4) provides a means to compare potential biases in various survey designs.

EXAMPLE 4.4 Potential Bias of Self-Selection Web Surveys

Regular surveys of Statistics Netherlands are all based on probability sampling. Their response rates are around 70%. This means the absolute maximum bias is equal to $0.65 \times S_y$.

One of the largest self-selection Web surveys in The Netherlands was *21minuten.nl*. Within a period of six weeks in 2006 about 170,000 people completed the Web questionnaire. The target population of this survey was not defined, as everyone could participate. If it is assumed the target population consists of all Dutch from the age of 18, the average response propensity is equal to $170{,}000/12{,}800{,}000 = 0.0133$. Hence the absolute maximum bias is equal to $8.61 \times S_y$.

It can be concluded that the bias of the large Web survey can be a factor 13 larger than the bias of the smaller probability survey. Apparently, when in a random sample a reasonable response rate is achieved, the bandwidth of a possible bias is reduced.

Web surveys can suffer from nonresponse. A Web survey questionnaire is a form of self-administered questionnaire. Therefore Web surveys have a potential of high nonresponse rates. An additional source of nonresponse problems are technical problems of respondents having to interact with the Internet; for example, see Couper (2000), Dillman and Bowker (2001), Fricker and Schonlau (2002), and Heerwegh and Loosveldt (2002).

Respondents need a browser to open and complete a Web survey questionnaire. There are many different browsers available. Examples are Internet Explorer, Firefox, Safari, Opera, and Google Chrome. And there are many more. These browsers do not behave exactly the same way. Therefore a questionnaire may behave differently in different browsers. A feature may not even work in a browser, preventing respondents from recording answers to the questions. As a result respondents may drop out.

The Internet is in continuous development. Specific features may work in the most recent version of a browser but not in earlier versions. Unfortunately, not all people have the latest version of their browser installed. Again, as a result a specific feature in the questionnaire may or may not work.

Some questionnaire features (e.g., use of animation and video) require an Internet connection with substantial bandwidth. It is clear from Table 4.2 that not every Internet user has a broadband Internet connection. So these features will not work properly on their computers.

Slow modem speeds, unreliable connections, high connection costs, low-end browsers, and incompatible browsers may frustrate respondents. This often results in prematurely interrupting completion of the questionnaire. To keep the survey response up to an acceptable level, every measure must be taken to avoid these problems. This requires a careful design of Web survey questionnaire instruments. Self-selection may be an effective way to boost the number of respondents, but it does not help increase response rates, let alone nonresponse bias.

Designing a questionnaire for a Web survey is to some extend similar to designing it for a mail survey. At first sight, one could say that a Web survey questionnaire is nothing more than a paper form displayed on a computer screen. There are, however, a lot of differences that may affect response rates. Here is a list of some issues:

- A paper questionnaire page usually does not fit on the computer screen. This means that the respondent has to scroll to see all parts of the page. Failing to do so may mean that some (not visible) questions are skipped.
- The questionnaire designer has the choice to display just one question per screen, or to put more questions on the screen. One question per screen may be more appropriate if the questionnaire is to contain extensive routing instructions (skip patterns). However, this will increase the perceived length of the questionnaire, possibly resulting in an interrupted completion.
- The respondents are completely free in the way they complete a paper questionnaire. They can answer any question they like in any order. Web questionnaires can be designed such that routing is enforced as in CAPI or CATI surveys. This sees to it that only relevant questions are answered. However, respondent may be frustrated by this forced approach.
- CAPI and CATI surveys questionnaires may contain extensive edit checks. These checks help detect and correct inconsistencies in the answers of respondents. Such checks are not possible in mail surveys, and this may have a negative effect on the quality of the collected data. The designer of a Web survey questionnaire has a choice to include edit checks or not. There is a risk that many (unfriendly) error messages may scare away respondents. This leaves the designer with the dilemma to either focus on data quality (with a risk of a lower response rate) or on high response rates (with a risk of lower data quality).

There are many more design issues that may affect response to a Web survey. A detailed description is given by Couper (2008).

Couper (2000) notes that the advantages of the Internet have made it possible for many organizations to conduct surveys. Several of these organizations are not really survey organizations. They are not aware of potential methodological risks involved in conducting Web surveys. As a consequence there are many bad surveys

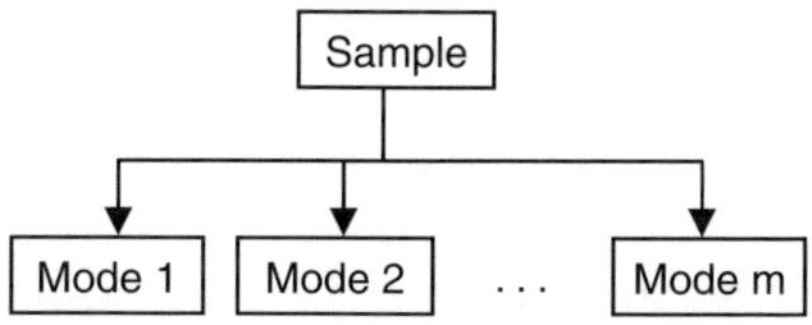

FIGURE 4.1 **Concurrent mixed mode data collection.**

on the Internet. For respondents it is not always easy to distinguish the good from the bad. This abundance of Web surveys can have a negative effect on response rates. Potential respondents are overcrowded and so will refuse to participate. The effect is similar to that of telephone surveys, where the survey climate is spoiled by telemarketing activities.

4.2.5 MIXED MODE SURVEYS

Every mode of data collection has advantages and disadvantages. The choice of the mode is therefore often a compromise between quality and costs. When high-quality data and a high response rate are required, the best choice will be a face-to-face survey. This is expensive. Consequently the sample size may need to be limited. If the objective is to obtain a large sample, and quality is not so important, a mail or Web survey could be considered.

Instead of choosing one mode of data collection, it is also possible to combine various modes of data collection in one survey. Budget cuts, on the one hand, and demands for more and more detailed information, on the other hand, while maintaining an acceptable level of data quality, have stimulated national statistical institutes to explore different approaches to data collection. One such approach is the *mixed mode survey*.

De Leeuw (2005) describes two mixed mode approaches. The first is using different modes concurrently. The sample is divided into groups and each group is approached by a different mode. Concurrent mixed-mode data collection is illustrated in Figure 4.1

Need to obtain a high response rate can be a reason to implement a survey with a mixed mode design. Nonresponse research has shown that the response of certain groups may depend on the mode of data collection. For example, young people may prefer the Web for completing a questionnaire, whereas the elderly appreciate being visited by interviewers. This could imply a Web survey for the young and a CAPI survey for the elderly. Of course, this approach requires that the age of each sample person to be known in advance.

EXAMPLE 4.5 The Basic Question Approach

Statistics Netherlands re-approached in 2005 a sample of nonrespondents from the Dutch Labor Force Survey (LFS). The objective was to obtain more insight in the nonresponse. Two techniques were used: the basic-question approach and an intensive follow-up. A concurrent mixed mode design was implemented for the basic-question approach using CATI and a combination of a Web and a mail survey. Persons with a listed telephone number were approached by CATI, whereas persons without a listed number received a letter with a paper questionnaire and a login to a secured website from Statistics Netherlands, where they could also fill in the questionnaire.

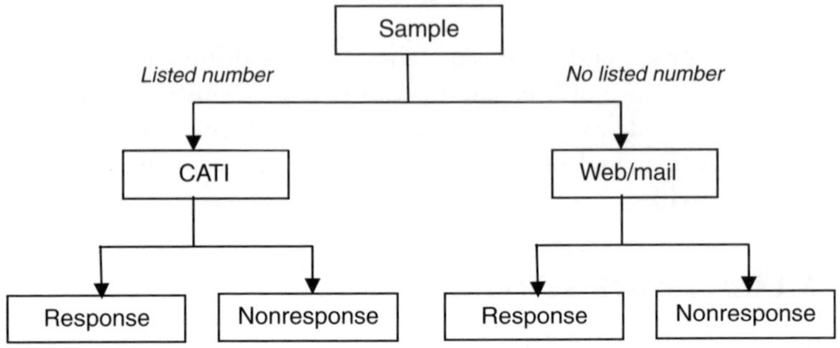

A second mixed mode approach described by De Leeuw (2005) is the sequential approach. All sample persons are approached by one mode. The nonrespondents are then followed up by a different mode than the one used in the first approach. This process can be repeated for a number of modes. The sequential mixed mode approach is illustrated in Figure 4.2.

If the main objective is to keep survey costs as low as possible, a sequential mixed mode survey could start with a mail questionnaire or a questionnaire on the Web. Nonrespondents are followed up by CATI. Nonrespondents remaining after CATI could be followed up by CAPI. So the survey starts with the cheapest mode and ends with the most expensive one.

If quality and response rates are of vital importance, one could think of a mixed mode design starting with CAPI. The nonresponse is then followed up by CATI. Finally, remaining nonrespondents could be asked to complete the questionnaire on the Web.

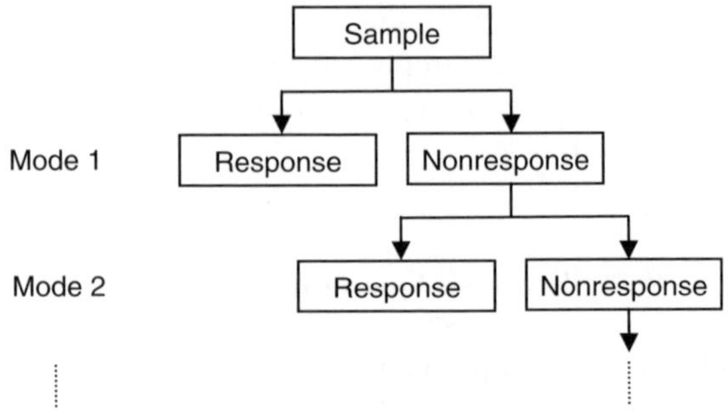

FIGURE 4.2 Sequential mixed mode data collection

EXAMPLE 4.6 The Safety Monitor Pilot

Statistics Netherlands carried out a number of mixed mode pilot surveys in 2005 to 2007. The objective of this research was to determine an optimal mixed mode design for surveys of The Netherlands; see Janssen et al. (2007) for the preliminary results. The ultimate goal was to reduce the costs of data collection while preserving the quality of the data. In particular, the possibility of using a Web survey as one of the data collection modes was investigated. Pilot surveys were conducted for three different kinds of surveys. One was the Safety Monitor. This survey measures, among other things, peoples' feelings about neighborhood security and insufficient police presence.

The first mode was a Web survey. The sample was selected from the population register. Each selected person was sent a letter with a login that gave access to the questionnaire on the Internet.

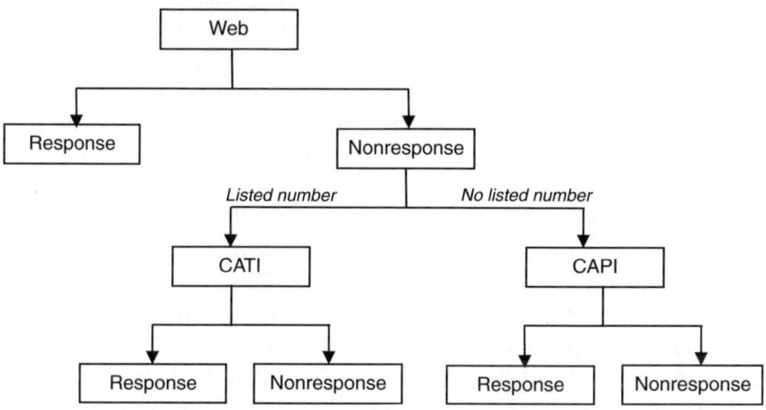

All nonrespondents of the Web survey were re-approached. Those persons with a listed telephone number were contacted by telephone in a CATI survey. Those without a listed telephone number were included in a CAPI survey.

Surveys can suffer from mode effects. Mode effects occur if the same question produces a different answer when asked in a different mode. Biemer and Lyberg (2003) suggest two types of mode effects:

- *Pure mode effect* A difference in survey outcomes originates solely from the choice of data collection mode.
- *Design mode effect* A difference in survey outcomes is caused by the combined effect of all differences in the survey design for the different modes. For example, if one mode is a CAPI survey and the other mode is a mail survey,

there will be differences in the questionnaire used (electronic vs. paper), the response rate, and the composition of the response.

Pure mode effects are much more difficult to assess than design mode effects. We will therefore consider only design mode effects, and refer to them as *mode effects*. Differences in survey outcomes primarily can be caused by the error sources described in Chapter 1 (Section 1.4). According to Roberts (2007) and Ariel et al. (2008) the choice of data collection mode can influence three sources of nonsampling error: coverage error, nonresponse bias, and measurement error. Ariel et al. (2008) studied the literature on mode-specific measurement errors and concluded that measurement errors can be serious and should be accounted for.

De Leeuw (1992) used a meta-analysis to study the data quality of face-to-face and telephone surveys. She concluded that differences in data quality between well-conducted face-to-face and telephone surveys are small. This conclusion is in line with that of Groves (1989), who found that the most consistent characteristic in studies comparing responses in face-to-face and telephone surveys is the lack of difference between the two modes. Snijkers (2002) compared CATI to CAPI with respect to cost and quality. The advantages of CATI are the efficient use of hardware and software (in case of centralized telephone interviewing), the immediate availability of the data, reduced costs because interviewers do not have to travel, and the smaller number of interviewers required for the same number of interviews.

But there are important differences between a CAPI and a CATI survey in the means of communicating. Groves (1989) calls these differences the "channel capacity" and the "intimacy of the social interaction." A face-to-face survey is capable of carrying more types of nonverbal messages, whereas telephone calls received from strangers tend to be short interactions (more frequently used for business purposes) and socially less personal. Snijkers (2002) observes that in the use of the telephone as a communication medium, telephone interviewing is only adequate for simple questions that can be answered instantaneously and that need little time to think out the answers. The telephone questionnaire should consist of questions with simple wordings, have a short list of response categories in closed-ended questions, and be of shorter duration.

De Leeuw (2005) provides an extensive overview of the advantages and disadvantages of mixed mode survey designs, and distinguishes between the use of multiple modes in the contact phase and in the actual data collection phase. De Leeuw (2005) considers the differences of modes in terms of *cognitive burden*, namely the burden of the processing of information related to the decision-making. She thereby applies the stimulus–response model (see also Tourangeau and Rasinski, 1988, Snijkers, 2002, Ariel et al., 2008), according to which the cognitive process of survey response can be divided into four phases. In each of the phases, the respondent has to perform cognitive tasks that can be different across modes.

The first phase concerns the interpretation and comprehension of the question. This depends on question wording, syntax, length, order, perceived sensitivity, and the response categories. The information has to be retrieved from the respondents' memory or from other sources. The second phase then depends on the type of information that is being asked, whether it is factual, attitudinal or proxy. Phase three concerns judgment. In this phase the respondent uses the retrieved information to form an internal answer. The retrieved information is integrated and evaluated. The judgment of the answer depends on the consistency with prior answers, or with social desirability. Together these two steps—information retrieval and judgment—can be seen as the information transmission phase. Last, the respondent reports the answer by comparing the internal answer with the response options or by reporting the answer in the requested quantity. For this phase it is crucial that the respondent fully understand the wording of the response options.

Using a similar stimulus–response model, Pierzchala (2006) identifies attributes of surveys that affect cognition and response: presentation, transmission of response, segmentation of the questionnaire, dynamic or passive medium, type of administration, who/what determines the pace of the interview, and the channel of communication. De Leeuw (2005) had grouped these attributes or factors into three classes: medium-related factors (presentation, transmission, segmentation, and dynamic/passive), factors concerning the information transmission (administration, pace and channel of communication) and interviewer effects, with interviewer effects being medium-related and information transmission factors. Pierzchala (2006) added three more mode effects. These are an aural versus a visual presentation, self-administered versus interviewer-administered, and dynamic versus passive nature of the questionnaire. Questionnaires for computer-assisted interviewing are usually dynamic, and paper questionnaires are passive.

In addition to these three mode effects, Pierzchala (2006) introduced degrees of *disparity* for the modes, by which modes differ by degree of risk of higher nonresponse. See Table 4.3.

CAPI and CATI are similar in presentation, administration, and in the dynamic nature of the questionnaire. Accordingly the degree of disparity of the other modes is indicated with respect to both CAPI and CATI. Web surveys also have a dynamic questionnaire. But the presentation is visual, and Web surveys are self-administered. Therefore the degree of disparity between Web surveys and CAPI/CATI surveys is 2. Mail surveys share none of the characteristics with CAPI/CATI, the degree of disparity therefore is 3. Mail surveys and Web surveys are similar in self-administration and visual presentation; however, mail surveys have a passive questionnaire. Their degree of disparity is 1.

It can be concluded from Table 4.3 that the largest mode differences can be expected for a combination of mail surveys and CAPI or CATI, followed by Web surveys and CAPI or CATI surveys. A combination of mail and Web surveys has

TABLE 4.3 Degree of disparity between data collection modes

Mode combination	CAPI/CATI	Mail	Web
CAPI/CATI	N/A	3	2
Mail	3	N/A	1
Web	2	1	N/A

Note: N/A stands for not applicable.

a reduced risk of mode effects. Combining CAPI and CATI is the safest option to avoid mode effects.

There are other factors to consider when choosing the modes to mix. Roberts (2007) identifies budget restrictions, timeliness, and the available infrastructure. Biemer and Lyberg (2003) discuss the desired level of data quality, the available budget and time, and the specific content of the survey (types of questions, length of survey, complexity of the questions, need for visual aids). They also point out that choosing an optimal design is especially difficult in situations where there are a lot of options.

Some mode effects are now discussed in some detail. Of course, the most important mode effect is whether the interviewer is present. As was noted earlier in this chapter, the presence of interviewers leads to more *socially desirable answers*, particularly for questions that could potentially embarrass the respondent. A social desirability bias can occur as in phase three of the response–stimulus model, if respondents change their "internal" answer to an answer that they perceive to be more socially correct. This behavior leads to underreporting of perceived socially undesirable behavior such as alcohol consumption, and to overreporting of perceived social behavior such as voting.

Satisficing is caused by respondents that are not motivated to think carefully about the meaning of the question or the answer. They just give a quick answer to be able to continue and finish the questionnaire. According to Krosnick (1991) such respondents reduce the cognitive burden of optimal question answering by short-cutting the response process. The degree of satisficing is related to ability and motivation of the respondents, and also to task difficulty. The amount of satisficing may differ from mode to mode. For example, there is less satisficing in face-to-face surveys than in telephone surveys. Apparently the physical presence of interviewers motivates the respondents to answer the questions more carefully. The fact that people more often answer "don't know" in a telephone survey can be seen as supporting evidence for this effect.

If interviewers are present, they are in control of presenting the questions to the respondents in CAPI and CATI surveys. They can see to it that the respondents hear and understand every word of it. When necessary, additional explanation may be provided. This is different for self-completion surveys like mail and Web surveys. There is no guarantee that questions are carefully read and clearly understood.

The presence of interviewers can also cause *acquiescence*. This is the tendency to agree with statements by interviewers. It is easier to agree than to disagree.

There can be mode effects to answering closed questions. Research suggests that respondents in mail or Web surveys more often choose the first answer option (*primacy effect*), while there is a preference for the last answer option on CAPI and CATI surveys (*recency effect*).

Another mode effect that needs to be mentioned is the treatment of "don't know." This option is often not offered explicitly in CAPI or CATI surveys, but the interviewers have a facility to record this answer if the respondents insist they really do not know the answer. For self-administered surveys, this option may or may not be explicitly offered. If "don't know" is clearly one of the answer options in a mail or Web survey, more respondents may select this option as the easy way out.

Mode effects can also be caused by coverage errors. If a sample element does not appear in the sampling frame, it can never be selected in the sample. This is called *undercoverage*. If an element does not belong to the target population, but it appears in The sampling frame, it may be wrongfully selected in the sample. This is called *overcoverage*. Mode effects may occur if different modes require different sampling frames and these sampling frames suffer from different coverage problems.

EXAMPLE 4.7 Mode Effects Caused by Coverage Errors

The target population of the surveys of Statistics Netherlands is the population that can be found in the population register, with the exception of the institutionalized population. The population register is the sampling frame for all surveys. Hence CAPI and mail surveys have a low risk of coverage errors. For CATI and Web surveys, however, sample elements need a telephone or access to the Internet. The sampling frame for a telephone survey is obtained by matching the records from the population register with the register from the Dutch telephone company (KPN). Originally KPN distributed all landline telephones. However, nowadays there are also other providers of landline telephones (mostly digital). Therefore the sample is also matched with the telephone directory, where numbers from other providers are listed too. As a consequence sample elements without a listed landline number (i.e., mobile only, an unlisted telephone, or no telephone at all) do not appear in the sampling frame for the telephone survey.

Undercoverage problems of telephone surveys may be avoided by applying random digit dialing (RDD). This method randomly generates telephone numbers from the frame of all possible telephone numbers. Lepkowski (1988) provides a review of RDD and other telephone survey sampling methods. The advantage of RDD compared to sampling from a telephone directory is that it provides coverage of both listed and unlisted numbers (and possibly mobile numbers as well). However, there are some disadvantages. A large problem of RDD is the number of ineligible listings. Numbers can belong to businesses, or households occur in the sample more than once because they own more than one number. It is thus unknown whether the generated telephone numbers correspond to eligible sample elements. With respect to adjustment methods, a disadvantage of RDD is that there is no auxiliary information at all available on the nonrespondents. Sometimes there is no difference between nonexisting telephone numbers and nonrespondents, in which case nonexisting numbers are wrongfully regarded as nonrespondents.

EXAMPLE 4.8 Undercoverage in Telephone and Web Surveys

Data collection for Web surveys at Statistics Netherlands takes the form of sending an advance letter to sample elements, with a login to a secured website where the questionnaire can be found. Sample elements are asked to fill in the form on the Internet. If they have no access to the Internet, they cannot respond to the survey. CATI and Web surveys hence have to do with undercoverage of the part of the population that does not have a listed landline telephone, or internet access. The number of listed landline telephones in The Netherlands is decreasing as the number of households that have access to the Internet increases.

Another problem that may occur in surveys is that of overcoverage. This is the phenomenon the persons are included in the sample that do not belong to the target population. Cases of overcoverage should be removed from the survey data set because they can affect estimates. However, it is not easy to detect overcoverage if respondents for a Web survey are recruited by an advance letter.

There may be ways around the undercoverage problems in Web surveys due to lack of Internet access at home. For example, persons can have access to the Internet at the public library or at their workplace. Another solution is provided in the LISS-panel of the Tilburg University in The Netherlands. This is a Web panel. If persons selected for this panel have no Internet access, they are offered a simple computer with Internet access; see Scherpenzeel (2008) for more information about the LISS panel.

It should be clear that mixed mode surveys can suffer from mode effects. There are two ways to reduce these effects. One is to develop separate questionnaires for different modes. A specific question may be defined differently in different modes as long as what is measured is the same. The different versions of the question should be cognitively equivalent. This is not very easy to realize, as it may take substantial research and experimentation.

Dillman (2007) proposes what he calls the *unimode approach*. This is a set of guidelines used to define questions in ways that minimize the mode effects. Some examples of his suggestions are:

- Keep the wording of a question the same across modes.
- Keep the number of answer options to a closed question as small as possible.
- Keep the wording of possible answers to a closed question the same across modes.
- Randomize as much as possible the order of the answer options to a closed question for different respondents.
- Include all answer options in the text of a closed question.
- Develop equivalent instructions for skip patterns.

One may wonder whether it is possible to develop a questionnaire that completely satisfies all unimode guidelines. Particularly for attitudinal questions, it may turn out to be necessary to prepare different mode-dependent versions.

Case management is a point of concern for mixed mode surveys. A case-managed system ensures that selected persons are assigned to the proper mode at the proper moment. Duplication must be avoided and so must dropouts.

4.3 Application

National statistical institutes are facing a challenge to reduce data collection costs, on the one hand, and to publish more detailed statistical information, on the other. New ways of data collecting are being considered in which Web surveys are expected to initiate the far-reaching change of replacing expensive CAPI and CATI surveys. In the meantime a viable option is to use mixed mode surveys. Statistics Netherlands is experimenting with mixed mode surveys. One such experiment is described here. A far more detailed account of this experiment is given by Beukenhorst and Wetzels (2009).

The Dutch Safety Monitor is a mixed mode survey that measures how the Dutch feel about safety and security in their neighborhoods Questions are asked about feelings of at-home security, quality of life, and level of neighborhood crime. The sample for this survey was selected from the Dutch population register. Only persons aged 15 years and older were selected. All sample persons received a letter in which they were asked to complete the survey questionnaire on the Internet. The letter also included a postcard that could be used to request a paper questionnaire. Two reminders were sent to those that did not respond by Web or mail. If still no response was obtained, nonrespondents were approached by means of CATI if a listed telephone number was available. If not, these nonrespondents were approached by CAPI. This four-mode survey is denoted by SM4.

To be able to compare this four-mode survey with a traditional survey, also a two-mode survey was conducted for an independent sample. Sampled persons where approached by CATI if there was a listed telephone number, and otherwise, they were approached by CAPI. This survey is denoted by SM2.

The response rate for SM4 turned out to be 59.7%. The response rate for SM2 was 63.5%. So there is not much difference. Table 4.4 shows the composition of response in both surveys.

More than half of the responses (58%) was obtained in the SM4 with a self-administered mode of data collection (CAWI or PAPI). The conclusion can be drawn that the four-mode survey did not increase the response. The costs of the survey were, however, much lower because interviewers where deployed in only 42% of the cases. Focusing on just interviewer costs, and ignoring all other costs (which are much lower), Beukenhorst and Wetzels (2009) found that the costs of SM4 were only 60% of the costs of SM2.

TABLE 4.4 Composition of the response of both safety monitor surveys

Data collection mode	SM4	SM2
Web	41.8%	
Mail	16.2%	
CATI	30.5%	71.6%
CAPI	11.5%	28.4%
Total	100.0%	100.0%

TABLE 4.5 Ranges of the response rates (%) in the categories of auxiliary variables

Auxliary variable	Categories	Range in SM4	Range in SM2
Household size	2	52–64	54–67
Gender	2	60–62	63–65
Age	7	55–66	52–70
Ethnicity	4	41–64	44–67
Degree of urbanization	5	52–66	53–71
Average		60	64

The quality of the response is not determined only by the response. The composition of the response is more important. It the response composition differs substantially from the sample composition, estimates may be biased. To get insight in possible composition differences, response rates in the categories of auxiliary variables can be compared. Table 4.5 shows the results.

As the table shows, there is little difference in the response rates. Only for age and degree of urbanization the range seems to be smaller for SM4. This is an indication that the composition improved. Chapter 7 discusses more advanced indicators for measuring the representativity of the response to a survey.

A phenomenon that repeats itself in many general population surveys is the low response rate of specific groups. Well-known examples are people in highly urbanized areas and ethnic minority groups. Beukenhorst and Wetzels (2009) investigated whether the four-mode survey reduces this problem. This turned out not to be the case for the urbanized areas because responses were very low on the Web and in the Mail mode. The same conclusion can be drawn for ethnic groups.

Analysis of mode effects in such studies is not straightforward because it is not possible to separate mode effects from selection effects. The group addressed by one mode may have a different composition than the group doing another mode. For example, the people completing a CATI questionnaire in SM4 are the nonrespondents from the Web and mail mode, and the Web and mail nonrespondents may differ from the Web and mail respondents. If it is assumed there are no substantial selection effects, it is possible to explore the existence of mode effects in the four-mode survey. Figure 4.3 shows the response by mode to the statement "I feel at home with the people here in the neighborhood."

There are two striking phenomena in this graph. The first is that more people agree with this statement in the CATI and CAPI mode. This is a well-known phenomenon called *acquiescence*. It is the tendency to agree with statements made by the interviewer. Apparently most people prefer to avoid contradicting the interviewer. CATI and CAPI are both interviewer-assisted modes of data collection. Therefore one can expect that acquiescence occurs only for these modes, and not for the Web and mail modes.

A second striking phenomena is that much less people choose for the neutral category in CAPI and CATI. This may be caused by the tendency of people to give more *socially desirable* answers when interviewers are present. Not having a clear

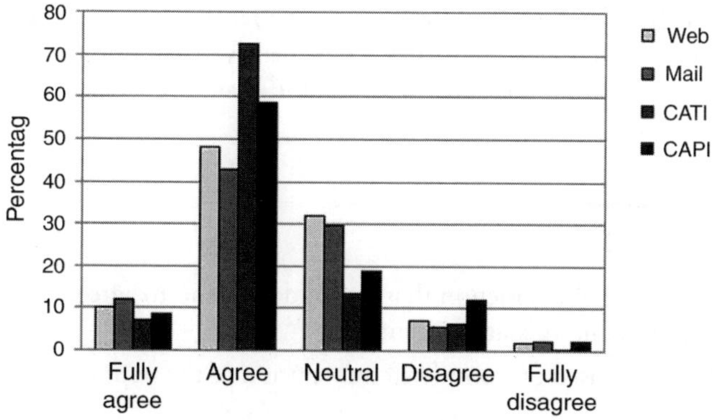

FIGURE 4.3 **Response in SM4 to the question "I feel at home with the people here in the neighborhood".**

opinion is not considered socially desirable. Therefore they may express an opinion they may not have.

The results of this experiment show that this mixed mode case (with four modes) was not very successful with respect to the quality of the outcomes. The response rate did not increase substantially, and also the composition of the response did not improve much. The hope that introducing new modes would increase the response rate in low-response groups, was not fulfilled.

The experiment showed that a mixed mode survey can lower survey costs, and that the four-mode survey was substantially less expensive than the two-mode survey.

4.4 Summary

A researcher has a choice of the mode of data collection. Some modes are expensive (e.g., CAPI and CATI) but produce good quality data. Other modes (mail, Web) are less expensive but may have negative effects on data quality.

For good response rates, it is best to use an interviewer-assisted mode of data collection. This means face-to-face or telephone interviewing, and also their computer assisted analogues CAPI and CATI. The presence and prodding of interviewers can lead to higher response rates.

Although mail and Web surveys are not as good as the interviewer-assisted modes, responses are more honest with respect to sensitive questions. Respondents tend to answer such questions more truthfully if no interviewers are present.

Web surveys are growing in popularity, but their response rates are still disappointingly low. This is partly due to the self-administered nature of this mode of data collection. Nonresponse can also be caused by technical problems, like slow modems (no broadband connection), and old or incompatible browsers.

A mixed mode survey can sometimes increase the response rate of a survey. The idea is that every respondent can choose his or her favorite mode. Results of a recent experiment suggest that response rates do not increase. An advantage of mixed mode surveys is nevertheless that they less expensive than, for example, a single mode CAPI survey while maintaining the same level of quality.

4.5 KEY TERMS

Acquiescence The phenomenon that respondents tend to agree with statements in questions if interviewers are present.

Census A survey based on complete enumeration of the population. There is no sampling.

Computer-assisted interviewing (CAI) Interviewing electronically by way of a questionnaire that is not printed on paper. The questions are answered using a computer program.

Computer-assisted personal interviewing (CAPI) Face-to-face interviewing whereby interviewers use a laptop computer to ask the questions and to record the answers.

Computer-assisted self-administered questionnaires (CASAQ) Data collection whereby respondents complete the questionnaires on their own computers. See also CASI.

Computer-assisted self-interviewing (CASI) Data collection whereby respondents complete the questionnaires on their own computers. See also CASAQ.

Computer-assisted telephone interviewing (CATI) Telephone interviewing whereby interviewers use a computer to ask the questions and to record the answers.

Computer-assisted Web interviewing (CAWI) Self-interviewing whereby respondents complete the questionnaires on the Internet. This is also called a Web survey.

Face-to-face interviewing Interviewing where an interviewer visits the homes of respondents (or a location convenient for the respondent). Together, the interviewer and the respondent complete the questionnaire.

Mail survey Data collection where paper questionnaire forms are sent to the respondents. Completed questionnaires are returned to the research organization.

Mixed mode survey A survey where various modes of data collection are combined. Modes can be used concurrently (different groups are approached by different modes) or sequentially (nonrespondents of a mode are re-approached by a different mode).

Mode effect The same question produces a different answer when asked in a different mode.

Primacy effect Respondents show a preference for the first option in the list of answers of a closed question. This typically happens in face-to-face and telephone surveys.

Purposive sampling Nonprobability sampling where the selection of the sample is based on the judgment of the researcher as to which elements best fit the criteria of the study.

Probability sampling Sampling where selection of elements is a random process. Each element must have a positive and known probability of selection.

Quota sampling Purposive sampling where elements are selected from the population in such a way that the distribution of some auxiliary variables matches the population distribution of these variables.

Random digit dialing (RDD) Sample selection for a telephone survey where random telephone numbers are generated by a computer algorithm.

Recency effect Respondents show a preference for the last option in the list of answers of a closed question. This typically happens in mail and Web surveys.

Representative method A methods proposed by Anders Kiaer in 1896 to select a sample from a population that forms a “miniature” of the population.

Satisficing The phenomenon that respondents do not think very carefully about the meaning of the question or the answer. They just give a quick answer to be able to continue and finish the questionnaire.

Socially desirable answer The phenomenon that respondents do not give the true answer but an answer that is socially desirable.

Telephone interviewing Interviewing whereby interviewers call selected persons by telephone. If contact is made with the desired person, and this person wants to cooperate, the interview is started and conducted over the telephone.

Unimode questionnaire A questionnaire that satisfies Don Dillmans's unimode guidelines. Such questionnaires do not suffer from mode effects.

Web survey Data collection whereby respondents complete the questionnaires on the Internet.

4.6 REFERENCES

Ariel, A., Giessen, D., Kerssemakers, F., and Vis-Visschers, R. (2008), *Literature Review on Mixed-Mode Studies.* Technical paper. Statistics Netherlands, Heerlen.

Armstrong, J. (1975), Monetary Incentives in Mail Surveys. *Public Opinion Quarterly* 39, pp. 111–116.

Bethlehem, J.G. (1988), Reduction of the Nonresponse Bias through Regression Estimation. *Journal of Official Statistics* 4, pp. 251–260.

Bethlehem, J.G. (1997), Integrated Control Systems for Survey Processing. In: Lyberg, L, Biemer, P., Collins, M., De Leeuw, E., Dippo, C., Schwarz, N., and Trewin, D. (eds.), *Survey Measurement and Process Control.* Wiley, New York, pp. 371–392.

Bethlehem, J.G. (2009), *Applied Survey Methods: A Statistical Perspective.* Wiley, Hoboken NJ.

Bethlehem, J.G., and Hofman, L.P.M.B. (2006), *Blaise—Alive and Kicking for 20 Years.* Proceedings of the 10th Blaise Users Meeting. Statistics Netherlands, Voorburg/Heerlen, pp. 61–88.

Beukenhorst, D., and Wetzels, W. (2009), *A Comparison of Two Mixed-mode Designs of the Dutch Safety Monitor: Mode Effects, Costs, Logistics.* Technical paper DMH 206546. Statistics Netherlands, Methodology Department, Heerlen.

Biemer, P.P., and Lyberg, L.E. (2003), *Introduction to Survey Quality.* Wiley, New York.

Bowley, A.L. (1906), Address to the Economic Science and Statistics Section of the British Association for the Advancement of Science. *Journal of the Royal Statistical Society* 69, pp. 548–557.

Bowley, A.L. (1926), Measurement of the Precision Attained in Sampling. Bulletin of the International Statistical Institute 12 (Book 1), pp. 6–62.

CBS (1987), Automation in Survey Processing. *CBS Select*, Staatsuitgeverij, the Hague.

Couper, M.P. (2000), Web Surveys: A Review of Issues and Approaches. *Public Opinion Quarterly* 64, pp. 464–494.

Couper, M.P. (2008), *Designing Effective Web Surveys.* Cambridge University Press, Cambridge.

Couper, M.P., Baker, R.P., Bethlehem, J.G., Clark, C.Z.F., Martin, J., NichollsII W.L., and O'Reilly, J.M. (eds.) (1998), *Computer Assisted Survey Information Collection.* Wiley, New York.

Couper, M.P., and Nicholls, W.L. (1998), The History and Development of Computer Assisted Survey Information Collection Methods. In: Couper, M.P., Baker, R.P., Bethlehem, J., Clark, C.Z.F., Martin, J., Nicholls, W.L., and O'Reilly, J. (eds.), *Computer Assisted Survey Information Collection.* Wiley, New York.

De Bie, S.E., Stoop, I.A.L., and Vries, K.L.M. de (1989), *CAI Software, an Evaluation of Software for Computer Assisted Interviewing.* VOI Uitgeverij, Amsterdam.

De Leeuw, E.D. (1992). *Data Quality in Mail, Telephone, and Face to Face Surveys.* PhDthesis. T.T.-publikaties, Amsterdam.

De Leeuw, E.D. (2005), To Mix or Not to Mix Data Collection Modes in Surveys. *Journal of Official Statistics* 21, pp. 233–255.

De Leeuw, E.D. (2008), Choosing the Method of Data Collection. In: De Leeuw, E.D., Hox, J.J., and Dillman, D.A. (eds.), *International Handbook of Survey Methodology.* Erlbaum Associates, New York, pp. 113–135.

De Leeuw, E., Callegaro, M., Hox, J., Korendijk, E., and Lensvelt-Mulders, G. (2007), The Influence of Advance Letters in Telephone Surveys. *Public Opinion Quarterly* 71, pp. 413–443.

Dillman, D.A. (2007), *Mail and Internet Surveys: The Tailored Design Method.* Wiley, Hoboken, NJ.

Dillman, D.A., and Bowker, D. (2001), The Web Questionnaire Challenge to Survey Methodologists. In: Reips, U.D., and Bosnjak, M. (eds.), *Dimensions of Internet Science.* Pabst Science Publishers, Lengerich, Germany.

Dillman, D.A., Smyth, J.D., and Christian, L.M. (2008), *Internet, Mail and Mixed-mode Surveys: The Tailored Design Method.* Wiley, Hoboken, NJ.

Dillman, D.A., Sinclair, M.D., and Clark, J.R. (1993), Effects of Questionnaire Length, Respondent-friendly Design and a Difficult Question on Response Rates for Occupant-Addressed Census Mail Surveys. *Public Opinion Quarterly* 57, pp. 289–304.

Eurostat (2008), Internet usage in 2008—Households and Individuals. *Data in focus 46/ 2008*. Eurostat, Luxembourg.

Fox, R., Crask, M., and Kim, J. (1988), Mail Survey Response Rate: A Meta-analysis of Selected Techniques for Inducing Response. *Public Opinion Quarterly* 52, pp. 467–491.

Fricker, R., and Schonlau, M. (2002), Advantages and Disadvantages of Internet Research Surveys.: Evidence from the Literature. *Field Methods* 15, pp. 347–367.

Goyder. J. (1987), *The Silent Minority: Nonrespondents on Sample Surveys*. Westview Press, Boulder, CO.

Groves, R.M. (1989), *Survey Errors and Survey Costs*. Wiley, New York.

Groves, R.M., and Couper, M.P. (1998), *Nonresponse in Household Interview Surveys*. John Wiley, Ney York.

Heerwegh, D., and Loosveldt, G. (2002), An Evaluation of the Effect of Response Formats on Data Quality in Web Surveys. Paper presented at the International Conference on Improving Surveys, Copenhagen 2002.

Hox, J.J., and De Leeuw, E.D. (1994), A Comparison of Nonresponse in Mail, Telephone and Face-to-face Surveys. Applying Multilevel Modeling to Meta-analysis. *Quality and Quantity* 28, pp. 329–344.

Janssen, B. and Schrooten, M., and Wetzels, W. (2007), Mixed-Mode Enquêtering bij Personen en Huishoudens 2005-2007: Een Overzicht (in Dutch). Technical paper SOO -2007-H231. Statistics Netherlands, Heerlen.

Kiaer, A. N. (1895), Observations et Expériences Concernant des Dénombrements Représentatives. Bulletin of the International Statistical Institute (Book 2) pp. 176–183.

Kiaer, A. N.(1997 reprint), Den Repräsentative Undersökelsesmetode. *Christiania Videnskabsselskabets Skrifter. II. Historiskfilosofiske klasse,* Nr 4 (1897). English translation: The Representative Method of Statistical Surveys, Statistics Norway.

Krosnick, J.A. (1991), Response Strategies for Coping with the Cognitive Demands of Attitude Measures in Surveys. *Applied Cognitive Psychology*, 5, pp. 213–236.

Kuusela, V., Vehovar, V., and Callegaro, M. (2006), Mobile phones—Influence on Telephone Surveys. Paper presented at the Second International Conference on Telephone Survey Methodology, FL.

Lepkowski, J.M. (1988), Telephone Sampling Methods in the United States. In: Groves, R. M., Biemer, L.P., Lyberg, L.E., Massey, J.T., NichollsII W.L., and Waksberg, J. (eds.), *Telephone Survey Methodology.* Wiley, New York, pp. 73–98.

Lyberg, L., Biemer, P., Collins, M., De Leeuw, E., Dippo, C., Schwarz, N., and Trewin, D. (1997), *Survey Measurement and Process Quality.* Wiley, New York.

Neyman, J. (1934), On the Two Different Aspects of the Representative Method: the Method of Stratified Sampling and the Method of Purposive Selection. *Journal of the Royal Statistical Society* 97, pp. 558–606.

Nicholls, W.L., and Groves, R.M. (1986), The Status of Computer Assisted Telephone Interviewing. *Journal of Official Statistics* 2, pp. 93–134.

Pierzchala, M. (2006), *Disparate Modes and Their Effect on Instrument Design.* Paper presented at the 10th International Blaise Users Conference, Papendal, The Netherlands.

Roberts, C. (2007), *Mixing Modes of Data Collection in Surveys: A Methodological Review.* NCRM review paper, 8. ESRC National Centre for Research Methods, Swindon, UK.

Saris, W.E. (1998), Ten Years of Interviewing without Interviewers: The Telepanel. In: Couper, M.P., Baker, R.P., Bethlehem, J.G., Clark, C.Z.F., Martin, J., NichollsII W.L., and O'Reilly, J.M. (eds.), *Computer Assisted Survey Information Collection*. Wiley, New York, pp. 409–430.

Scherpenzeel, A. (2008), An Online Panel as a Platform for Multi-disciplinary Research. In: Stoop, I., and Wittenberg, M. (eds.), *Access Panels and Online Research, Panacea or Pitfall?* Aksant, Amsterdam, pp. 101–106.

Snijkers, G.J.M.E. (2002). Cognitive Laboratory Experiences: *On Pre-testing Computerised Questionnaires and Data Quality*. PhD thesis. Utrecht University, Utrecht.

Tourangeau, R., and Rasinski, K.A. (1998), Cognitive Processes Underlying Context Effects in Attitude Measurement. *Psychological Bulletin* 103, pp. 299–314.

Wagner, J. (2008), *Adaptive Survey Design to Reduce Nonresponse Bias*. PhD thesis. University of Michigan, Ann Arbor.

4.7 EXERCISES

Exercise 4.1 What is quota sampling?

(a) A form of probability sampling.

(b) A form of purposive sampling.

(c) A form of sampling that is only used in opinion polls.

(d) A form of sampling introduced by Arthur Bowley in 1906.

Exercise 4.2 Why is random digit dialing (RDD) sometimes preferred for a telephone survey instead of random sampling from a telephone directory?

(a) RDD sampling provides more auxiliary information about nonrespondents.

(b) Response rates are lower in an RDD survey.

(c) RDD guarantees full coverage of the population.

(d) An RDD sample is less expensive than a sample form directory.

Exercise 4.3 What does acquiescence mean?

(a) This is the tendency to not answer sensitive questions.

(b) This is the tendency to give more extreme answers.

(c) This is the tendency to disagree with what the interviewers say.

(d) This is the tendency to agree with what the interviewers say.

Exercise 4.4 How can the primacy effect in closed questions in web surveys be reduced?

(a) By randomizing the order of the answer options.

(b) By putting the answer options in the reverse order.

(c) By reducing the number of answer options.

(d) By increasing the number of answer options.

Exercise 4.5 Which of the following options would not reduce a possible mode effect in a mixed mode survey?

(a) Reduce the number of answer options of closed questions.

(b) Include all answer options also in the text of a closed question.

(c) Keep the text of a question the same across modes.

(d) Avoid routing instructions in self-administered modes.

Exercise 4.6 Which of the following options is not an advantage of computer-assisted interviewing (CAI) as compared to traditional modes of data collection?

(a) Data quality is higher due to included checks.

(b) The software is in charge of routing through the questionnaire.

(c) CAI leads to higher response rates.

(d) Data can be processed quicker.

Exercise 4.7 What is the effect of the mode of data collection on an opinion question with the possible answers as follows: strongly disagree, disagree, neutral, agree, and strongly agree?

(a) More people will select the neutral category in mail and Web surveys.

(b) More people will select the neutral category in CAPI and CATI surveys.

(c) More respondents will select the extreme options strongly disagree and strongly agree.

(d) The mode of data collection does not influence the answer patterns.

CHAPTER FIVE

Analysis of Nonresponse

5.1 Introduction

Nonresponse has the potential to introduce a bias. It is therefore important to carry out a nonresponse analysis on the data that have been collected in a survey. Such an analysis should reveal whether or not the response to the survey is selective, and if so, how the distribution of the characteristics of the sampled elements is affected by the selective nonresponse.

An example of such a nonresponse analysis is provided in this chapter. First, a number of techniques that can be used to analyze nonresponse are presented. Data from the General Population Survey are then used to illustrate these different techniques.

5.2 Theory

5.2.1 HOW TO DETECT A BIAS?

How to detect that the nonresponse is selective? The available data with respect to the target variables will not be of much use because these data are only available for the respondents and not for the nonrespondents. So it cannot be established whether respondents and nonrespondents differ with respect to these variables. The way out of this problem is to use auxiliary variables; see Figure 5.1.

Auxiliary variables are, in general, variables that are available prior to sampling, either from the sampling frame or from a different source. Usually values of

Handbook of Nonresponse in Household Surveys, By Jelke Bethlehem, Fannie Cobben, and Barry Schouten

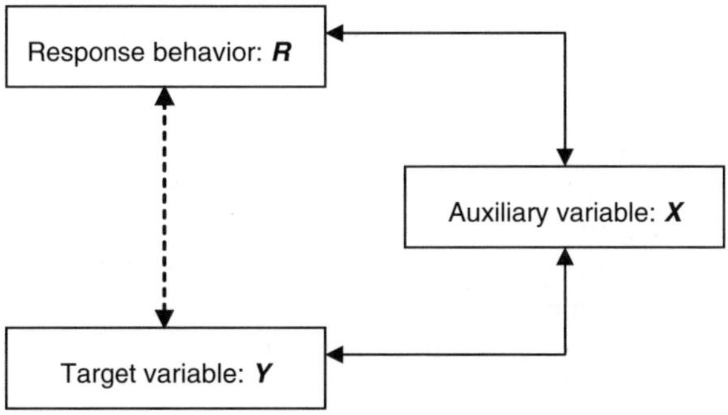

FIGURE 5.1 Relationships between target variable, response behavior and auxiliary variable.

auxiliary variables are available for every element in the sample or the population. If such detailed information is not available, a variable that is measured in the survey, and for which the distribution in the population (or in the complete sample) is known, can also be used as an auxiliary variable. In this situation it can also be established whether there is a relationship between this variable and the response behavior. Use of sampling information may lead to somewhat weaker conclusions because sampling causes small deviations between the sample and population distribution.

To better understand this situation, let us recall the three different response mechanisms discussed in Chapter 2. The first is the response mechanism called *missing completely at random* (MCAR). The occurrence of nonresponse (R) is completely independent of both the target variable (Y) and the auxiliary variable (X). The response is not selective. Estimates are not biased. There is no problem.

In MCAR the response behavior (R) and an auxiliary variable (X) are unrelated. If there also is a strong relationship between target variable (Y) and auxiliary variable (X), this situation indicates that there is no strong relation between target variable (Y) and response behavior (R), and thus that estimators do not have a severe bias.

Then again, if there is no strong relationship between the auxiliary variable (X) and the target variable (Y), analysis of the relationship between the auxiliary variable (X) and the response behavior will give us no information about a possible bias of estimates.

The second response mechanism is *missing at random* (MAR). This situation occurs when there is no direct relation between the target variable (Y) and the response behavior (R), but there is a relation between the auxiliary variable (X) and the response behavior (R). The response will be selective, but this can be solved by applying a weighting technique using the auxiliary variable.

In MAR the response behavior (R) and an auxiliary variable (X) are related. If there also is a strong relationship between target variable (Y) and auxiliary variable (X), this situation indicates that there is an (indirect) relation between target variable (Y) and response behavior (R), and thus that estimators may be biased.

The third response mechanism is *not missing at random* (NMAR). There is a direct relationship between the target variable (Y) and response behavior (R), and this relationship cannot be accounted for by an auxiliary variable. Estimators are biased. Correction techniques based on use of auxiliary variables may not be able to reduce such a bias.

In all three mechanisms the relationship between auxiliary variables and response behavior should be analyzed, and also the relationship between target variables and auxiliary variables. If such relationships exist, there is a risk of biased estimates which suggests the need for some nonresponse correction technique. It should be noted that surveys usually measure several target variables Y. The relationship between every Y and the set of X-variables may be different, and likewise different response mechanisms may apply.

5.2.2 WHERE TO FIND AUXILIARY VARIABLES?

To analyze the effects of nonresponse, auxiliary variables are needed. The auxiliary variables have to be available for the respondents and some kind of information about the distribution of these variables in the population (or in the complete sample) is also needed. In the ideal situation the auxiliary variables are available for every element in the population, and hence for both respondents and nonrespondents.

One obvious source of auxiliary information is the sampling frame itself. For example, if a sample from a population register is selected, it is possible to find in there variables like age (computed from date of birth), gender, marital status, household composition, and geographical location (e.g., neighborhood). Note that these frame variables are available for both respondents and nonrespondents.

A survey data set may sometimes be linked to a register or another source of administrative information. For example, Statistics Netherlands is allowed to link its survey data to data obtained from tax authorities and the social security administration. Records are matched using personal identification numbers.

Another way to obtain information on respondents and nonrespondents is by recording data about the process of data collection. This type of information is usually referred to as *paradata.* Paradata are always restricted to the sample level, meaning there are no paradata available on the population. Examples of paradata are the number of contact attempts, or the sequence of outcomes of contact attempts. Also interviewer observations about the location of the selected persons can be regarded as paradata. Examples of interviewer observations are the neighborhood, type of house, and age of house. A special type of paradata is the classification of response types (e.g., unprocessed, noncontact, not-able, refusal, or response).

The availability of this type of data depends on the mode of data collection. In interviewer-assisted surveys, in general, more information is available about the selected people and their behavior when contacted. But in self-administered surveys, no information about the location of the selected people, or about their response behavior, is available. Exceptions are Web surveys and CASI surveys, in which a different type of paradata (so-called audit trails) becomes available about, for instance, the number of changes made to answers, the sequence in which questions are answered, and the time spent on each question; see Heerwegh (2003).

5.2.3 METHODS OF ANALYSIS

To investigate the relationship between the auxiliary variables and response behavior, and the auxiliary variables and the target variables, different techniques can be used. A global analysis can be performed first by doing bivariate tests of independence. For this purpose, Cramér's V statistic can be used (e.g., see Agresti, 2002). Cramér's V is a χ^2-based measure of association that accounts for the different degrees of freedom of the variables (a different number of categories) and the dimension of the dataset:

$$V = \sqrt{\frac{\chi^2}{N \times \min(r-1,\ c-1)}}, \tag{5.1}$$

where r is the number of categories from the auxiliary variable (in rows) and c is the number of categories of the response type or the target variable (in columns). For a binary response indicator or target variable, Cramér's V reduces to $\sqrt{\chi^2/N}$. $V=0$ when the auxiliary variable and the response indicator or target variable are completely independent. $V=1$ when the auxiliary variable and the response indicator or target variable are completely dependent.

EXAMPLE 5.1 Cramér's *V* Statistic

In the General Population Survey (GPS), it is useful to look at the relationship between response and auxiliary variables to see whether a nonresponse was selective. Take the auxiliary variable *marital status*. This variable has four categories—being married, divorced, unmarried, and widowed—and hence $r=4$. The response indicator has two categories—response and nonresponse—and hence $c=2$. See the table below.

	Response	
Marital status	No	Yes
Not married	4015	4559
Married	7180	12006
Widowed	995	1151
Divorced	1037	1076

The dimension of the data set is $N=32{,}019$. The value for χ^2 is 302.42. Hence the value for Cramér's V becomes

$$V = \sqrt{\frac{302.42}{32019 \times \min(1,3)}} = \sqrt{\frac{302.42}{32019}} = 0.07$$

This indicates that the relationship between response and marital status is weak.

In addition the bivariate relationship between the auxiliary variables and the response behavior or the target variable can be displayed in a graph. Examples will be given later.

Remember, in a bivariate analysis, the effects of other auxiliary variables on the relationship between the variable considered and the response indicator or target variable are not accounted for. So a multivariate analysis has to be performed.

Usually, in household surveys, the target variables are *qualitative variables.* Qualitative variables divide the population or sample into groups. The values denote categories. Sample elements in the same category belong to the same group. Examples of qualitative variables are ethnicity, religion, and employment status.

Indicator variables measure whether or not a sample element has a certain property. They can only assume the values 0 and 1. The value is 1 if an element has the property; otherwise, it is 0. The response indicator is an indicator variable. If a sample element responded to the survey, the response indicator takes the value 1. Otherwise, its value is 0. However, when a distinction must be made between the different types of responses, the response indicator is described by a qualitative variable as well.

Suppose that the target population U consists of N elements, $k = 1, 2, \ldots, N$, that can be observed and made measurements on. Assume, U to be finite and N to be known. Assume further that one target variable, denoted Y, is in the survey. Every element k of the population has a fixed, nonrandom value for the target variable, denoted by Y_k. The N-vector of values for the target variables is denoted by $Y = (Y_1, Y_2, \ldots, Y_N)'$.

Now let there be P auxiliary variables. Every person k can be associated with a p-vector of values for the auxiliary variables $X_k = (X_{k1}, X_{k2}, \ldots, X_{kp})'$. The auxiliary information for the entire population is denoted by the $N \times p$-matrix X.

Under the *random response model* (see Chapter 2) every element k of the population U has an unknown, nonzero probability of response denoted by ρ_k. If element k is selected in the sample, a random mechanism is activated that results with probability ρ_k in response and with probability $(1 - \rho_k)$ in nonresponse. R is a vector of response indicators, $R = (R_1, R_2, \ldots, R_N)'$ where $R_k = 1$ if the corresponding element k responds; otherwise, $R_k = 0$. So $P(R_k = 1) = \rho_k$, and $P(R_k = 0) = 1 - \rho_k$.

For indicator variables, a multivariate analysis can be performed by way of a logistic regression model:.

$$\log\left(\frac{\rho(X_k)}{1-\rho(X_k)}\right) = \text{logit}(\rho(X_k)) = X'_k\beta, \tag{5.2}$$

where β is a p-vector of coefficients that correspond to the different auxiliary variables in X_k. The response indicator (or target variable) is the dependent variable, and the auxiliary variables are selected as independent variables. Because all the variables are considered in this model, the effects that the auxiliary variables have on each other have been accounted for in the final model output.

For qualitative variables with more than two categories, a multinomial logistic regression model can be used; see, for instance, Johnston and DiNardo (1997) or Greene (2003).

5.3 Application

The methods for nonresponse analysis introduced in this chapter are applied to a real survey data set in this section. This data set is based on a Dutch survey that has been carried out by Statistics Netherlands. It is called the General Population Survey (GPS). A detailed description of the GPS is given in Chapter 1.

The results from this analysis confirm to a large extent the results in other studies on nonresponse, in both Dutch surveys and international surveys. For a literature overview of the determinants of nonresponse, see Cobben (2009).

Persons were selected by means of a stratified two-stage sample. In the first stage, municipalities were selected within regions with probabilities proportional to the number of inhabitants. In the second stage, an equal probability sample was drawn in each selected municipality. Sampling frames were the population registers of the municipalities. The sampling design was such that each person had the same probability of being selected (a so-called self-weighting sample). In Table 5.1 the survey results for the GPS are displayed.

Ultimately about 59% of the sampled persons responded.

The composition of the nonresponse is displayed in Figure 5.2. It is clear that refusal is by far the largest cause of nonresponse (59%). In 14% of the cases, no contact could be established with the sampled persons. And 19% of the cases were not processed in the field. Reasons for this type of nonresponse are lack of capacity (high workload of the interviewer) and interviewer not available (illness, holiday). Only 8% of the nonresponse is caused by people not able to answer the questions due to illness, handicap, or language problems.

5.3.1 BIVARIATE ANALYSIS

A first step in the analysis of nonresponse would be to explore possible relationships between auxiliary variables and response behavior, and auxiliary variables and the

TABLE 5.1 Fieldwork results of the General Population Survey

Result	Frequency	Percentage
Sample size	32,019	100.0 %
Response	18,792	58.7%
Nonresponse	13,227	41.3%
Unprocessed cases	2,456	7.7%
Noncontact	1,847	5.8%
Not-able	1,034	3.2%
Refusal	7,890	24.6%

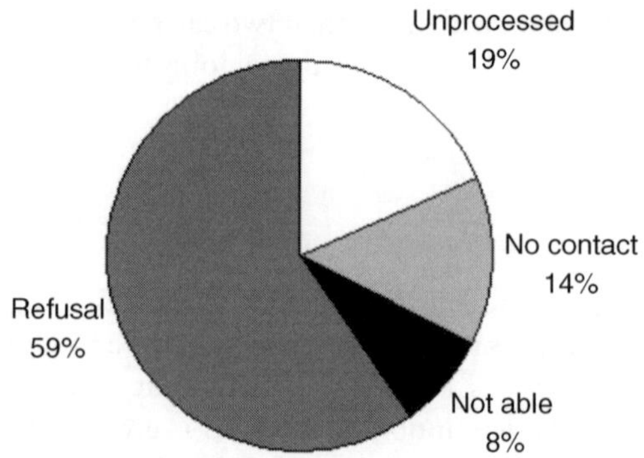

FIGURE 5.2 Composition of the nonresponse.

target variables. For the GPS, bivariate tests of independence as described in Section 5.2.3 have been performed. The results are discussed in Section 5.3.1.1. In addition a number of graphs were made for the most interesting variables. These are provided in Section 5.3.1.2.

5.3.1.1 Bivariate Tests of Independence Cramér's *V* statistic was calculated to test the independence between the auxiliary variables and the response indicator. SPSS was used to perform the tests. The results are displayed in Table 5.2.

In general, the relationships found are not very strong. The region of the country, the degree of urbanization, the percentage of nonnatives, and the average house value in the neighborhood show the strongest relation to the response behavior. A remarkable result is the strong relationship between having a listed phone number and response. Remember, the GPS is only partly a CATI survey. Households that have a listed phone are notably easier to reach. These results are also known from previous research; see Cobben (2009).

Suppose that the auxiliary variables have a strong relationship with the response behavior. Then the survey runs a risk of a nonresponse bias for survey variables that are related to those variables. For example, persons from a different region are not equally well presented in the response. If these persons give different answers to the survey questions, this will result in a nonresponse bias. Unfortunately, it is not possible to check this since the nonrespondents do not answer the survey questions. It is possible, though, to check the relationship between the auxiliary variables and the survey target variables for the respondents. In Table 5.3 Cramér's *V* statistic for the test of independence between the target variables and the region of residence is displayed, based on the respondents' data.

The results in Table 5.3 should be interpreted with care. After all, these relationships hold for the respondents but not necessarily for the nonrespondents. The relationship between religious denomination and region is high. Also whether the respondent owns a house would suggest a strong connection to the region.

Of course, the relationships can differ depending on the auxiliary variable for which the relationships are investigated. For instance, if a relationship with regard

TABLE 5.2 Cramér's V statistic for bivariate tests of independence between the auxiliary variables and the response behavior

Auxiliary variable	V
Region of the country	0.163
Degree of urbanization	0.153
Has listed phone number	0.150
Percentage nonnatives in neighborhood	0.138
Percentage nonwestern nonnatives in neighborhood	0.133
Average house value in neighborhood	0.115
Type of nonnative	0.112
Type of household	0.106
Size of the household	0.099
Marital status	0.097
Is married	0.097
Is nonnative	0.087
Has social allowance	0.077
Age in 13 classes	0.061
Has an allowance	0.061
Children in household	0.056
Has a job	0.037
Age in 3 classes	0.030
Has disability allowance	0.021
Gender	0.011
Has unemployment allowance	0.000

TABLE 5.3 Cramér's V statistic for bivariate tests of independence between the target variables and the region of residence (for the respondents only)

Target variable	V
Religious denomination	0.247
Owns house	0.182
Has newspaper subscription	0.094
Is active in a club	0.068
Level of education	0.064
PC in household	0.055
Wants to move within a year	0.049
Job level	0.045
Is interested politics	0.041
Employment situation	0.032
General health condition	0.031

TABLE 5.4 Cramér's V statistic for bivariate tests of independence between the target variables and the ethnic background (for the respondents only)

Target variable	V
Religious denomination	0,365
Has newspaper subscription	0,193
Owns house	0,163
Is active in a club	0,101
PC in household	0,088
Is interested politics	0,073
Level of education	0,066
Wants to move within a year	0,053
General health condition	0,048
Job level	0,045
Employment situation	0,030

to ethnic background is desired (see Table 5.4), different relationships can be discovered, with the relationships to the target variables being slightly stronger.

From these results it is not possible to say something about the nature of the relationship between the auxiliary variables and the response behavior, for instance, what categories of the variables show the strongest relationship, and the direction of that relationship. A bivariate graph will give more insight. In the next section a graphical representation of the different relationships is therefore provided.

5.3.1.2 Bivariate Graphs In the previous section the relationship between the auxiliary variables and the response behavior was tested with Cramér's *V* statistic. The strongest relationships (see Table 5.2) are found for the following variables: region of the country, degree of urbanization, having a listed phone number, percentage of nonnatives in the neighborhood, average house value in the neighborhood, type of nonnative (or ethnic background), and type of household. These are all categorical variables.

Some of the variables are ordered: degree of urbanization, percentage of nonnatives, and average house value. Having a listed phone is an indicator variable, that being either no (0) or yes (1). The other variables are unordered categorical variables. These are region, ethnic background, and type of household.

Two series of graphs are presented. The first series examines the relationship between each auxiliary variable and response/nonresponse. In addition one may wonder what causes the nonresponse. As outlined in Chapter 1, it is important to distinguish between the different causes of nonresponse. These causes here are unprocessed cases, not being able to participate either due to illness or due to language problems, noncontact, and refusal. The second series show the relationship between each auxiliary variable and the fieldwork result, with categories response, refusal, no contact, not-able, and unprocessed.

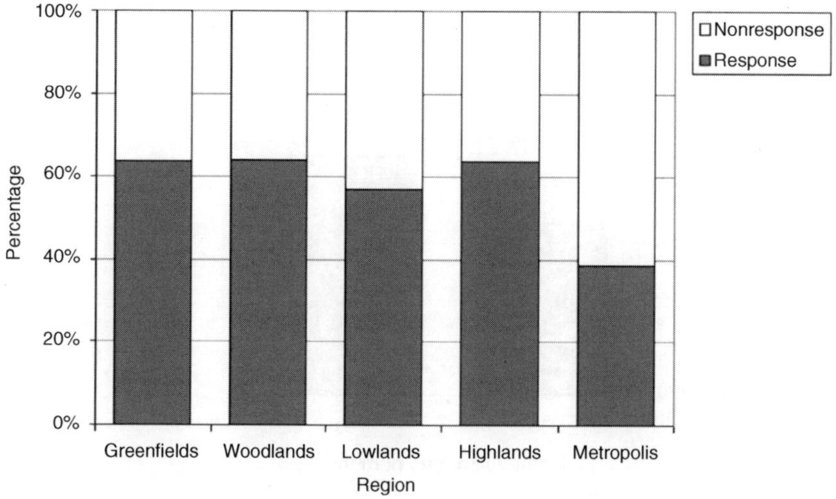

FIGURE 5.3 Relationship between response and region of the country.

From Figure 5.3 it is clear that in Greenfields, Woodlands, and Highlands the response is almost the same. In Lowlands it is approximately 5% lower. The most striking result in this graph, however, is the low response in the urbanized region of Metropolis. The response is more 20% lower than in the other regions. Apparently persons in Metropolis respond less to surveys.

Figure 5.4 shows a clear relationship between the degree of urbanization and response. The more urbanized the area, the lower is the response. This confirms the strong negative relationship between response behavior and the Metropolis region,

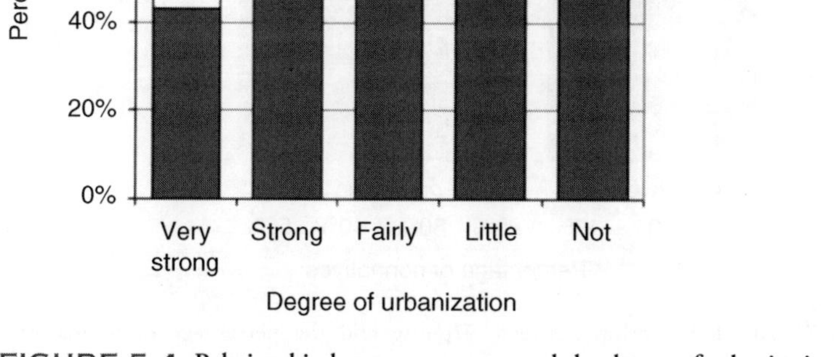

FIGURE 5.4 Relationship between response and the degree of urbanization.

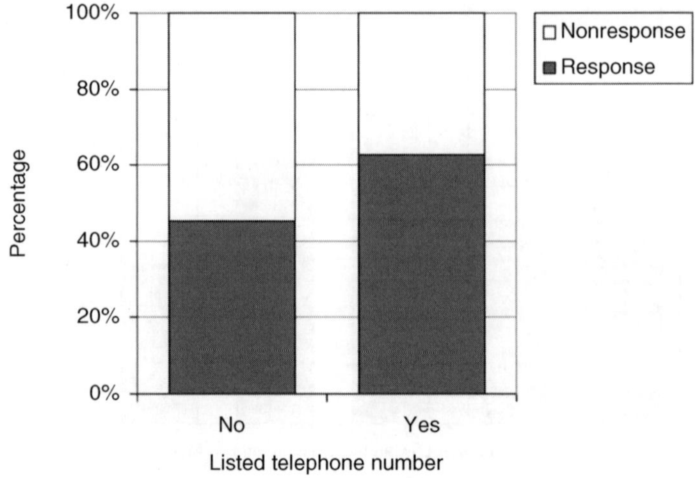

FIGURE 5.5 Relationship between response and having a listed phone number.

since these regions are more densely populated than the other regions of the country.

In Figure 5.5 it can be observed that persons with a listed phone number respond more to the survey than persons without a listed phone number. This latter group actually consists of three different groups: persons with a nonlisted phone, persons with a mobile phone only, and persons without a phone. It was not possible, though, to distinguish among these three groups.

From Figure 5.6 it becomes clear that the higher the percentage of nonnatives in the neighborhood, the lower is the response. The difference between

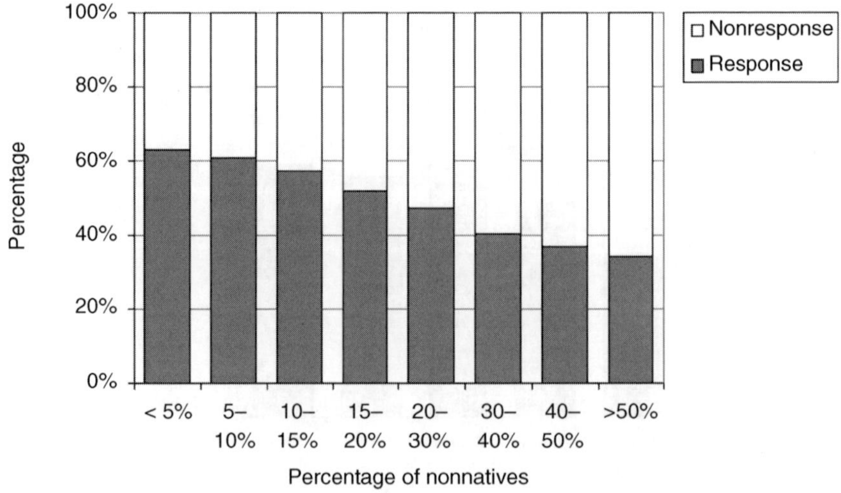

FIGURE 5.6 Relationship between response and the percentage of nonnatives in the neighborhood.

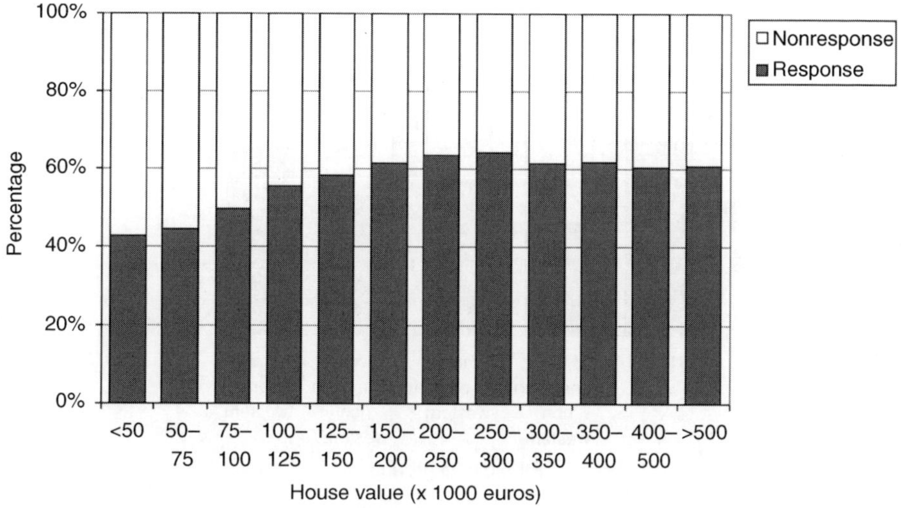

FIGURE 5.7 **Relationship between response and the average house value in the neighborhood.**

neighborhoods with less than 5% nonnatives and neighborhoods with more than 50% nonnatives is well over 20%.

Figure 5.7 shows the relationship between response/nonresponse and the average house value in the neighborhood. The relationship is more complicated here. It appears that in neighborhoods with a lower average house value the response is lower. However, when the average house value exceeds 300 thousand euros, this relationship is reversed. In neighborhoods with an average house value of 300 thousand euros and more, the response seems to decrease with an increasing average house value.

In Figure 5.8 native persons can be observed to have the highest response percentage, followed by second generation western nonnatives. About a 10% lower response is obtained in the group of first generation western nonnatives. In general, western nonnatives have a higher response than nonwestern nonnatives. First-generation western nonnatives seem to respond more than second-generation western nonnatives. The response is lowest for the first-generation nonwestern nonnatives.

Figure 5.9 shows that the highest response percentage is obtained in the group of couples with children, followed by couples. Singles and single parents have a lower response, and other types of households have the lowest response percentage.

With regard to the type of household, one may wonder what causes the nonresponse. Intuitively it makes sense that couples with children have the highest response percentage. However, is this caused by their at-home behavior, meaning they are more often at home? Or are they more inclined to participate in surveys than persons without children due to their link to society because of the children?

In order to answer those questions, it is informative to distinguish the nonresponse by cause. Consider, once again, the different reasons for nonresponse:

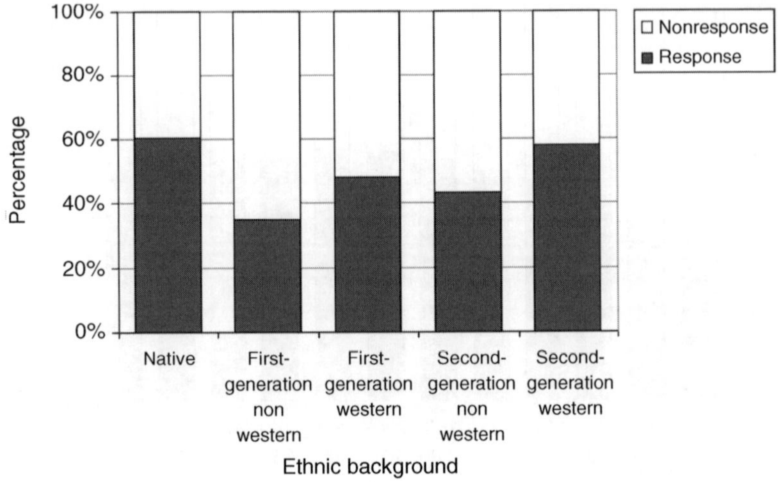

FIGURE 5.8 Relationship between response and the ethnic background.

unprocessed cases, not being able to participate either due to illness or due to language problems, noncontact, and refusal. As shown in Figure 5.2, the largest cause of nonresponse is refusal. However, also a large proportion of nonresponse is due to unprocessed cases and noncontact. The smallest group of nonresponse is the group that is not able to participate. Notice that these types of nonresponses are nested. In order to refuse participation, the case has to be processed in the field, contact has to be made, and the person has to be able to participate.

From Figure 5.3 it is known that the response was lowest in the Metropolis region of the country. From Figure 5.10 it becomes clear that this is mostly due to

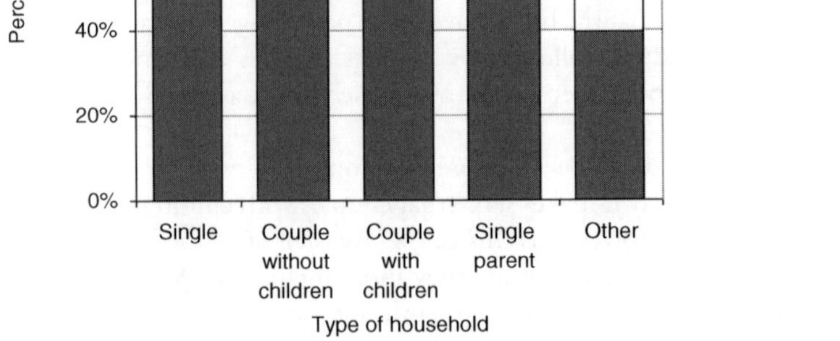

FIGURE 5.9 Relationship between response and the type of household.

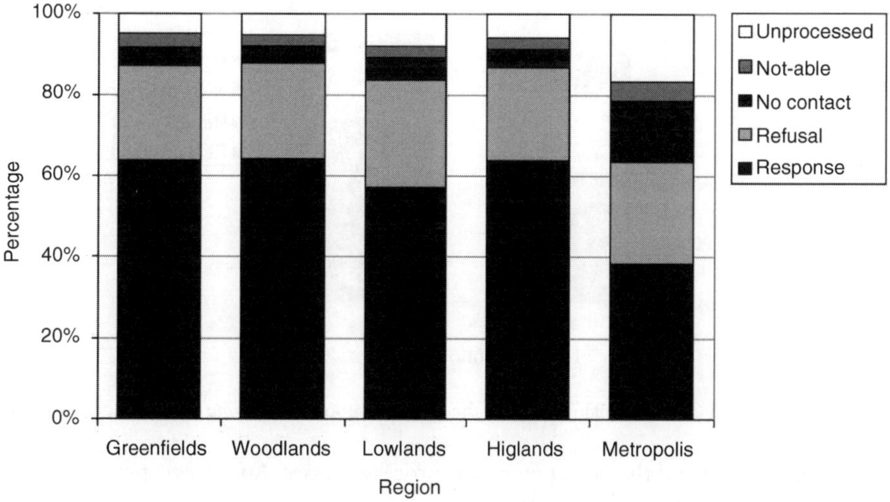

FIGURE 5.10 Relationship between the type of response and the region of the country.

more unprocessed cases, and a higher percentage of noncontacts. The percentage of refusals in the Metropolis region does not differ much from the other regions. The percentage not-able is somewhat higher in Metropolis. This is mainly caused by language problems.

In Figure 5.11 the same pattern as for the region of the country can be observed for the degree of urbanization. The very strongly urbanized areas have a higher percentage of unprocessed cases, and also a higher percentage of noncontacts, as was found for the urbanized Metropolis region of the country. In addition it can be seen in Figure 5.11 that the lower the degree of urbanization, the lower is the percentage

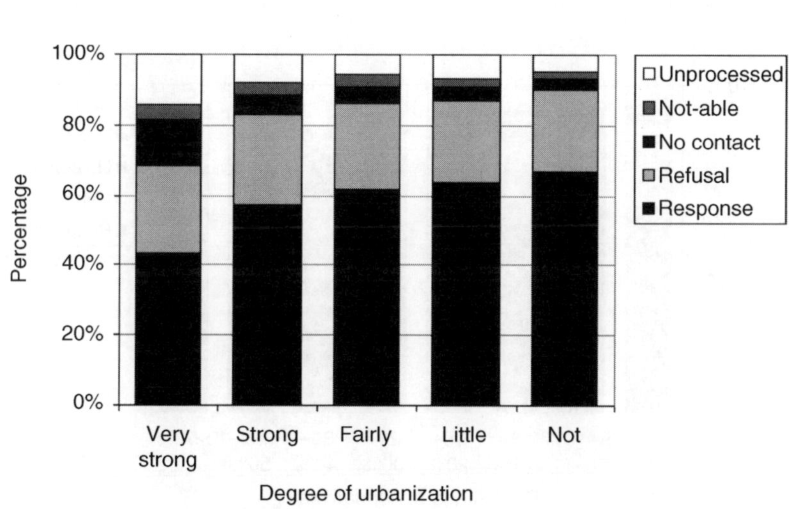

FIGURE 5.11 Relationship between the type of response and the degree of urbanization.

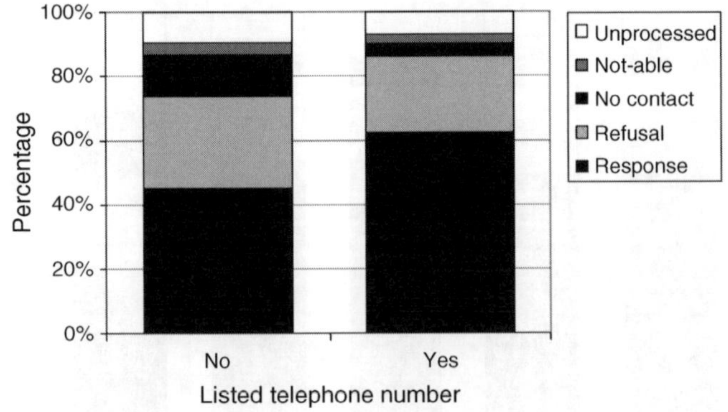

FIGURE 5.12 Relationship between the type of response and having a listed phone number.

of noncontacts and the percentage of unprocessed cases. Apparently persons in less urbanized areas are easier to reach.

From Figure 5.12 it can be concluded that the lower response for the persons without a listed phone number is apparently caused by a higher percentage of noncontacts and to a lesser extent by a higher refusal rate. Persons without a listed phone number are more difficult to reach than persons with a listed phone. The other causes of nonresponse show no large differences between having and not having a listed phone number.

Figure 5.13 shows a clear relationship between the type of response and the percentage of nonnatives in the neighborhood. The higher the percentage of nonnatives, the lower is the percentage of refusals and the higher is the percentage of unprocessed cases. Also the percentage of noncontact and not able to participate seem to increase with the percentage of nonnatives. The nonresponse that is caused by not being able to participate is for a large part be attributable to language problems.

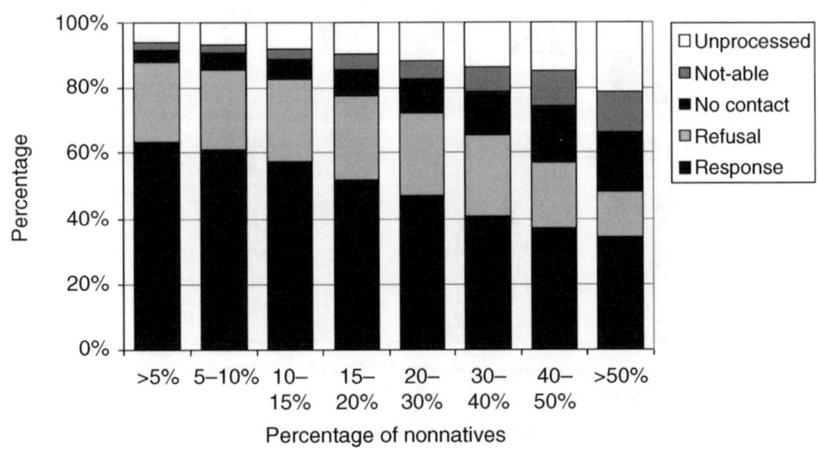

FIGURE 5.13 Relationship between the type of response and the percentage of nonnatives.

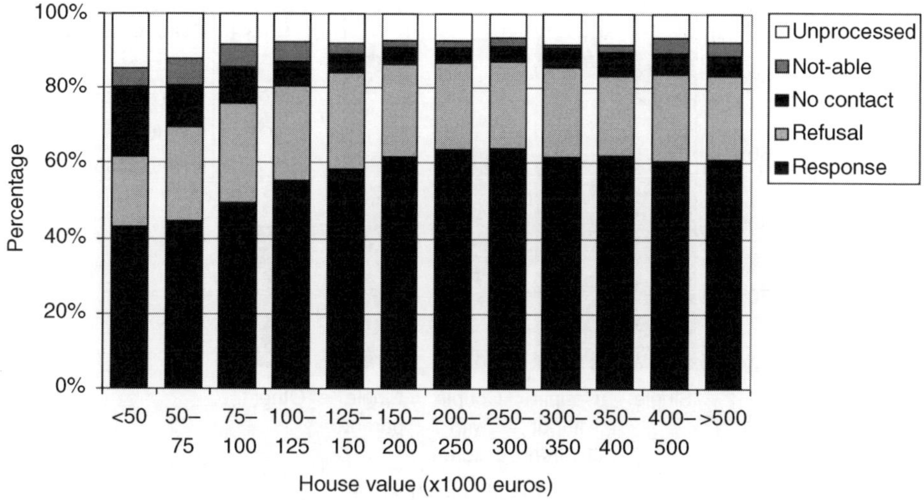

FIGURE 5.14 Relationship between the type of response and the average house value in the neighborhood.

In Figure 5.14 a mixed pattern can be observed. In neighborhoods with an average house value up to 300 thousand euros, the percentage of refusals seems to increase with an increasing average house value. The percentage of unprocessed cases and noncontacts seems to decrease with an increasing average house value. For neighborhoods with an average house value of 300 thousand euros and more, the composition of the response does not seem to differ much.

In Figure 5.15 the relationship between the type of response and the ethnic background of the sampled person is displayed. For persons from the first generation of nonnatives, the percentage that is not able to participate in the survey is large. This especially holds for nonwestern nonnatives and is probably

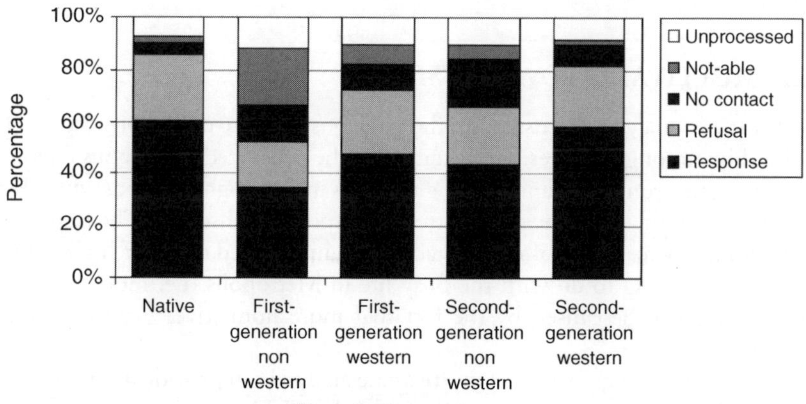

FIGURE 5.15 Relationship between the type of response and the ethnic background.

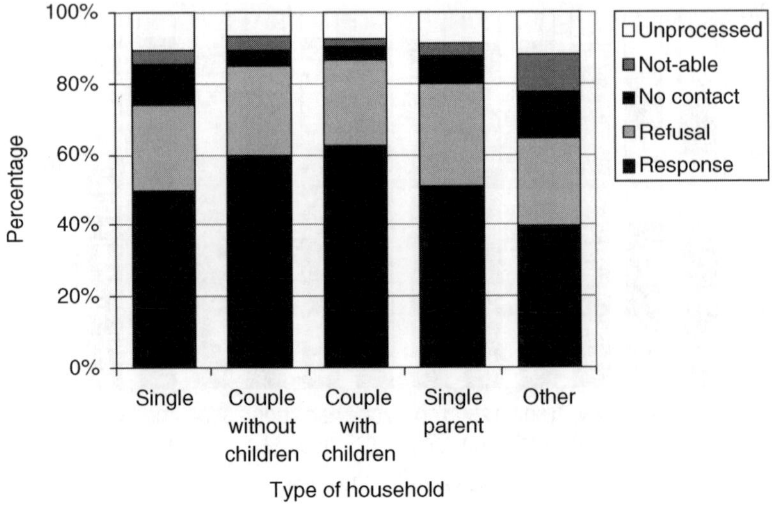

FIGURE 5.16 **Relationship between the type of response and the type of household.**

caused by language problems. Also noncontact in these two groups seems to be a more frequent cause of nonresponse than it is in the other groups. The percentage of noncontacts in the second-generation nonwestern nonnatives is also remarkable high. The percentage of refusals is lower in these two groups than in the other groups.

It can be observed in Figure 5.16 that single households are more difficult to reach than households that consist of a couple (with or without children). This makes sense, since the probability of finding someone at home simply increases with an increasing number of persons in the household. Households consisting of a single parent seem to refuse more frequently. Other households, for instance, households that take care of an elderly person, seem to be more frequently not able to participate.

5.3.2 MULTIVARIATE ANALYSIS

The bivariate analysis discussed in the previous section does not account for relationships among variables. For instance, in the urbanized Metropolis region of the country, in strongly urbanized areas as well as in the neighborhoods with a high percentage of nonnatives, the same pattern of response is observed. The response is lower, and the percentages of noncontacts and unprocessed cases are higher. Does this have something to do with the busy life in Metropolis (i.e. more urbanized areas), or could it be caused by the fact that more nonnatives live in the more urbanized areas?

To answer those questions, a multivariate analysis can provide useful insights. With a multivariate analysis the net effect of each variable on the response or target variable is revealed, adjusted for the effect of all the auxiliary variables included in the analysis.

Section 5.3.2.1 provides an outline of how the models are built. A logistic regression model for the response indicator is built and the resulting model is discussed in Section 5.3.2.2. Section 5.3.2.3 describes how these models can be extended to include different response types.

5.3.2.1 Model Selection and Interpretation All analyses were run in Stata. Based on Cramér's *V*, the strongest candidate for the model is selected. Stepwise, the logistic regression model with one additional variable is evaluated. The final model is found if none of the additional variables has a significant relation with the independent variable (the response indicator in Section 5.3.2.2 and the target variable(s) in Section 5.3.2.3).

The relative significance (i.e., adjusted for the effect of the other variables in the model) is evaluated by the p-value of a Wald test for joint significance of all the categories of a variable. The null hypothesis of the Wald test is that all coefficients corresponding to the categories of a variable are equal to zero, against the alternative hypothesis that at least one of the coefficients is not equal to zero. If $p > 0.05$, the variable is not added to the model.

The estimated coefficients β with the corresponding standard errors (se) are reported for the final model, for all categories except the reference category. Also the corresponding χ^2-value for the Wald test of joint significance of all categories of a variable is presented. An asterisk indicates that a category is significant on a 10% level; two asterisks indicate significance on a 5% level. The sign of the coefficient β determines whether the value of the dependent variable is increasing or decreasing with respect to the reference category.

For the general model fit the Nagelkerke pseudo R^2, the total χ^2, and the degrees of freedom (df) of the model are reported. The Nagelkerke pseudo R^2 (Nagelkerke, 1991) is expressed as

$$\text{Pseudo } R^2 = \frac{1-\left\{L(0)/L(\hat{\beta})\right\}^{2/n}}{1-L(0)^{2/n}}, \tag{5.3}$$

where $L(0)$ denotes the likelihood of the null model and $L(\hat{\beta})$ the likelihood of the fitted model. The pseudo R^2 can be interpreted as the proportion of explained variation by the model, like the R^2 in classical regression analysis. Furthermore the pseudo R^2 takes on values between 0 and 1.

5.3.2.2 A Logistic Regression Model for Response From the bivariate analysis of response and the auxiliary variables described in Section 5.2, the variables with the strongest relationship with response are known. These are region of the country, degree of urbanization, having a listed phone number, percentage of nonnatives in the neighborhood, percentage of nonwestern nonnatives in the neighborhood, average house value in the neighborhood, ethnic background, and type of household. In Table 5.5 the results from the multivariate analysis are presented, and in Table 5.6 the final logistic regression model.

TABLE 5.5 Multivariate analysis of response behavior

Variable	Wald χ^2					
Region	817.31	190.83	159.02	174.30	163.31	159.96
Degree of urbanization		89.90	50.00	22.31	19.97	16.36
Having a listed phone			415.61	344.05	274.83	251.62
Average housevalue				96.00	73.83	37.47
Ethnic background					69.58	83.89
Type of household						116.56
Size of household						
Marital status						
Has a social allowance						
Has a job						
Age (3 categories)						
Gender						
Pseudo R^2	0.019	0.022	0.031	0.033	0.035	0.038
χ^2	842.55	932.59	1347.74	1443.56	1514.46	1631.31
df	4	8	9	20	24	28
Region	159.17	160.22	162.57	163.70	163.62	164.79
Degree of urbanization	15.56	15.38	15.10	16.04	15.93	16.28
Having a listed phone	251.18	244.48	233.54	238.21	241.78	242.04
Average housevalue	34.43	30.87	26.84	24.64	25.71	25.29
Ethnic background	96.79	107.71	92.42	90.91	96.25	93.65
Type of household	31.06	16.25	16.74	15.34	14.08	23.61
Size of household	52.50	52.19	52.78	52.58	52.38	52.58
Marital status		49.91	51.30	65.01	76.66	74.12
Has a social allowance			14.76	8.08	8.50	8.62
Has a job				29.32	17.89	23.70
Age (3 categories)					13.05	10.97
Gender						14.72
Pseudo R^2	0.039	0.040	0.040	0.041	0.041	0.042
χ^2	1683.96	1733.62	1748.45	1777.79	1790.89	1805.62
df	32	35	36	37	39	40

The contributions of the different auxiliary variables to the model are displayed in Table 5.5. The first variable to be included is the region. This variable showed the strongest bivariate relation to the variable and was hence included first. In each following column, another variable is added to the model. In the last column, the final model is presented.

Notice that in general, the model does not fit well. The value for Nagelkerke's pseudo R^2 is low; only 4% of the total variance can be explained by all the auxiliary

TABLE 5.6 Logistic regression model for the response behavior

Variable	Category	β
Intercept		−0.2030
Region	Woodlands	0.0109
(Reference Greenfields)	Lowlands	−0.2752**
	Highlands	−0.0083
	Metropolis	−0.7210**
Urbanization	Strong	−0.0792
(Reference very strong)	Fairly	−0.0003
	Little	−0.0084
	Not	0.0825
Listed phone	Yes	0.4595**
(Reference no)		
Average house value	50–75 thousand	−0.0389
(Reference < 50thousand)	75–100 thousand	−0.0091
	100–125 thousand	
	125–150 thousand	0.0965
	150–200 thousand	0.1144
	200–250 thousand	0.1478
	250–300 thousand	0.1749
	300–350 thousand	0.1860
	350–400 thousand	0.0938
	400–450 thousand	0.1217
	450–500 thousand	0.0849
	> 500 thousand	0.2209
Ethnic background	First-generation nonwestern	−0.6336**
(Reference native)	First-generation western	−0.1871**
	Second-generation nonwestern	−0.2440
	Second-generation western	0.0056
Type of household	Couple without children	−0.3231
(Reference single)	Couple with children	−0.4391
	Single parent	−0.3410
	Other	−0.7442*
Size of household	2	0.3415
(Reference 1)	3	0.4114
	4	0.5972
	5 or more	0.7391*
Marital status	Married	0.3229**
(Reference not married)	Widowed	0.0867
	Divorced	0.1763**
Has social allowance	Yes	−0.1884**

(Continued)

TABLE 5.6 (*Continued*)

Variable	Category	β
(Reference no)		
Has a job	Yes	0.1408**
(Reference no)		
Age	35–54 years	−0.1086**
(Reference 18–34 years)	55 years and older	−0.1074*
Gender	Female	0.0932**
(Reference male)		

variables. As will be shown in Chapter 7, this may be a good thing. The variable degree of urbanization appeared to be very strongly related to the response behavior in the bivariate analysis. However, in Table 5.5 it can be seen that in addition to the variable region, the contribution of the region is greatly reduced. It is still significant, however. Another result to consider is that the variable type of household has a large contribution when first added to the model, with a Wald χ^2 of 116.56. However, when the variable size of the household is added next, the contribution of the household type reduces to 31.06. In the final model the largest contributions come from the variables having a listed phone, region of the country, ethnic background, and marital status.

The results for the final logistic model are shown in Table 5.6. The nature of the relationship between the categories of the variables and the response behavior can be deduced from this table. Having a listed phone, compared to not having one, largely increases the probability of response. In the Lowlands and Metropolis the probability of response is lower than in Greenfields. First-generation nonnatives have a lower response probability than natives, and this especially holds for first-generation nonwestern nonnatives. Both married and divorced persons have a higher response probability than nonmarried persons. Having a social allowance implies a lower response probability; having a job a higher. The response probability is highest for persons aged 18 to 34 years. Women have a slightly higher response probability than men.

5.3.2.3 Different Response Types The bivariate relationships between the different types of response and the auxiliary variables are presented in Section 5.2. Ideally also the multivariate relationships between response types and auxiliary variables should be considered.

When analyzing the different causes of nonresponse, a different method than the logistic regression model has to be used because the dependent variable (response yes/no) is not binary anymore (type of response). In addition, since the causes of nonresponse are nested, this nesting has to be taken into account.

One way to do so is by fitting separate logistic regression models for the subset of observations for which a cause of nonresponse is observed. Suppose that there are

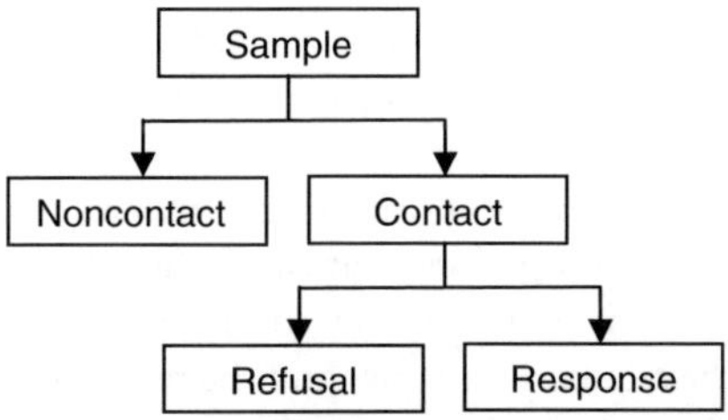

FIGURE 5.17 Nested representation of the response process.

only two causes of nonresponse: noncontact and refusal. The response process can be displayed as in Figure 5.17.

For the entire sample consisting of n observations, the objective is to observe whether the sample elements are contacted ($C_i = 1$) or not ($C_i = 0$). However, whether sample elements refuse participation ($P_i = 0$) or respond to the survey request ($P_i = 1$) can only be observed for the subset of sample elements that have been contacted. So C_i is defined for all elements i, whereas P_i is only defined for those sample elements for which $C_i = 1$. In other words, participation is censored.

This process can be described by a so-called nested logit model. The first step of response, obtaining contact, is observed for every sample element and can be described as

$$\log\left(\frac{\gamma(X_i)}{1-\gamma(X_i)}\right) = \text{logit}(\gamma(X_i)) = X'_i\beta^C \quad \text{for } i = 1, \ldots, n, \tag{5.4}$$

where $\gamma(X_i)$ is the propensity to be contacted: $\gamma(x) = P(C_i = 1|X_i = x)$. Only sample elements for which $C_i = 1$ can be asked to participate in the survey. For these elements the logit model for the participation propensity $\theta(X_i)$ is defined as

$$\log\left(\frac{\theta(X_i)}{1-\theta(X_i)}\right) = \text{logit}(\theta(X_i)) = X'_i\beta^P \quad \text{for } i = 1, \ldots, n^c, \tag{5.5}$$

where n^c is the number of contacted sample elements, and $\theta(x) = P(P_i = 1|X_i = x)$.

Because the different types of response are nested, this model is referred to as the nested logit model. The assumption underlying the nested logit model is that the different response types are not correlated. As a consequence equations (5.4) and (5.5) can be estimated separately.

A detailed description of methods for analysis of nonresponse that distinguish between different causes of nonresponse is provided by Cobben (2009).

5.4 Summary

This chapter describes how bivariate and multivariate methods can be used to analyze response behavior with the use of auxiliary variables. The bivariate methods can be used to evaluate the relationship between one auxiliary variable and the response behavior. Both the statistical measure and the graphical representation of a bivariate relationship is illustrated using a real survey data set, the GPS (General Population Survey).

Bivariate relationships can be misleading, however, because there may be relationships among auxiliary variables that cause the relationship to the response behavior. Therefore a multivariate analysis should be performed to evaluate the relationship between the auxiliary variables and the response behavior, while taking into account other relationships among variables as well. The logit model can be used for this purpose. An illustration of the logit model used to analyze nonresponse is provided by an application to the nonresponse results of the GPS.

In addition the chapter distinguishes among the different causes of nonresponse using bivariate graphs to illustrate the different types of responses. Also shown is how this information can help the survey researcher recognize at different stages of the response process the connections to auxiliary variables. This information can be used for the reduction of nonresponse as well as for the building of weighting models to adjust for nonresponse bias. An extension to the multivariate analysis method to account for different types of responses is provided.

Traditionally fieldwork reports are made to monitor fieldwork and interviewer performance. Use of this type of information in a nonresponse analysis requires systematic record keeping. It is important that fieldwork information becomes a standard part of the survey data file.

The response mechanism can be split in two (or more) sequential phases. The first phase is that of the contact attempt. The second phase is that of the cooperation attempt, once contact has been established. These two phases usually contribute most to the nonresponse rate.

Different models and different auxiliary variables are used in analyses of the two phases. However, in practical survey situations it is not easy to separate the two mechanisms. Future survey designers should take heed and attempt to find a better way to do this. Of course, it also remains important to distinguish among other types of nonrespondents.

An analysis of nonresponse in the GPS shows that it helps to have the following auxiliary information available for both respondents and nonrespondents:

- Whether or not a person has a listed phone number.
- Demographic variables, like the region of the country.
- The ethnic background of persons.
- The marital status of persons.
- The composition of the household, like the type and size of the household.

5.5 KEY TERMS

Bivariate analysis An analysis of the relationship between two variables.

Logistic regression model A multivariate regression model for a binary dependent variable.

Multivariate analysis An analysis of the relationship between more than two variables, in which relationships among variables (i.e., both dependent and independent) are explored and accounted for.

Cramér's *V* statistic A statistic that says something about the strength of the relationship between two variables.

Nested logit model A model used to describe a sequence of logistic regression models that are nested within each other. It is assumed that the different models are uncorrelated. Hence the models can be estimated separately.

Paradata Data that has been recorded about the process of collecting data.

5.6 REFERENCES

Agresti, A. (2002), *Categorical Data Analysis*. Wiley, New York.

Cobben, F. (2009). *Nonresponse in Household Surveys. Methods for Analysis and Adjustment.* PhD thesis. University of Amsterdam, Statistics Netherlands, Amsterdam.

Greene, W.H. (2003), *Econometric Analysis*, 5th ed. Prentice-Hall, Upper Saddle River, NJ.

Johnston, J., and DiNardo, J. (1997), *Econometric Methods*. McGraw-Hill, New York.

Heerwegh, D. (2003), Explaining Response Latencies and Changing Answers Using Client-Side Paradata from a Web Survey. *Social Science Computer Review* 21, pp. 360–373.

Nagelkerke, N.J.D. (1991), Miscellanea. A Note on a General Definition of the Coefficient of Determination. *Biometrika* 78, pp. 691–692.

5.7 EXERCISES

Exercise 5.1 Which statement about the relationship between response and region is not correct?

(a) In Greenfields persons are difficult to contact for the GPS.

(b) In large cities, persons more often refuse participation than in other regions.

(c) In large cities, persons are easier to contact than in other regions.

(d) Persons in large cities are overrepresented in the GPS.

Exercise 5.2 Which statement about the relationship between response and the ethnic background is not correct?

(a) Western nonnatives have a higher response than nonwestern nonnatives.

(b) First-generation nonnatives have a lower response than second-generation nonnatives;

(c) Because of language problems, first-generation nonnatives are more often not able to participate in surveys than natives.

(d) First-generation nonnatives less often refuse than natives and second-generation nonnatives.

Exercise 5.3 Which statement about the relationship between response and having a listed phone is not correct?

(a) Persons with a listed phone number respond more to the survey than persons without a listed phone number.

(b) Persons with a listed phone number are more difficult to reach than persons without a listed phone.

(c) Persons with a listed phone respond better in all types of surveys, not just telephone surveys but also face-to-face surveys and Web surveys.

Exercise 5.4 What should you do with unprocessed cases in a survey?

(a) You can ignore them. They just lead to a smaller sample size.

(b) You should increase the interviewer capacity.

(c) You should investigate a possible bias due to unprocessed cases.

(d) You should treat these cases as refusals.

Exercise 5.5 Why are response rates low in areas with a high percentage of non-natives?

(a) There is a high number of unprocessed cases.

(b) It is difficult to contact people.

(c) People are unable to participate because of language problems.

(d) All three reasons mentioned above.

Exercise 5.6 Why should you split the response mechanism into two sequential steps (a contact attempt followed by a cooperation attempt)?

(a) The bias due to noncontact is usually smaller than the bias due to refusal.

(b) The group of successful contacts does not resemble the complete sample.

(c) Analysis of these steps may require different models and auxiliary variables.

(d) It is easier to compute the response rate.

CHAPTER SIX

An International Comparison of Nonresponse

6.1 Introduction

The response rate of a survey and the composition of the nonresponses is determined by many different factors. One of these factors may be the country in which a survey is conducted. It is, however, not so easy to detect country effects. Such effects can only be measured if identical surveys from different countries are compared. This implies that every aspect of these surveys should be the same: the same sampling design, the same questionnaire, the same data collection method, the same fieldwork procedures, and so forth. These conditions are almost never satisfied in practical situations.

De Leeuw and De Heer (2002) and Groves and Couper (1998) attempted a multilevel approach of factors having an influence on the size and composition of the nonresponse. The levels and factors within levels they used are summarized in Table 6.1.

Notice that the *macro level* consists of factors affecting nonresponse at the country level. Societal characteristics belong to the macro level. For example, if people recognize the legitimacy of societal institutions in their country, they may feel more socially responsible and therefore more inclined to cooperate.

Response rates can also be affected by the general survey climate. If there are many surveys in a country, there will be an increased response burden. The survey climate may also be determined by the amount of telemarketing activities in the

Handbook of Nonresponse in Household Surveys, By Jelke Bethlehem, Fannie Cobben, and Barry Schouten

TABLE 6.1 Factors affecting nonresponse

Macro level	Meso level	Micro level
Societal characteristics	Mode of data collection	Interviewer characteristics
Survey taking climate	Fieldwork organization	Household characteristics
Demographic characteristics	Use of incentives	Physical impediments
Economic situation	Reminders	
Regional characteristics	Saliency of the topic	
	Name of the survey	
	Survey organization	
	Length of questionnaire	

country. Typically people are called around dinner time (when the contact probability is high) in attempts to sell products or services. This spoils the survey climate and has caused people to develop a hostile attitude toward interviewers.

Demographic characteristics may vary from country to country. Examples are the age distribution or the percentage of nonnatives living in the country. Contact rates are usually low among young people, and refusal rates are relatively high among the elderly. Typically response rates are lower among nonnatives because of language problems.

The economic situation in the country can influence survey participation. For example, a weak economy usually is accompanied by high unemployment, and nonresponse tends to be higher among the unemployed.

An important regional effect is the degree of urbanization. It is the general case that response rates are low in highly urbanized areas worldwide, even if response rates are corrected for other demographic variables. Apparently, the type of housing in urban areas and the population density reduce the response rate. Response rates are typically lowest in neighborhoods with a low socioeconomic status. As a consequence more urbanized countries have lower response rates.

The *meso level* consists of factors affecting nonresponse due to the survey's design and the organization of the fieldwork. The survey organization that controls most of these factors may desire a high response rate but may find that to be too expensive. So the actual survey will be a compromise between costs and quality.

As we saw in Chapter 4, the mode of data collection can affect the response rate. Recall that response rates are highest in face-to-face surveys, and somewhat lower in telephone surveys. Mail and Web surveys have the lowest response rates. There are large differences in the coverage of modes and the acceptance of modes as survey instruments across countries.

The organization of fieldwork is important. Factors influencing response rates are training of interviewers and use of incentives and reminders. Fieldwork needs to be planned and supervised. Small details are important, like the layout and contents of advance letters. Even the color of the envelope of the advance letter can have an effect. Use of blue envelopes caused a lower response rate in a

Dutch survey because the envelopes looked like the envelopes used for sending income tax forms.

Some aspects are survey dependent and hard to control. Examples are the saliency of the survey topic, name recognition of the organization conducting the survey, and the length of the questionnaire. There are examples of small changes in survey designs that haved raised the response rates. An interesting case is the family planning survey in The Netherlands. Responses increased when the name was changed (keeping all other aspects the same) from "Fertility Survey" into "Family Planning Survey."

The *micro level* consists of factors affecting nonresponse at the level of the people involved in the survey. These are the sample persons or households, and the interviewers. Interviewers should recognize that the first impression is a lasting one for the potential respondent. The interviewer must make it clear what the purpose of the visit is. The persons in the households must be convinced the interviewer is not selling something, and that there is no risk of physical harm. The interviewer must appear as someone to be trusted.

Interviewer characteristics play a role in obtaining participation. More experienced interviewers get cooperation whereas inexperienced ones fail. There is evidence that interviewers (both in face-to-face and telephone surveys) get more responses if they speak the dialect of the respondents.

Some households may be easier to interview than others. It is easier to establish contact with larger families than with small ones. It is difficult to contact young people if the sample is selected from a sampling frame consisting of registered landline telephone numbers, whereas establishing contact with the elderly via landlines is more probable.

Sociodemographic characteristics of households have an effect on response rates. Some types of households are more inclined to cooperate than others. Experiences with participation in previous surveys may determine future participation. If the perceived burden of completing the questionnaire is high, refusal may be likely. There are also temporary conditions affecting response, such as the current socioeconomic situation of the household. A family happily living together may be more inclined to cooperate.

The result of a contact attempt is largely determined by the interaction between the respondent and the interviewer. Therefore interviewers should tailor their behavior to that of the household. For example, an interviewer should not wear expensive clothes in a poor neighborhood. It is also wise not to drive in an expensive car in a poor neighborhood.

Finally, physical impediments may prevent some households from being contacted. In case of face-to-face surveys, locked apartment buildings, gated housing complexes, doormen, no-trespassing enforcements, intercoms, and other devices can limit contact with households. In case of telephone surveys, contact attempts may fail due to caller IDs, call blocking, or answering machines.

It is very difficult to compare cross-country response rates. Differences at the macro level can only be distinguished if there are no differences at the meso level or the micro level, or it must be possible to correct for the differences at the meso or micro level. Some such approaches will be described in this chapter.

6.2 Theory

Depending on the type and amount of data available, several theoretical approaches are possible to investigate nonresponse differences between countries. The ideal situation is that where a survey has been carried under the same conditions in several different countries. This implies that the factors at the meso level are the same for all countries involved. Consequently observed differences in nonresponse can be attributed to factors at the macro level or the micro level. Such data can be analyzed using a technique called correspondence analysis. This technique is described in Section 6.2.1.

Another situation to be considered here is that where data are available for surveys that have been conducted in several countries for several years. By including year and survey as explanatory variables in the model, it is possible to detect survey effects and trends over time. This approach is described in Section 6.2.2.

6.2.1 CORRESPONDENCE ANALYSIS

Correspondence analysis is used to identify and explore existing relationships in contingency tables. The strength of these associations can be expressed by the chi-square statistic. Through correspondence analysis insight can be obtained both in the degree and the type of association. The technique bears some resemblance to principal components analysis (PCA): a contingency table is decomposed into factors. Contrary to principal component analysis, however, correspondence analysis explains the strength of association as expressed by the chi-square statistic instead of by the variance. More information about correspondence analysis can be found in Greenacre (1984) and Bethlehem (1988). The correspondence analysis plots in this chapter were made with a slightly adapted version of the R-package by Nenadic and Greenacre (2007).

Correspondence analysis assigns scale values to rows and columns in such a way that the correlation coefficient between the scaled row and column variables is maximized. The first factor consists of this set of scale values. For each following factor, scale values are determined that maximize the correlation coefficient for that part of the association not accounted for by the previous factors.

The scale values can be used for a graphical investigation of the associations in the table. The *thermo-plot* compares scale values of rows and columns of a single factor, thereby offering insight in the type of association explained by the particular factor. The *bi-plot* is a two-dimensional plot, where the *x*-axis and *y*-axis are thermo-plots of the first two factors. The bi-plot offers the possibility to explore the structure of the association of two factors simultaneously.

EXAMPLE 6.1 Fieldwork Result by Household Type in the GPS

The table below is an example of a contingency table of the two variables Household type and Fieldwork result, as measured in the GPS. The row variable Household type has five categories, and the column variable Fieldwork result has five categories.

	Response	No-contact	Refusal	Not-able	Unprocessed	Total
Single	2695	628	1318	233	560	5434
Couple with children	6311	510	2649	390	709	10569
Couple	8849	547	3394	321	1016	14127
Single parent	829	126	462	61	139	1617
Other	108	36	67	29	32	272
Total	18792	1847	7890	1034	2456	32019

The value of the chi-square statistic for the test for independence is equal to 806.855. Since the *P*-value is close to 0, the hypothesis of independence is rejected. So there is a relationship between the two variables. The test provides no insight as to the nature of the relationship.

We will assume, without loss of generality, that the number of columns does not exceed the number of rows, meaning $r \geq c$. If this is not the case, the table can always be transposed. The observed frequency in row i (the ith category of the row variable) and column j (the jth category of the column variable), that is, the frequency in the cell (i, j) of the table, is denoted by n_{ij}. The total over all cells is denoted by n. Therefore

$$\sum_{i=1}^{r} \sum_{j=1}^{c} n_{ij} = n. \tag{6.1}$$

The first step in the analysis is to transform the frequencies in the table into relative frequencies, by dividing all cell frequencies by the total n. The relative frequency corresponding to the absolute frequency n_{ij} is denoted by $f_{ij} = n_{ij}/n$. The relative frequencies add up to 1:

$$\sum_{i=1}^{r} \sum_{j=1}^{c} f_{ij} = 1. \tag{6.2}$$

Furthermore row and column totals of the relative frequencies are introduced. The row totals of the relative frequencies are defined by

$$f_{i+} = \sum_{j=1}^{c} f_{ij} \tag{6.3}$$

and column totals by

$$f_{+j} = \sum_{i=1}^{r} f_{ij}. \tag{6.4}$$

EXAMPLE 6.2 Fieldwork Result by Household Type in the GPS—Relative Frequencies

The table below contains the relative cell frequencies and corresponding marginal totals for the sample data form Example 6.1.

	Response	No-contact	Refusal	Not-able	Unprocessed	Total
Single	0.0842	0.0196	0.0412	0.0007	0.0175	0.1697
Couple with children	0.1971	0.0159	0.0827	0.0122	0.0221	0.3301
Couple	0.2764	0.0171	0.1060	0.0100	0.0317	0.4412
Single parent	0.0259	0.0004	0.0144	0.0002	0.0004	0.0505
Other	0.0004	0.0001	0.0002	0.0001	0.0001	0.0008
Total	0.5869	0.0577	0.2464	0.0323	0.0767	1.0000

We further assume that all marginal totals are strictly positive, which implies there are no empty rows or columns in the table. The strength of the association between the row variable and the column variable can be expressed in the chi-square statistic. In terms of relative frequencies, and omitting a factor n, the adjusted chi-square statistic G is defined by

$$G = \sum_{i=1}^{r} \sum_{j=1}^{c} \frac{\left(f_{ij} - f_{i+} f_{+j}\right)^2}{f_{i+} f_{+j}}. \tag{6.5}$$

If G is multiplied by the absolute total n, the chi-square statistic for the original table is obtained. If there is no association in the table, the chi-square statistic has the value 0. The larger the value of this statistic is, the stronger the association in the table will be. The chi-square statistic for the table in Example 6.1 is equal to 806.855, and thus G = 806.855/32019 = 0.025. Clearly, there is some (but not much) association in the table.

Let m be the minimum of the number of rows and the number of columns. Then correspondence analysis decompose the chi-square statistic G into $m-1$ components:

$$G = \sum_{k=1}^{m-1} G_k^2. \tag{6.6}$$

The values G_1, G_2, ..., G_{m-1} are sorted in decreasing order, so $G_1 \geq G_2 \geq \ldots \geq G_{m-1}$.

To each component belong two sets of *scale values*: a set for the row variable and a set for the column variable. The set of scale values for the row variable consists of r (the number of rows) values, and the set for the column variable of c (the number of columns) values. Together, the two sets of scale values are called a *factor*.

The set of scale values for the rows of the first factor is denoted by

$$g_{1,1}, g_{1,2}, \ldots, g_{1,r}, \tag{6.7}$$

and the set of scale values for the columns of the first factor is denoted by

$$h_{1,1}, h_{1,2}, \ldots, h_{1,c}. \tag{6.8}$$

Suppose that the qualitative row variable is treated as a numerical variable with the row numbers l, 2, . . ., *r* as possible values. And likewise suppose that the column variable is also treated as a numerical variable with the column numbers l, 2, . . . , *c* as possible values. Then correlation coefficient between the row and column variable can be calculated. Of course, row numbers and column numbers do not have a real meaning. If the rows or columns were in a different order, a different correlation coefficient would be obtained. Now the two sets of scale values (6.7) and (6.8) are constructed in such a way that they have the nice property that if they are interpreted as the values of the row and column variable (instead of the row numbers 1, 2, . . ., *r* and column numbers 1, 2, . . ., *c*), the correlation coefficient obtains its maximum value (over all possible sets of real valued scale values), and this maximized correlation is equal to the first component G_1.

If the maximum correlation coefficient is equal to 1, there is a perfect relationship between the row and the column variable. If the correlation coefficient is equal to 0, there is no relationship.

A linear transformation is applied to both sets of scale values such that the means of the corresponding continuous variables are equal to 0, and that their variances are equal to G_1^2.

EXAMPLE 6.3 Fieldwork Result by Household Type in the GPS—The First Factor

The first component of the table in Example 6.1 is equal to $G_1 = 0.151$. Both sets of scale values are given in table below.

Category	Scale value
Single	0.288
Couple + children	−0.039
Couple	−0.107
Single parent	0.131
Other	0.527
Response	−0.088
Noncontact	0.495
Refusal	0.008
Not-able	0.262
Unprocessed	0.162

A maximized correlation coefficient of 0.151 does not indicate a very strong association. Still there is a relationship of some form that is worth investigating more thoroughly.

Generally, the first factor is not able to explain all association in the table completely. As is clear from (6.6), the quantity G_1^2 is only one component of the adjusted chi-square statistic G. The amount of association explained by the first factor is indicated by means of the quantity t_1, defined by

$$t_1 = \frac{G_1^2}{G}. \tag{6.9}$$

A value of t_1 close to 1 implies that the first factor is able to explain almost all association in the table. The smaller t_1, is the less the factor explains.

What about the association not explained by the first factor? If a substantial amount of unexplained association remains, a second factor can be computed, and so on. In general, $m-1$ factors can be computed. Each factor tries to explain as much as possible of the association not explained by the previous factors. To each component G_k, there are two corresponding sets of scale values (one for the rows, and one for the columns), and each G_k is the maximized correlation coefficient for that part of the table not explained by the previous components. The row values for factor k are denoted by

$$g_{k,1}, g_{k,2}, \ldots, g_{k,r}, \tag{6.10}$$

and the column values for factor k by

$$h_{k,1}, h_{k,2}, \ldots, h_{k,c}. \tag{6.11}$$

A linear transformation is applied to both sets of scale values such that the means of the corresponding continuous variables are equal to 0 and their variances are equal to G_k^2. The components are ordered in descending order of importance. The first factor can be interpreted as the one that gives the most information about how the association in the table deviates from independence. The second factor contains most information about the deviation that is not accounted for by the first factor, and so forth. The quantity t_k, defined by

$$t_k = \frac{G_k^2}{G} \quad \text{for } k = 1, 2, \ldots, m-1 \tag{6.12}$$

measures the relative importance of a factor. The more t_k approaches 1, the more important the factor is in explaining the relationship.

EXAMPLE 6.4 Fieldwork Result by Household Type in the GPS—The First two Factors

The value of t_1 for the table in Example 6.1 is $(0.151)^2/0.025 = 0.908$. This means that the first factor is able to explain almost 91% of the (modest) association in the table. The table below gives information about the first two factors.

Category	Factor 1	Factor 2
Correlation	0.151	0.042
Explained	0.908	0.070
Explained (cumulative)	0.908	0.978
Single	0.288	−0.039
Couple with children	−0.039	0.045
Couple	−0.107	−0.026
Single parent	0.131	0.021
Other	0.527	0.245
Response	−0.088	−0.007
Non-contact	0.495	−0.051
Refusal	0.008	0.019
Not-able	0.262	0.199
Unprocessed	0.162	−0.052

It turns out that when combined, the first two factors explain 98% of the association in the table. Thus nearly all association can be analyzed by investigating the first two factors. Note that the contribution of the second factor is only small. Its correlation coefficient is only 0.042.

In the case of a perfect relationship between the two variables, the scale value of one variable can be predicted exactly from the scale value of the other variable. The *thermo-plot* is a graphical tool used to investigate the relationship between the two variables that is accounted for by one factor.

The thermo-plot derives its name from the thermometer, which often has two scales: one measures the temperature in degrees Fahrenheit and the other one in degrees Celsius. Likewise the thermo-plot has two scales: one for the row variable and one for the column variable.

The scale values of the two variables are plotted on the same vertical axis. Labels corresponding to the row variable are printed to the left of the axis and the labels of the column variable to the right. Drawing correct conclusions from such a plot requires careful interpretation. Here are some guidelines:

- If two categories of one variable are close to each other, the behavior of the people in those categories is similar with respect to the other variable.
- If a category is close to the origin, the people in that category have a behavior that does not differ from the sample as a whole.
- If categories of both variables are opposed to each other, and are far away from the origin, this indicates a positive relationship.

- If categories of both variables are far away from the origin, but with opposite signs. this indicates a negative relationship.

EXAMPLE 6.5 Fieldwork Result by Household Type in the GPS—The Thermo-plot

The example of a thermo-plot below contains the thermo-plot for the first factor of the sample table. The black dot represents the origin.

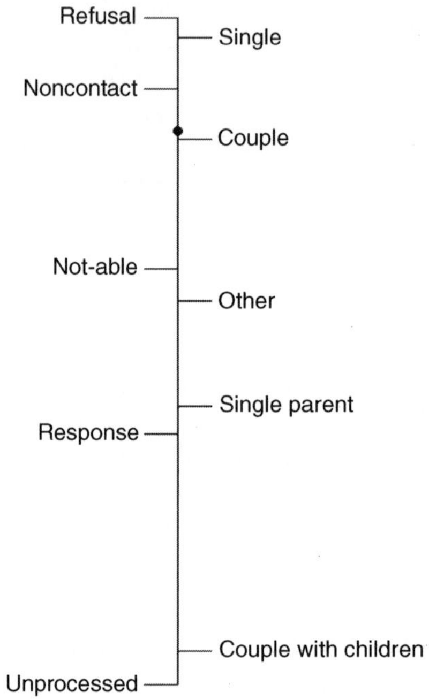

It may be concluded that refusal (and to a lesser extent noncontact) particularly occurs among singles. Unprocessed cases occur more than on average among couples with children. The behavior of couples without children does not differ from the average behavior in the sample. Refusal is much lower than average among couples with children. The fraction of unprocessed cases is lower than average among singles.

The *bi-plot* tool offers the possibility to graphically investigate the association explained by two factors simultaneously. Correspondence analysis assigns $m-1$ scale values to every row and column. Hence the categories of the variables represent points in a $(m-1)$-dimensional space. Since it is rather difficult to make and interpret an $(m-1)$-dimensional plot, only two-dimensional projections are used.

The bi-plot of two factors plots each category as a point in two-dimensional space. The x-coordinate of a category is the scale value in one factor, and the y-coordinate is the scale value in the other factor. Attention should be paid to three distinct relationships: (1) distances of points from the origin, (2) the direction of the vectors from the origin to points, and (3) distances between points.

The origin of the bi-plot represents the marginal distributions of both the row and column variables. For a point close to the origin, the distribution within the corresponding category resembles the sample distribution. The behavior of elements within that row or column does not differ from the general behavior in the sample. The larger the distance from the origin, the more different the behavior of the category will be.

Points corresponding to categories of different variables that point in the same direction indicate a positive relationship between the categories. If categories of different variables point in opposite directions, it means that there is a negative relationship. Since the vertical scale may differ from the horizontal scale, direction vectors with a 90-degree angle do not necessarily indicate independence.

If the distance between two points corresponding to categories of the same variable is small, then the distributions within those categories resemble each other. This does not necessarily mean that the elements within the two categories have a lot in common. It only means that their distributions with respect to the other variable are nearly the same. The distance between categories of different variables cannot be interpreted in this way.

EXAMPLE 6.6 Fieldwork Result by Household Type in the GPS—The Bi-plot

The figure below gives the bi-plot of the first two factors of the sample table. As can be seen in the table of Example 6.4, the two factors together explain 97.8% of the association in the table.

The origin is the point where the two dotted lines cross. The behavior of elements in a category close to the origin does not differ from the average behavior of the elements in the sample. So the response behavior of couples without children does not deviate from the average behavior.

Single parents point in the direction of unprocessed cases. This means there is a higher than average proportion of unprocessed cases for single parents. Singles point in the direction of refusal. So more singles on average refuse. Response and couples with children point in the same direction. Therefore responses are particularly high among couples with children.

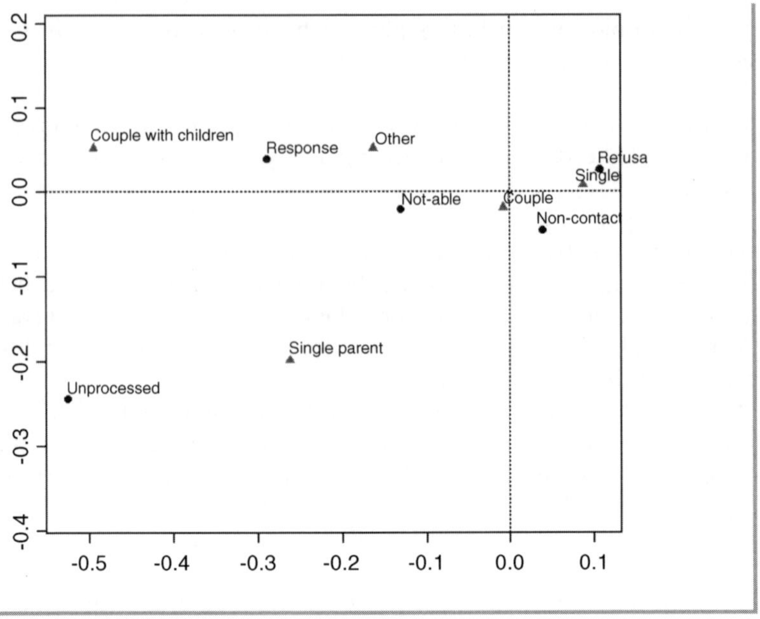

The application of correspondence analysis in Example 6.6, where one variable is the fieldwork result and the other variable is the type of household, can be used for international comparisons. If a survey is carried out in several countries, one variable is the fieldwork results and the other variable is the country where the survey was conducted. Section 6.3 provides such an application.

6.2.2 MULTINOMIAL MULTILEVEL MODELING

International comparisons of response behavior in surveys often take the form of a meta-analysis. The analysis does not take place at the level of the individual person or household but at the level of the study. One reason to conduct a meta-analysis is in circumstances where individual data are simply not available. Only data at the level of the survey and higher can be used in an analysis.

This type of data can be analyzed with a multinomial multilevel model. A multinomial model is appropriate because the dependent variable is the result of the fieldwork. This is a categorical variable with categories response, noncontact, refusal, and so forth. Note that if the dependent variable is a dummy variable with just the values 1 (response) and 0 (nonresponse), the obvious model would be a binary dependent variable model, like a logit or a probit model. Of course, distinguishing among various types of nonresponses gives more insight into the response behavior.

The model used should be a multilevel model because there are at least two levels in the available data: data at the country level and data at the survey level. Furthermore surveys should be nested within countries.

EXAMPLE 6.7 Nonresponse in Official Statistics

The International Workshop on Household Survey Nonresponse took an initiative in 1990 to collect data on nonresponse in official statistics surveys. A questionnaire was designed to collect information about nonresponse in surveys conducted by governmental survey agencies. These questionnaires were completed by key informants in those agencies. See De Heer (1999) for more information about this project. Data was collected up to 1997. The data came from 16 countries (15 European countries and the United States) and 10 different surveys (the Labor Force Survey, the Family Expenditure Survey, etc.) that were carried out in several countries.

De Leeuw and De Heer (2002) have analyzed the data with multinomial multilevel models Three models were fitted: one for the overall response, one for noncontact, and one for refusal.

	Response	Noncontact	Refusal
Fixed effects			
Intercept	1.48**	−2.72**	−2.31**
Random effects			
Country (variance)	0.33**	0.18**	0.64**
Survey (variance)	0.17**	−0.01	0.44**

Significance at the 95% confidence level of coefficients is indicated by **. Almost all coefficients, with the exception of −0.01 (the variance for the variable survey in case of noncontact) are significant. Therefore one can conclude there are real country effects. Moreover within countries there were significant differences from survey to survey.

There are surveys, particularly for official statistics purposes, that are repeated on a yearly basis. This makes it possible to include time as a variable in the multinomial multilevel model, so, trends over time can be explored. Some complications such as when the time series of surveys are not the same each year, may be solved by cross-classifying the variables survey and year.

EXAMPLE 6.8 Nonresponse in Official Statistics

The International Workshop on Household Survey Nonresponse initiative collected data on nonresponses in official statistics surveys. There were data on a series of years for eight surveys. De Leeuw and De Heer (2002) analyzed these data by including the variable year in the multinomial multilevel models of Example 6.7. The table below summarizes the results.

	Response	Noncontact	Refusal
Fixed effects			
Intercept	1.35**	−2.56**	−2.14**
Year	−0.02**	0.03**	0.03**
Random effects			
Country (variance)	0.39**	0.24**	0.57**
Survey (variance)	0.16**	0.003	0.46**
Year (variance)	0.001**	0.001	0.002**

Significance at the 95% confidence level of coefficients is indicated by **. The coefficient of the variable year in the category response is negative. This means there is a decreasing response over time. Both noncontact and refusal increase over time.

These results show again the country differences and also survey differences within countries. It should be recognized that an analysis like this assumes that the definitions of response, noncontact, and refusal are the same in all countries and all surveys.

6.3 Application

The European Social Survey (ESS) is an academically driven research instrument, funded jointly by the European Commission, the European Science Foundation, and scientific funding agencies in 30 European countries. The survey is conducted bi-annually. In each participating country approximately 2000 respondents are asked to answer a questionnaire in a one-hour-long face-to-face interview and to complete a short supplementary questionnaire either face-to-face or on paper.

The ESS has two objectives. The first is to explore and explain the interaction among Europe's changing institutions, its political and economic structures, and the attitudes, beliefs, and behavior of its people. The second objective is to show that it is possible to conduct a high-quality cross-country social survey.

A major challenge of the ESS is collecting data that allow for country comparisons. This can only be accomplished by applying fundamental methodological principles of survey sampling. The ESS attempts to achieve this by enforcing consistent, standardized procedures and protocols in all participating countries. These include sampling designs, questionnaire definitions, and fieldwork procedures.

A major objective of the ESS is to obtain a response rate of at least 70% and a noncontact rate of at most 3%. An important feature of the ESS is the use of detailed contact forms. These forms enable standardized, systematic reports to be made documenting all call contact attempts and reasons for refusal. The forms collect the following information:

- Identification of the interviewer and (potentially) the respondent.
- Information about the selection procedure.
- Date, month, day of the week, and time of day of visit.
- Mode of visit (telephone or face-to-face).
- Result of each visit.
- Outcome when there was no interview.
- Reason of refusal.
- Interviewer's assessment of future cooperation.
- Reason why an outcome was ineligible.
- Information about the neighborhood.

With these contact forms, analyses and comparisons of the fieldwork results are vastly simplified across countries. See Stoop et al. (2003) for more information about the development of the contact forms.

The fieldwork results of the second round of the ESS are analyzed here. The fieldwork took place in the second half of 2004 and the first half of 2005. The number of participating countries was 26. The fieldwork results for these countries are summarized in Table 6.2.

Sample sizes vary between 1129 (for Iceland) and 5633 in (Germany). Where possible, the data in the table were compiled from the contact forms. See Billiet and Pleysier (2007) for more details.

The next step is to transform the data in Table 6.2 into percentages. The result is Table 6.3. The data in this table form the basis for the first exploration of the response and the nonresponse in the participating countries.

An exploratory analysis of a large amount of data is more effectively presented using graphs. The Chinese proverb that one picture is worth a 1000 words certainly applies. As a first step in a graphical exploratory analysis, Figure 6.1 shows a bar chart of the data in Table 6.3. Countries have been ordered by their response rate.

Notice the large differences in response rates. France has the lowest response rate (43.6%), followed by Switzerland (48.6%). The highest response rates are obtained by Estonia (79.1%) and Greece (78.8%). The target response rate of the ESS is 70%. Only 6 out of the 26 countries succeeded in achieving this response rate: Estonia, Greece, Poland, Portugal, Finland, and Slovenia.

Figure 6.2 contains the noncontact rates of the countries participating in the second round of the ESS. Only a few countries managed to keep the noncontact rate below the target of 3%: Poland, Norway, Finland, Sweden, The Netherlands, and Portugal. The noncontact rates are more than 10% in Ireland (10.2%), the Czech Republic (10.9%), and Turkey (13.5%). Billiet and Pleysier (2007) see only one possible explanation for the high noncontact rate in the Czech Republic: the proper procedures and strategy were not followed during the fieldwork.

TABLE 6.2 Fieldwork results for the second round of the ESS

Country	Code	Response	Noncontact	Refusal	Not-able	Sample
Austria	AT	2256	251	1076	32	3615
Belgium	BE	1778	101	768	259	2906
Czech Republic	CZ	3026	598	607	1243	5464
Denmark	DK	1487	114	573	143	2.317
Estonia	EE	1989	85	285	156	2515
Finland	FI	2022	59	650	128	2859
France	FR	1806	363	1653	322	4144
Germany	DE	2870	393	1848	522	5633
Greece	GR	2406	109	504	36	3055
Hungary	HU	1482	128	364	274	2248
Iceland	IS	579	52	441	57	1129
Ireland	IE	2286	347	817	207	3657
Italy	IT	1529	153	590	305	2577
Luxembourg	LU	1635	232	1135	259	3261
Netherlands	NL	1881	78	852	113	2924
Norway	NO	1760	44	701	155	2660
Poland	PL	1716	20	452	141	2329
Portugal	PT	2052	78	538	215	2883
Slovakia	SK	1511	143	547	209	2410
Slovenia	SI	1442	210	315	86	2053
Spain	ES	1663	216	762	390	3031
Sweden	SE	1948	70	654	308	2980
Switzerland	CH	2234	98	2025	243	4600
Turkey	TU	1856	495	880	430	3661
Ukraine	UA	2031	191	490	338	3050
United Kingdom	UK	1897	296	1263	290	3746

There are strict rules in the ESS with respect to the fieldwork. These rules are meant to keep data collection similar in all countries, and therefore should enable comparability of the results. Among the rules are these six:

- An interviewer has to make at least four visits to an address before abandoning it.
- Each visit has to be preceded by an advance letter announcing the visit.
- At least one visit has to be made in the evening.
- At least one visit has to be made on the weekend.
- Visits have to be spread over a period of at least two weeks.
- The fieldwork period of the survey has to cover at least 30 days.

TABLE 6.3 Response and nonresponse rates (%) in the second round of the ESS

Country	Code	Response rate	Noncontact rate	Refusal rate	Not-able rate
Austria	AT	62.4	6.9	29.8	0.9
Belgium	BE	61.2	3.5	26.4	8.9
Czech Republic	CZ	55.3	10.9	11.1	22.7
Denmark	DK	64.2	4.9	24.7	6.2
Estonia	EE	79.1	3.4	11.3	6.2
Finland	FI	70.7	2.1	22.7	4.5
France	FR	43.6	8.8	39.9	7.8
Germany	DE	50.9	7.0	32.8	9.3
Greece	GR	78.8	3.6	16.5	1.2
Hungary	HU	65.9	5.7	16.2	12.2
Iceland	IS	51.3	4.6	39.1	5.0
Ireland	IE	62.5	9.5	22.3	5.7
Italy	IT	59.3	5.9	22.9	11.8
Luxemburg	LU	50.1	7.1	34.8	7.9
Netherlands	NL	64.3	2.7	29.1	3.9
Norway	NO	66.2	1.7	26.4	5.8
Poland	PL	73.7	0.9	19.4	6.1
Portugal	PT	71.2	2.7	18.7	7.5
Slovakia	SK	62.7	5.9	22.7	8.7
Slovenia	SI	70.2	10.2	15.3	4.2
Spain	ES	54.9	7.1	25.1	12.9
Sweden	SE	65.4	2.3	21.9	10.3
Switzerland	CH	48.6	2.1	44.0	5.3
Turkey	TU	50.7	13.5	24.0	11.7
Ukraine	UA	66.6	6.3	16.1	11.1
United Kingdom	UK	50.6	7.9	33.7	7.7

Given these fieldwork rules and the results seen in Figure 6.2, it is clear that in some countries people are harder to contact than people in other countries. To improve the contact rate, Billiet and Philippens (2004) suggest that the minimum number of visits be increased, or the number of calls in the evening or on the weekends.

Figure 6.3 shows the refusal rates for the 26 countries in the ESS. The rates vary between 11.1% and 44.0%. Refusal is the main cause of nonresponse in almost all countries. There is only one exception: the not-able rate is higher than the refusal rate in the Czech Republic. There are large differences in refusal rates. The two countries with the lowest refusal rates are the Czech Republic (11.1%) and Estonia

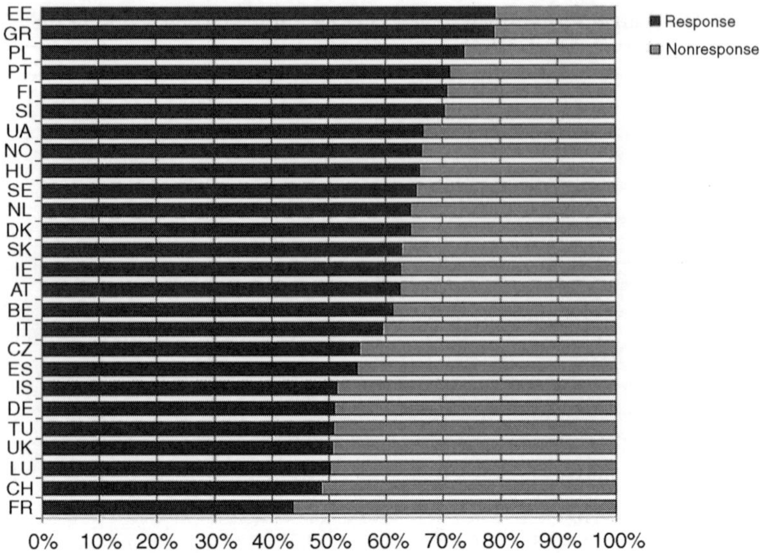

FIGURE 6.1 Response rates (%) in the second round of the ESS.

(11.3). The highest refusal rates are found in Iceland (39.1%), France (39.9%), and Switzerland (44.0%).

The countries participating in the ESS are advised to attempt refusal conversion. Initial refusers should be re-approached and encouraged to participate. At first sight, such an attempt may seem not very meaningful. Many refusals are, however, of a temporary nature. They may be due to a situation or mood of a person at the moment of approach. So an attempt at a different time can be successful. See also Chapter 3 on ways to reduce nonresponse.

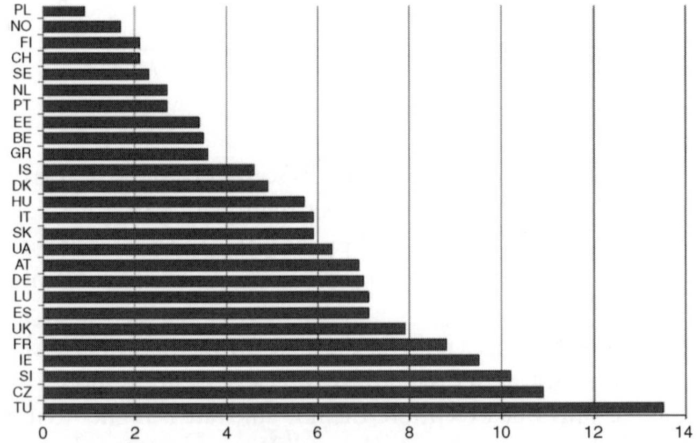

FIGURE 6.2 Noncontact rates (%) in the second round of the ESS.

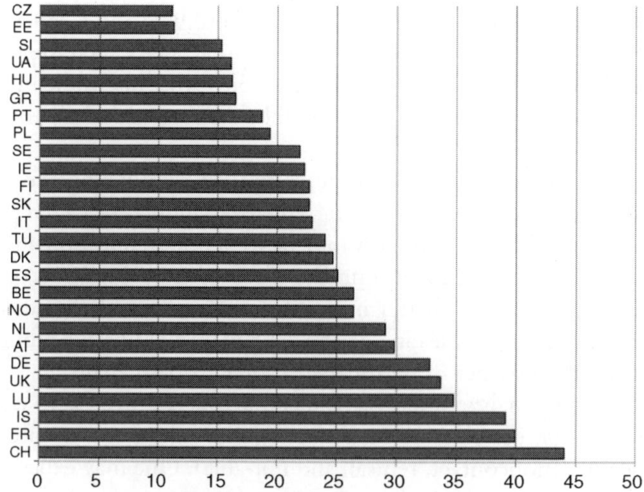

FIGURE 6.3 Refusal rates (%) in the second round of the ESS.

The successes of refusal conversion strategies appear to be country dependent. Billiet and Pleysier (2007) report that in Greece and Switzerland less than 10% of the refusal conversion attemps was successful. The success rates were much higher in other countries: 70% in Slovenia, 42% Estonia, and 38% in The Netherlands.

It should be noted that the chance of success in the second attempt partly depends on the effort put into the first attempt. If a refusal is accepted rather easily in the first attempt, the chances of success are higher in the second attempt. If almost everything is tried in the first attempt, the chances of success will be low in the second attempt.

Figure 6.4 contains the not-able rates for the 26 countries in the ESS. There are large differences in not-able rates. They vary between 0.9% and 22.7%. The two

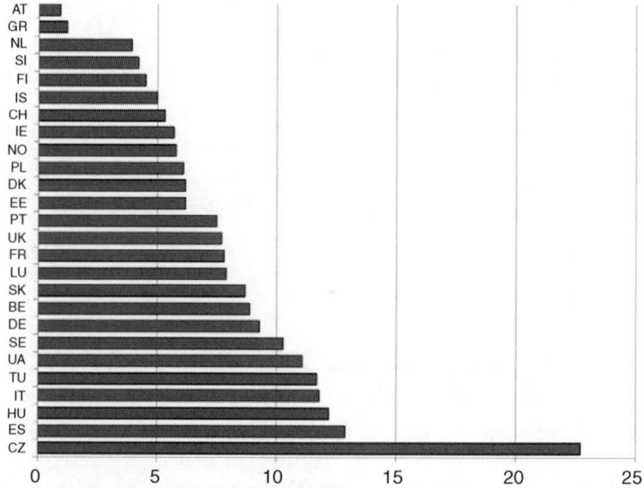

FIGURE 6.4 Not-able rates (%) in the second round of the ESS.

countries with the lowest not-able rates are Austria (0.9%) and Greece (1.2). The not-able rates of 20 of the 26 countries are below 10%. The Czech Republic has by far the highest rate: 22.7%. This is so unusually high that one is inclined to think that fieldwork procedures in the Czech Republic differ substantially from those in other countries.

The refusal rates in Figure 6.3 were computed by dividing the number of refusals by the sample size. It should be recognized that only contacted persons can refuse. Therefore a different approach may be to define a conditional refusal rate. This rate is obtained by dividing the number of refusals by the number of contacted persons. This may provide a better indication of the refusal behavior of people. Likewise a conditional not-able rate may be defined by dividing the number of not-able persons by the number of persons who agree to participate.

Another way to analyze response behavior across countries is to examine the fieldwork results for, correlations between response rates and the three different nonresponse rates (noncontact, refusal, and not-able). One may expect a negative correlation between response and nonresponse. Of course, correlations can differ for the three types of nonresponse. A graphical exploration can help identify countries with a different behavior.

Figure 6.5 contains the scatter plot of the response rate by the noncontact rate. The dotted line is the regression line. The correlation coefficient has a value of −0.49. So there is a reasonably strong negative relationship. The noncontact rate decreases as the response rate increases.

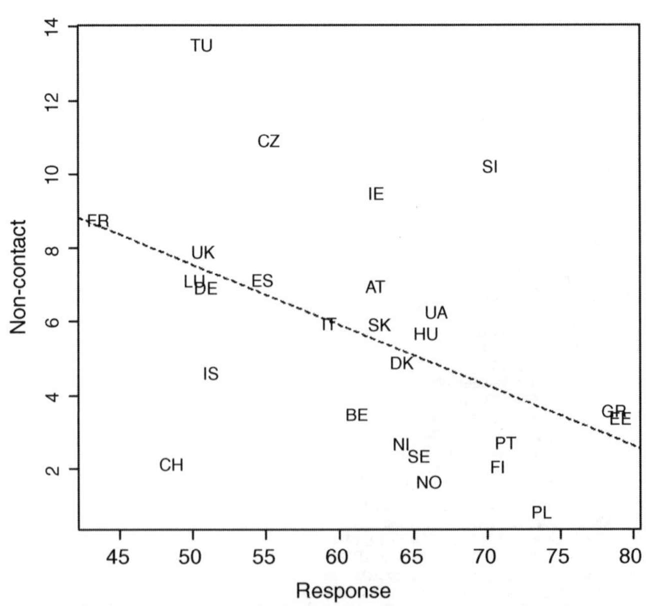

FIGURE 6.5 Scatterplot of the response rate by the noncontact rate in the second round of the ESS.

TABLE 6.4 First two factors of Table 6.2

Category	Factor 1	Factor 2
Correlation	0.214	0.193
Explained	0.492	0.401
Explained (cumulative)	0.492	0.893
Response	−0.040	0.152
Noncontact	−0.243	−0.355
Refusal	0.304	−0.173
Not-able	−0.459	−0.308
AT	0.208	0.082
BE	0.030	0.038
DK	0.043	0.094
FI	0.071	0.244
FR	0.219	−0.301
GR	0.021	0.388
HU	−0.220	0.074
IE	−0.029	0.026
IT	−0.108	−0.037
LU	0.149	−0.176
NL	0.180	0.134
NO	0.106	0.161
PL	−0.002	0.294
PT	−0.059	0.224
CH	0.397	−0.137
DE	0.092	−0.170
EE	−0.159	0.360
ES	−0.103	−0.131
SE	−0.059	0.110
SK	−0.049	0.042
CZ	−0.558	−0.228
IS	0.298	−0.113
SI	−0.120	0.160
TU	−0.159	−0.254
UA	−0.206	0.088
UK	0.128	−0.173

To interpret both factors separately, thermo-plots are drawn. Figure 6.8 shows the thermo-plot for the first factor. By this plot, response behavior is characterized by contrasting two types of nonresponse. The top end of the scale corresponds to active nonresponse (persons who refuse to participate), and the bottom end to more

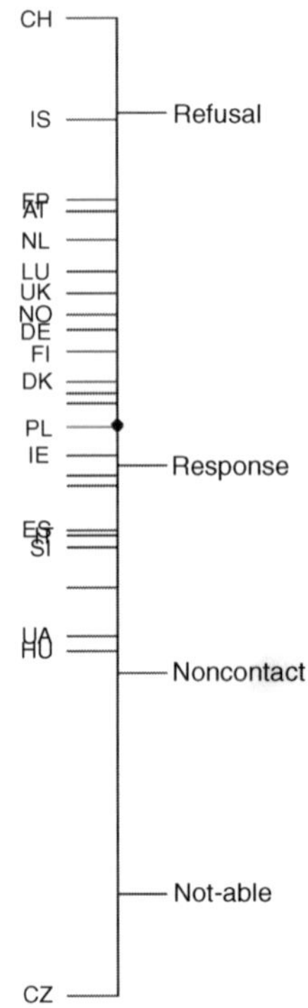

FIGURE 6.8 Thermo-plot of the first factor of Table 6.2.

passive nonresponse (noncontact and not-able). Response is near to the origin of the scale.

Countries like Switzerland (CH) and Iceland (IS) point in the direction of refusal, as there is more than average refusal in these countries. The Czech Republic (CZ) points in the direction of not-able, and to a lesser extent in the direction of noncontact. So there is a more than average amount of nonresponse due to not-able in the Czech Republic.

Switzerland and not-able point in opposite directions. So there is a lower than average amount of nonresponse due to not-able in Switzerland. Similarly there is lower than average refusal in the Czech Republic.

Figure 6.9 shows the thermo-plot for the second factor. This plot characterizes response behavior in terms of response and nonresponse. Countries on the top part of the scale point to response. So they have a more than average amount of response. Particularly Greece (GR) and Estonia (EE), and to a lesser extent Poland (PL) have high response rates. France (FR), Turkey (TU), and the Czech Republic (CZ) have a higher than average nonresponse rate.

Both thermo-plots are combined in the bi-plot in Figure 6.10. The first thermo-plot is obtained by projecting the points on the *x*-axis, and the second plot by projecting the points on the *y*-axis. The bi-plot gives a more comprehensive view

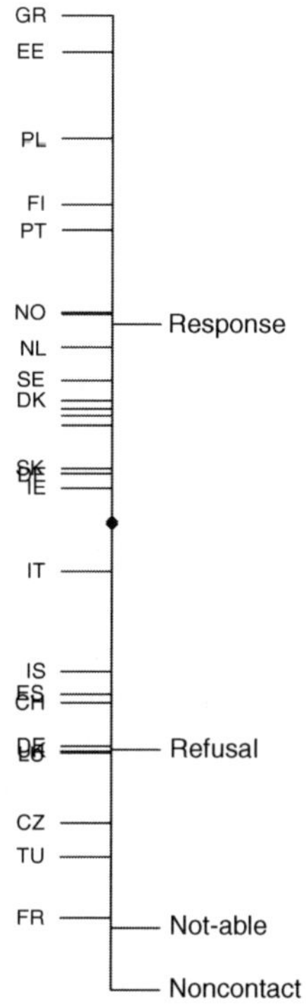

FIGURE 6.9 Thermo-plot of the second factor of Table 6.2.

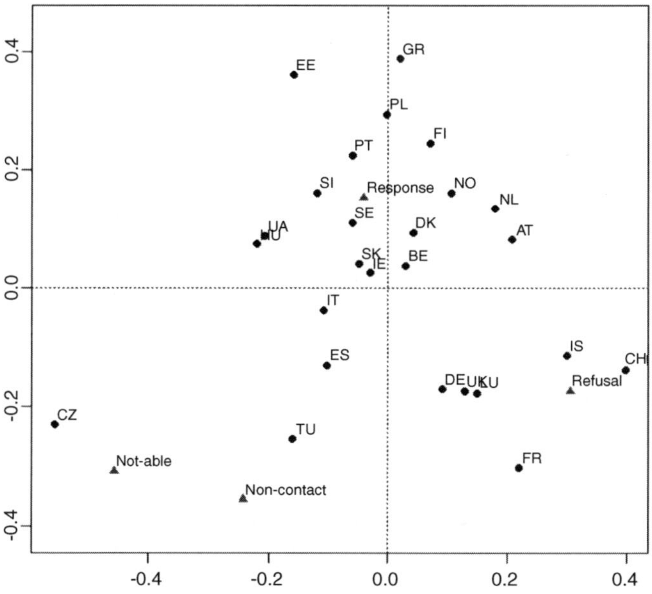

FIGURE 6.10 Bi-plot of the first two factors of Table 6.2.

of the nature of the association between the variables country and response behavior.

Countries that are far away from the origin have a behavior that differs from the average behavior. Greece (GR) and Estonia (EE) point in the direction of response, so they have a higher than average response.

The Czech Republic (CZ) points in the direction of not-able, so there is more than average nonresponse due to not-able in this country. Turkey (TU) points to noncontact, indicating there are problems with noncontacts.

Switzerland (CH), Iceland (IS) and France (FR) point to refusal. So these countries have higher than average refusal rates.

Countries like Belgium (BE), Ireland (IE), and Slovakia (SK) are close to the origin. It can be concluded that these three countries show an average response behavior.

The first round of the European Social Survey was carried out in 2002 to 2003 and the second round in 2004 to 2005. Twenty-two countries participated in both rounds. For these countries it is possible to compare response rates between rounds and so to establish whether nonresponse is caused by factors at the macro level or at the meso level. If, on the one hand, nonresponse is only caused by macro-level factors, it will be difficult to improve response rates. If, on the other hand, nonresponse is caused by meso-level factors, the response rates could be raised by improving the fieldwork of the survey.

Figure 6.11 shows a scatter plot of the response rates of the 22 countries in the first and second rounds of the ESS. The data were compiled by Billiet and Pleysier (2007). The dotted line represents the situation in where response rates

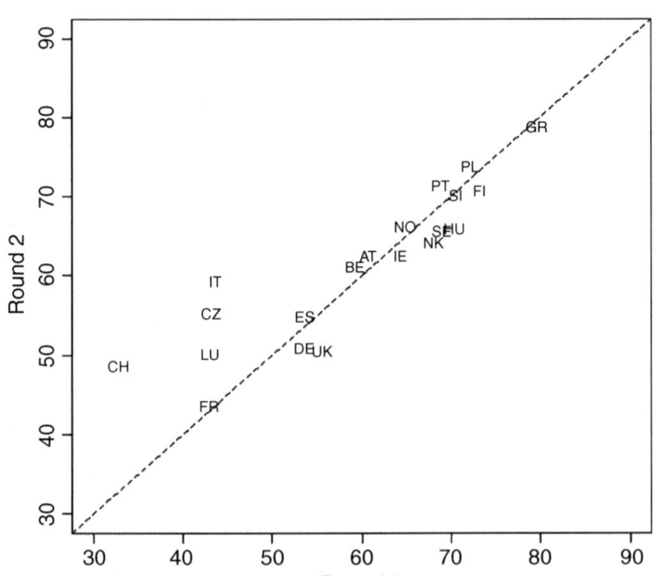

FIGURE 6.11 Scatter plot of the response rates in the first and second round of the ESS.

in both rounds are the same. Most countries are close to or on this line. This implies there were no changes in response rates between rounds.

There are four countries that lie above the line. Response rates for these countries are higher in the second round. Particularly Italy (IT) succeeded in improving its response rate. It went up from 43.7% to 59.3%. Switzerland (CH) realized a major increase of its response rate: from 33.0% to 48.6%. Also the Czech Republic (CZ) and Luxemburg (LU) increased their response rates. It should be noted, however, that the response rates of these four countries are still at the lower end of the spectrum.

A more extensive analysis of the nonresponse in the European Social Survey has been carried out by some of the researchers involved in this project. Examples are Billiet et al. (2007), Billiet and Pleysier (2007), Billiet and Welkenhuysen-Gybels (2004), Blom (2009), and Stoop (2005).

6.4 Summary

It is difficult to analyze and compare survey nonresponse in different countries. The response rate of a survey and the composition of the nonresponse is determined by many different factors. One of these factors may be the country in which a survey is conducted. It is, however, not so easy to detect country effects. Such effects can only be measured if identical surveys from different countries are compared. To do so, every aspect of these surveys should be the same: the same sampling design, the same questionnaire, the same data collection method, the same fieldwork procedures, and so forth. These conditions are almost never satisfied in practical situations.

Three levels of factors possibly affecting nonresponse are distinguished. The *macro level* is the highest level. It consists of factors affecting nonresponse at the country level. An example of such a factor is the perceived legitimacy of societal institutions in the participating countries.

The second level is the *meso level.* It consists of factors affecting nonresponse at the level of the survey. There are several aspects of the survey design and the organization of the fieldwork that may have an influence in response rates. The survey organization controls most of these factors and will attempt to conduct a survey with a high response rate.

The third, and final, level is the *micro level.* It consists of factors affecting nonresponse at the level of the people involved in the survey. These are the sample persons or households, and the interviewers. The characteristics and behavior can play an important role in obtaining participation. For example, more experienced interviewers are capable of getting cooperation where inexperienced interviewers fail. Also for some types of households it will be easier to obtain response than for others. It is easier to establish contact with larger families than with single adults. It is difficult to contact young people if the sample is selected from a sampling frame consisting of registered landline telephone numbers.

Two approaches are described for comparing survey results from different countries. The first approach is the ideal situation in which it can be assumed that the survey design and the fieldwork procedures in all countries are the same. Consequently all differences can be attributed to macro-level country effects. Some graphical tools, including correspondence analysis, are described to analyze such data.

If several surveys are conducted in several countries, and the survey specifications are not necessarily the same in each country, another type of analysis can be carried out. A multinomial multilevel model can be fitted. Such a model makes it possible to distinguish macro-level effects from meso-level effects.

It should be realized that once different nonresponse effects are detected in different countries (different response rates and/or different response compositions), it becomes almost impossible to properly analyze the survey data. Observed differences for a specific target variable may be caused by real country differences for that variable, but they may also be caused by different response compositions.

The R-indicator is introduced in Chapter 7 as an indicator for the representativity of the survey response. If sufficient auxiliary variables are available in all participating countries, the R-indicator can be used the compare representativity. If there are large differences, this is a clear warning that survey results may be incomparable.

6.5 KEY TERMS

Bi-plot One of the graphical tools of correspondence analysis whereby, in a contingency table, dependence is split into a number of independent factors. The bi-plot shows the correlation within two factors in a graphical way.

Correspondence analysis An exploratory analysis technique that displays the probable existing dependency between two categorical variables in a graphical way

Macro-level factors The set of factors affecting nonresponse at the country level. An example of such a factor is the perceived legitimacy of societal institutions in the participating countries.

Meso-level factors The set of factors affecting nonresponse at the level of the survey. There are several aspects of the survey design. For example, the organization of the fieldwork that can have an influence on response rates.

Micro-level factors The set of factors affecting nonresponse at the level of the people involved in the survey. These are the sample persons or households, and the interviewers.

Multinomial multilevel model A statistical model that attempts to predict the proportions in the categories of a (categorical) dependent variable, where some of the explanatory variables may be nested in other explanatory variables.

Thermo-plot One of the graphical tools of correspondence analysis. This analysis technique splits the dependence in a contingency table in a number of independent factors. The thermo-plot shows the correlation within one factor in a graphical way.

6.6 REFERENCES

Bethlehem, J.G. (1988), *The Program Coran 2.0 for Correspondence Analysis.* Report 332-88-M3. Netherlands Central Bureau of Statistics, Voorburg.

Billiet, J., and Philippens, M. (2004), *Data Quality Assessment in ESS Round 1. Between Wishes and Reality.* Onderzoeksverslag DA/2004-38. Departement Sociologie, Afdeling Data-verzameling en Analyse, K.U. Leuven, Belgium.

Billiet, J., Philippens, M., Fitzgerald, R., and Stoop, I. (2007), Estimation of Nonresponse Bias in the European Social Survey: Using Information from Reluctant Respondents. *Journal of Official Statistics* 23, pp. 135–162.

Billiet, J., and Pleysier, S. (2007), *Response Based Quality Assessment in the ESS-Round 2—An Update for 26 Countries.* Center of Sociological Research, K.U. Leuven, Belgium.

Billiet, J., and Welkenhuysen-Gybels, J. (2004), *Assessing Cross-national Construct Equivalence in the ESS: The Case of Six Immigration Items.* Paper presented at the European Conference on Quality and Methodology in Official Statistics, Mainz, Germany.

Blom, A.G. (2009), *Explaining Cross Country Differences in Survey Contact Rates: Application of Decomposition Methods.* Paper presented at the 20th International Workshop on Household Survey Nonresponse, Lausanne, Switzerland.

De Heer, W. (1999), International Response Trends: Results of an International Survey. *Journal of Official Statistics* 15, pp. 129–142.

De Leeuw, E., and De Heer, W. (2002), Trends in Household Survey Nonresponse: A Longitudinal and Internation Comparison. In: Groves, R.M. Dillman, D.A., Elting, J.L., and Little, R.J.A. (eds.), *Survey Nonresponse.* J Wiley, New York.

Greenacre, M.J. (1984), *Theory and Applications of Correspondence Analysis.* Academic Press, New York.

Nenadic, O., and Greenacre, M. (2007), Correspondence Analysis in R, with Two- and Three-dimensional Graphics: The ca Package. *Journal of Statistical Software* 20, pp. 1–13.

Stoop, I.A.L. (2005), *The Hunt for the Last Respondent: Nonresponse in Sample Surveys.* Social and Cultural Planning Agency, The Hague.

Stoop, I., Devacht, S., Billiet, J., Loosveldt, G., and Phillipens, M. (2003), The Development of a Uniform Contact Description Form in the ESS. Paper presented at the 14th International Workshop on Household Survey Nonresponse, Leuven, Belgium.

6.7 EXERCISES

Exercise 6.1 A survey is conducted in two different countries. The survey designs and the response rates are the same in both countries. The estimates for the mean of a target variable differ. What can one say?

(a) The population means of the target variables really differ.

(b) The population means of the target variables are equal.

(c) The conclusion depends on the composition of the response.

(d) The estimates should have been equal.

Exercise 6.2 A survey is conducted in two different countries. There are differences in the survey designs. There is no nonresponse in both surveys. The estimates for the mean of a target variable differ. What can one conclude?

(a) The difference is caused only by country differences.

(b) The difference is caused only by survey differences.

(c) The difference may be caused by both country and survey differences.

(d) The difference is not caused by country and/or survey differences.

Exercise 6.3 One survey is conducted in several different countries. There may be country and survey differences. Is it possible to distinguish country effects from survey effects by fitting a multinomial multilevel model?

(a) Yes, by nesting the survey in the countries.

(b) Yes, by nesting countries in the survey.

(c) No, this requires data on more than one survey per country.

(d) No, a single-level multinomial model should be used.

Exercise 6.4 If a category of the row variable and a category of the column variable of a contingency table are close together in a thermo-plot, this means that:

(a) People in one category are overrepresented in the other category.

(b) People in one category are underrepresented in the other category.

(c) The distribution of row variable within the category of the other variable differs from the sample distribution.

(d) It depends on how far both categories are from the origin of the thermo-plot.

Exercise 6.5 If a category of the row variable and a category of the column variable of a contingency table are far away from the origin in the bi-plot, and the located in opposing directions, this means that:

(a) People in one category are overrepresented in the other category.

(b) People in one categorie are underrepresented in the other category.

(c) The distribution of row variable within the category of the other variable does not differ from the sample distribution.

(d) No conclusion can be drawn.

Exercise 6.6 When comparing response rates in different countries, there always seems to be a strong correlation between the refusal rate and the response rate. The explanation is that:

(a) This is an obvious consequence of the fact that usually response and refusal together make up the largest part of the sample.

(b) Refusers cannot be respondents, and vice versa.

(c) Typically refusers in countries with high response rates are very hard to persuade to participate.

(d) The noncontact rates and not-able rates are also very high in countries with high refusal rates.

CHAPTER SEVEN

Nonresponse and Representativity

7.1 Introduction

Because the *response rate* is an important indicator of the quality of a survey, response rates are frequently used to compare surveys, and also surveys repeated over time. Nevertheless, it is often the case that response rates alone are poor indicators of any nonresponse bias; for example, see Curtin, Presser, and Singer (2000), Groves, Presser, and Dipko (2004), Groves (2006), Groves and Peytcheva (2006), Keeter et al. (2000), Merkle and Edelman (2002), and Heerwegh et al. (2007). Recently, some alternative quality indicators of response have been suggested by Schouten et al. (2009) that may be useful in assessing overall survey quality.

Nonresponse adjustments are usually based on weighting classes that vary in their response rates and their scores with regard to the key survey variables; see Chapters 8 and 9. Sometimes weighting classes are based on expected values under propensity models. Hence implicitly a variation in response rates over the classes is assumed to be an indicator of a nonresponse error in the adjustment for nonresponse.

The *concept of representative response* is introduced and discussed in this chapter. Generally, it is the degree of similarity between the set of respondents and the complete sample with certain sociodemographic and socioeconomic characteristics. From representative response it is a small step to quality indicators, called

Handbook of Nonresponse in Household Surveys, By Jelke Bethlehem, Fannie Cobben, and Barry Schouten

representativeness indicators or *R-indicators* for short. These indicators are defined and illustrated.

Part of the theory and some of the examples in this chapter were developed by the RISQ project, which was part of the 7th EU Research Framework Programme. Background, additional examples and papers can be found at www.risq-project.eu. The website also contains syntax codes in SAS and R that can be downloaded.

The concepts and ideas behind representative response and R-indicators are introduced and illustrated in Section 7.2. In Section 7.3 the theory is applied to the General Population Survey.

7.2 Theory

Representativity theory is introduced in five sections. In Section 7.2.1, the representative response is defined. In section 7.2.2, R-indicators are defined and illustrated. In section 7.2.3, the R-indicators are linked to nonresponse bias. In Section 7.2.4, additional indicators, called partial R-indicators, are defined. In section 7.2.5, guidelines are presented for the use of the these indicators in practical settings.

7.2.1 WHAT IS REPRESENTATIVE RESPONSE?

Individual response probabilities are defined by the random response model (see Chapter 2). Let $\rho_X(x) = E(R = 1|X = x) = P(R = 1|X = x)$ be the propensity to respond for a population unit carrying value x on the vector of auxiliary variables. Then the response propensity is the conditional distribution of the response indicator R on X. Furthermore let

$$\rho_X = (\rho_X(x_1), \rho_X(x_2), \ldots, \rho_X(x_N))^T \tag{7.1}$$

denote the vector of response propensities in the population (because in this chapter distinctions are made between different sets of auxiliary variables, the subscript is not omitted as in Chapter 2). Each set of auxiliary variables leads to different response propensities.

Let $\bar{\rho}$ be the average response propensity in the population:

$$\bar{\rho} = \frac{1}{N}\sum_{k=1}^{N} \rho_X(x_k). \tag{7.2}$$

The subscript X for $\bar{\rho}$ is omitted as the average response propensity always attains the same value irregardless of what vector X is employed.

EXAMPLE 7.1 Response Rate as a Quality Indicator

Literature warns us not to single-mindedly focus on response rates as an indicator of survey quality. This can easily be illustrated by an example from the 1998 Dutch survey POLS (short for Permanent Onderzoek LeefSituatie, or Integrated Survey on Household Living Conditions in English).

The following table contains the one- and two-month POLS survey estimates for the proportion of the Dutch population that receives a form of social allowance and for the proportion of nonnatives (i.e., persons that have at least one parent that was born outside the Netherlands). Both variables are taken from registry data and are artificially treated as survey items by deleting their values for nonrespondents The sample proportions are also given in the table. After one month, the response rate was 47.2%, while after the full two-month interview period, the rate was 59.7%. In the 1998 POLS, the first month was CAPI (computer-assisted personal interview). Nonrespondents after the first month were allocated to CATI (computer-assisted telephone interview) when they had a listed, land-line phone. Otherwise, they were allocated once more to CAPI. The second interview month gave another 12.5% of response. However, from the table it can be seen that after the second month, the survey estimates have a larger bias than after the first month.

From the example, it should be clear that the increased efforts led to a less representative response with respect to both auxiliary variables.

Variable	After 1 month	After 2 months	Sample
Receiving social allowance	10.5%	10.4%	12.1%
Nonnative	12.9%	12.5%	15.0%
Response rate	47.2%	59.7%	100%

When the literature is consulted, it turns out that the term "representative" is often used without explaining what it means. Kruskal and Mosteller (1979a, b, c) make an extensive inventory of the use of the word "representative" in the literature and identify nine interpretations. A number of interpretations they have found are omnipresent in the statistical literature. The statistical interpretations that Kruskal and Mosteller named "absence of selective forces," "miniature of the population," and "typical or ideal cases" relate to probability sampling, quota sampling, and purposive sampling. Below a definition of representative response is given that corresponds to the "absence of selective forces" interpretation. Why is this choice made?

The need for indicators for response quality comes from the need to monitor and compare the response to different surveys in topic or time. Because of these objectives it is neither appealing nor practical to define representative response as dependent on the survey topic, nor as dependent on the estimators used. Clearly, the survey topics influence the probability that households will participate in the survey, but the influence cannot be measured or tested. Hence this influence cannot be the input for assessing response quality. For this reason the composition of response is judged by predefined sets of variables that are observed outside the survey and can be employed for each survey under investigation. Respondent selection should be as close as possible to a "simple random sample of the survey sample," namely with as little relation as possible between response and characteristics that distinguish units from each other. The latter can be interpreted as having selective forces that are absent in the selection of respondents.

The definition of representative response adopted here was proposed by Schouten et al. (2009).

Definitaion 7.1 The response to a survey is *representative with respect to X* when the response propensities $\rho_X(x)$ are constant over x, meaning when they are equal to $\bar{\rho}$.

In other words, response leads to equal selection probabilities and can be considered as an additional phase in the sampling design. Its interpretation is straightforward as a result of that. It does not relate to a specific survey item, a specific estimator, or a specific model for response behavior other than that it is assumed that the random response model can be used.

The definition is dependent on the selected auxiliary variable vector X. Response may be representative with respect to one such vector but not representative with respect to another. Ideally one would like to assess whether response is representative with respect to some super auxiliary vector $\aleph$ that represents all relevant characteristics. However, this vector is not observed, so it only has a conceptual meaning. In order to compare the representativeness of surveys, one has to fix the auxiliary vector.

In the literature missing-data mechanisms (like nonresponse to a survey) are often termed *missing completely at random* (MCAR), *missing at random* (MAR), or *not missing at random* (NMAR), after the influential work of Little and Rubin (2002). The three mechanisms represent decreasingly strict assumptions about the missingness of data, with MCAR being the most favorable and NMAR being the least favorable setting. Chapter 2 describes in detail the ideas behind the three mechanisms. There is an essential conceptual difference between these mechanisms and the definition of representative response that is important to stress. This difference arises from the objectives behind the definitions. The missing-data mechanisms originate from the focus on estimation, while the definition of representative response comes from the focus on data collection.

Somewhat confusingly, in the literature the missing-data mechanisms are usually referred to without an explicit reference to what is missing. However, the mechanisms only have a meaning when they are connected to variables. A different mechanism may apply to different items Y in the same survey and different sets of auxiliary information X. MCAR(Y) means, in terms of response propensities, that $\rho_Y(y)$ is constant in y; that is, response is representative with respect to Y. MAR (X, Y) means that $\rho_{X,Y}(x,y) = \rho_X(x)$ for all y, while NMAR(X, Y) implies that $\rho_{X,Y}(x,y) \neq \rho_X(x)$ for at least one possible outcome y. The distinction between Y and X is deliberate. Y is a variable of interest in a survey and X is an auxiliary variable. The three mechanisms cannot be tested formally for any survey variable; they are underlying to models that attempt to adjust for the impact of nonresponse. MCAR means that no adjustment is needed. MAR means that the distribution of

Y is affected by nonresponse and parameters of that distribution like the mean may be biased as a result. Adjustment using X, if relations with Y and R are specified correctly, removes this bias. NMAR implies that X does not suffice to remove the bias of all parameters.

The definition of representative response aims at data collection, not a specific Y nor a specific parameter of any distribution. It can be used to assess whether data collection succeeded in obtaining a balanced response for a set of preselected variables X that is available before and during data collection. Of course, the selected variables may be of a general, wide interest when multiple surveys are compared, or the variables may be relevant for a particular survey when that survey is compared to itself. Hence representative response with respect to X is the same as MCAR(X), but nonrepresentative response does not conform to MAR or NMAR in any way. The most that can be said is that the more deviant it is from the representative response, the stronger one has to rely on MAR assumptions in the estimation of parameters of interest.

So far the focus has been on a univariate assessment of response quality. The definition of representative response cannot be used as the basis to provide insight into the impact of single variables conditionally on other variables. It is therefore worthwhile to translate representative response to a multivariate setting, which is termed conditionally representative response.

Definitaion 7.2 A response to a survey is *conditionally representative with respect to* X*given* Z when conditional response propensities given Z are constant for X; in other words, when $\rho_{X,Z}(x, z) = \rho_Z(z)$ for all x.

Hence Z may be a vector now that two variables are involved, X and Z. The representativeness of response for X is evaluated with respect to Z.

EXAMPLE 7.2 Representative and Conditionally Representative Response

From the fictitious population of Samplonia a sample of 1000 persons is selected with the following characteristics:

Gender	Watches sports two hours or less per week	Watches sports more than two hours per week	Total
Male	375	125	500
Female	450	50	500
Total	825	175	1000

Suppose that we are interested in the percentage of persons who watch sports for more than two hours a week, say y_{sport}. This percentage is equal to

$$y_{sport} = \frac{500}{1000} \times \frac{125}{500} + \frac{500}{1000} \times \frac{50}{500} = 17.5\%.$$

Now suppose that the survey has a response rate of 62.5%, with the following fieldwork results:

Gender	Nonresponse	Response	Total
Male	250	250	500
Female	125	375	500
Total	375	625	1000

The nonresponse does not affect sports-watching estimates for women and for men. It only affects the distribution of gender. Then, the estimate for the percentage of persons that watch sports for more than two hours a week would become

$$y_{sport} = \frac{250}{625} \times \frac{125}{500} + \frac{375}{625} \times \frac{50}{500} = 16.0\%.$$

Apparently the percentage of sports-lovers is underestimated because females are overrepresented among the respondents. The response is neither representative with respect to gender nor with respect to sports watching. However, the response is conditionally representative with respect to sports watching given gender.

Now suppose that the response rate is only 50%, with the following results:

Gender	Nonresponse	Response	Total
Male	275	225	500
Female	225	275	500
Total	500	500	1000

The estimate for the percentage of persons who watch sports for more than two hours a week would now be

$$y_{sport} = \frac{225}{500} \times \frac{125}{500} + \frac{275}{500} \times \frac{50}{500} = 16.75\%.$$

Again, response is not representative with respect to gender and sports watching and conditionally representative with respect to sports watching given gender. However, the response rate is much lower, but the estimate is better. This is because the composition of the response resembles that of the complete sample much better.

Example 7.2 clearly illustrates that defining *representative response* and *conditionally representative response* is not enough. Indicators are needed that measure the extent to which response is nonrepresentative. Such indicators must be based on definitions of representative response in order to avoid ambiguity.

7.2.2 INDICATORS FOR REPRESENTATIVE RESPONSE

In practice, response will hardly ever be fully representative with respect to all available auxiliary sociodemographic characteristics. Therefore, to evaluate the quality of the response, indicators are needed that measure how dissimilar the response data set is from the sample.

In this section, one such indicator for representative response is introduced. The indicator is proposed by Schouten et al. (2009) and is called a *representativeness indicator* or *R-indicator* for short. The indicator is equal to one minus two times the standard deviation of the response propensities:

$$R(\rho_X) = 1 - 2S(\rho_X), \tag{7.3}$$

with

$$S^2(\rho_X) = \frac{1}{N}\sum_{k=1}^{N}(\rho_X(x_k) - \bar{\rho})^2. \tag{7.4}$$

In (7.4), N is the population size and U is the set of population units.

The indicator exploits the idea of diversity among individual response probabilities and is based on the standard deviation of response propensities. The larger the standard deviation, the lower is the representativity of the response. Hence, when the standard deviation is zero, the indicator is 1. It is not difficult to prove that the standard deviation is maximally equal to 0.5.

The indicator R is motivated by the idea that the better the nonresponse mechanism can be explained (i.e., the better differences in response rates can be distinguished), the further away the response is from representativity. This is somewhat of a paradox. The better nonresponse can be explained, the better it can also be adjusted for. This is true in general. However, it is important to note that indications of an unbalanced or skewed response composition are sought. Such an indication is a first step toward fieldwork strategies that obtain a more balanced response; see Chapter 13. Attention needs to be focused to those groups in the population that become underrepresented in a survey. Of course, such groups can only be identified by using the auxiliary information at hand. And this is the same information that may be available for nonresponse adjustment. But it is not said that within groups formed by the auxiliary information, the answers to the survey variables are the same for nonrespondents and respondents. If "new" respondents need to be found, then it is best to seek them in those groups that are the most underrepresented.

The R-indicator (7.3) cannot be calculated directly, since the response probabilities are unknown. They need to be estimated using, for instance, a logistic or probit regression model. Furthermore the population variance (7.4) needs to be replaced by a sample variance. R-indicator (7.3) is replaced by an estimator $\hat{R}$ and the response propensities ρ_X by estimated response propensities $\hat{\rho}_X$:

$$\hat{R}(\hat{\rho}_X) = 1 - 2\hat{S}(\hat{\rho}_X), \tag{7.5}$$

with

$$\hat{S}^2(\hat{\rho}_X) = \frac{1}{n}\sum_{i=1}^{n} (\hat{\rho}_X(x_i) - \hat{\bar{\rho}})^2. \quad (7.6)$$

Estimator (7.6) can be used when the sampling design leads to equal inclusion probabilities. If this is not the case, then (7.6) needs to be replaced by

$$\hat{S}^2(\hat{\rho}_X) = \frac{1}{N}\sum_{i=1}^{n} d_i(\hat{\rho}_X(x_i) - \hat{\bar{\rho}})^2, \quad (7.7)$$

with d_i the inclusion or design weight of sample unit i. The inclusion weight is defined as $d_i = 1/\pi_i$, where π_i is the inclusion probability.

Chapter 11 explains in detail how to estimate response propensities from survey data. A discussion of the estimation is omitted here.

EXAMPLE 7.3 R-Indicators

R-indicators are computed for the setting of Example 7.2. The R-indicator values are presented in the following table:

	R-indicator	
X	50% response	62.5% response
Gender	90.0%	75.0%
Sports watching	98.0%	95.1%
Gender × sports watching	90.0%	75.0%

The R-indicator is lower for the setting with 62.5% response, both for X= gender and X= sports watching. The R-indicator for X= gender × sports watching is exactly equal to that of X= gender, because sports watching does not produce additional selectivity in the response. As a consequence the response propensities for X= gender and for X= gender × sports watching are the same.

7.2.3 WORST-CASE NONRESPONSE BIAS

It was shown in Chapter 2 that under the random response model the bias of the response mean $\bar{y}_R$ of some survey variable Y is equal to

$$B(\bar{y}_R) = \tilde{Y} - \bar{Y} = \frac{R_{\rho Y} S_\rho S_Y}{\bar{\rho}}, \quad (7.8)$$

with $R_{\rho Y}$ being the correlation between the response propensities and Y, and S_ρ and S_Y being the standard deviations of the response propensities and Y, respectively. Expression (7.8) implies that low-response rates have a negative impact on bias, but only when there is a linear relation between the survey variable and response behavior. Expression (7.8) also shows that a higher response rate does not necessarily lead to a smaller nonresponse bias. When, along with the response rate, also the correlation between response and survey variable increases, the bias may in fact be bigger. However, it can easily be shown that the maximal nonresponse bias of the response mean for categorical variables given the response rate $\bar{\rho}$ is always smaller than $1-\bar{\rho}$. Hence the response rate does provide an upper bound. In the following it will be shown that this bound can be made sharper based on the observed variance of response propensities and thus indirectly also on the R-indicator.

From (7.8) to the R-indicator $R(\rho_X)$ turns out to be a relatively small step. To enable R-indicators to be interpreted in terms of the impact of nonresponse on survey estimation, the *standardized bias* of the response mean $\bar{y}_R$ is considered:

$$\frac{|B(\hat{\bar{y}}_R)|}{S_Y} = \frac{|C_{\rho Y}|}{\bar{\rho} S_Y} = \frac{|R_{\rho Y}| S_\rho}{\bar{\rho}} \leq \frac{S_\rho}{\bar{\rho}} = \frac{1-R(\rho)}{2\bar{\rho}}, \tag{7.9}$$

with $C_{\rho Y}$ being the covariance between the response propensities and Y, and S_Y being the standard deviation of Y. Hence the R-indicator shows up as a component in the upper limit to the standardized bias. The upper limit in (7.9) can be viewed as the worst-case nonresponse bias of response means because it holds for any Y and it follows from maximal correlation with response. The upper limit in (7.9) is denoted by $B_m(\rho_X)$, when X is used to model response propensities,

$$B_m(\rho_X) = \frac{1-R(\rho_X)}{2\bar{\rho}}. \tag{7.10}$$

Clearly, the true response propensities are unknown and, again, need to be estimated. $B_m(\rho_X)$ is replaced by its estimator

$$\hat{B}_m(\hat{\rho}_X) = \frac{1-\hat{R}(\hat{\rho}_X)}{2\hat{\bar{\rho}}}. \tag{7.11}$$

$\hat{B}_m$ thus represents the estimated maximal absolute standardized bias under the scenario that nonresponse correlates maximally with the selected auxiliary variables. The maximal bias is interesting from the point of view that each survey carries multiple survey variables and that different surveys have different survey variables. As a consequence it is useful to remove the dependence on a specific Y and to move to a worst-case scenario.

EXAMPLE 7.4 Maximal Bias

R-indicators were computed in Example 7.3. The corresponding maximal bias is presented in the following table.

	Maximal bias	
X	50% response	62.5% response
Gender	10.0%	20.0%
Sports watching	2.0%	3.9%
Gender × sports watching	10.0%	20.0%

From the table it can be concluded that the maximal bias is bigger for the setting with 62.5% response, both for X= gender and X= sports watching. The maximal bias for X= gender × sports watching is exactly equal to that of X= gender, because sports watching does not produce additional selectivity in the response and leads to the same R-indicator values.

The R-indicator and the maximal bias provide means to evaluate the quality of response. Ideally one would like to bound the R-indicator from below, namely to derive values of the R-indicator that are acceptable and values that are not. If the R-indicator value is smaller than the threshold, then this is an incentive to enhance the distribution of response by lifting the response rates of underrepresented groups.

Response-representativity functions can be used for deriving lower bounds. They are a function of a specified threshold γ and the response rate $\bar{\rho}$. The threshold γ represents a quality level and may be provided by survey users or survey analysts beforehand.

Two functions are distinguished in this chapter, but many more are possible. The response-representativity functions are defined as

$$RR_1(\gamma, \bar{\rho}) = 1 - \frac{2}{\xi_{1-\alpha/2}}\gamma, \qquad \text{maximal variation in response propensities,} \tag{7.12}$$

and

$$RR_2(\gamma, \bar{\rho}) = 1 - 2\bar{\rho}\gamma, \qquad \text{maximal bias,} \tag{7.13}$$

with $\xi_{1-\alpha/2}$ being the $1-\alpha/2$ quantile of the standard normal distribution. The functions originate from setting a threshold γ to $\xi_{1-\alpha/2}S(\rho_X)$ and $B_m(\rho_X)$, respectively. A usual choice for α is 5%, leading to $1.96S(\rho_X)$.

The first function, RR_1, is the most general one. It is based on the idea that R-indicators present the quality of response regardless of the estimators that the survey researcher is going to use and the population parameters that the survey is aiming at. In that setting the concept of nonresponse bias has little meaning, since there is no estimator of which the bias can be derived, and a lower bound must be based on the distribution of response propensities alone. A quality threshold γ may be derived by demanding that a specified proportion of the response propensities must have a maximal distance to the mean response propensity. More specifically, it may be requested that $100(1-\alpha)\%$ of the probability mass of the response propensities should be within a distance γ to the response rate ρ. For example, if $\alpha = 0.05$ and $\gamma = 5\%$, 95% of the response propensities must be at most 5% away of the response rate. Clearly, the distribution underlying the response propensities is unknown. For the sake of simplicity it is therefore assumed that the propensities follow a normal distribution. Then the interval $[\rho-\xi_{1-\alpha/2}S(\rho_X),\ \rho+\xi_{1-\alpha/2}S(\rho_X)]$ contains $100(1-\alpha)\%$ of the probability mass. RR_1 follows easily from demanding that $\xi_{1-\alpha/2}S(\rho_X) \leq \gamma$. Note that (7.12) does not depend on the response rate $\bar{\rho}$. Hence it is constant with respect to the response rate.

RR_2 arises when it is demanded that the maximal bias must not exceed a prescribed threshold γ. In many surveys the population parameters of interest are population means or population totals. The maximal bias then gets a clear meaning; it reflects the quality of simple response means. RR_2 follows from $B_m(\rho_X) \leq \gamma$. For instance, $\gamma = 5\%$ implies that the maximal absolute bias may not be bigger than 5%.

EXAMPLE 7.5 Response-Representativity Functions for a Business Survey

In 2007 a business survey, the Short Term Statistics (STS) of Statistics Netherlands, was conducted for industry and retail businesses. The data collection lasted 90 days. Statistics were needed after 30 days.

The following two figures show RR_1 and RR_2 for the response to the STS after 15, 30, 45, and 60 days of data collection. The vertical axes represent the R-indicator and the horizontal axes the response rate. From top to bottom, the lines correspond to thresholds of 5% and 10% in the first figure, and to thresholds of 1%, 5%, and 10% in the second figure.

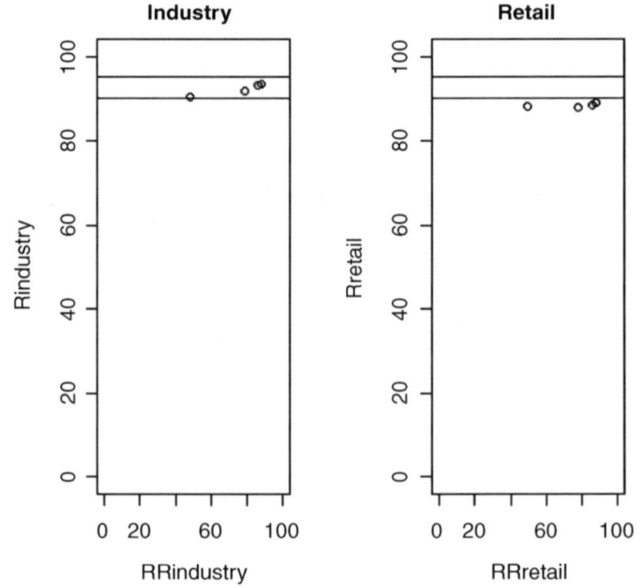

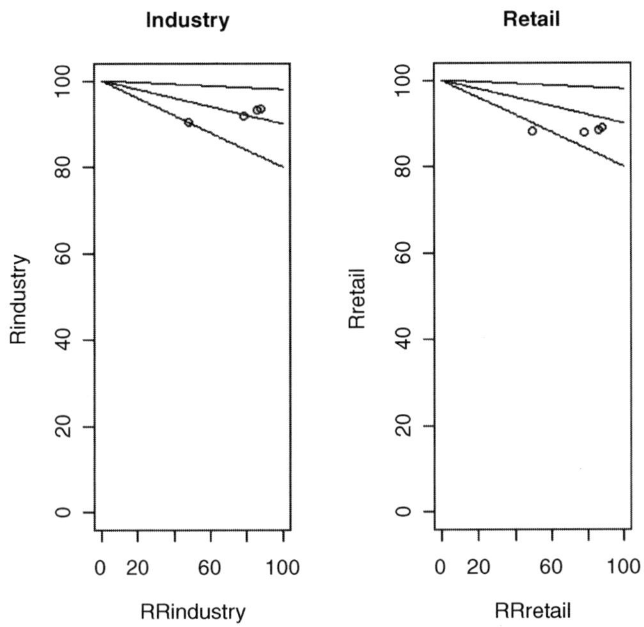

For industry the R-indicator meets the 10% threshold for RR_1 after 15 days, while for retail this threshold is approached only after 60 days. RR_2 gives a similar picture. Industry reaches the 5% threshold after 30 days; for retail this threshold is not attained even after 60 days of data collection.

It can be concluded that data collection for STS industry leads to acceptable response much sooner than for STS retail.

7.2.4 PARTIAL INDICATORS FOR REPRESENTATIVE RESPONSE

Two types of *partial R-indicators* are defined: unconditional and conditional. *Unconditional partial R-indicators* measure the contribution of single variables to a lack of representative response. *Conditional partial R-indicators* measure the contribution of single variables to a lack of representative response given other variables, meaning with respect to conditional representative response. Unconditional partial indicators are designed typically for comparisons of different surveys or surveys in time. Conditional partial indicators are especially suited for data collection monitoring.

Both unconditional and conditional partial R-indicators can be computed at the variable level and at the category level. The category level is only applicable when the variable of interest is categorical. The variable level indicators represent the impact of the variable as a whole, although the impact depend on the classification and categories that are used. If a categorical variable is recoded, the impact of the variable will change. The category level indicators correspond to the impact of single categories and are always smaller than the variable level indicators.

At the variable level the unconditional partial R-indicator P_u for a variable Z is defined as

$$P_u(Z) = S(\rho_Z), \tag{7.14}$$

or in other words, the standard deviation of the response propensity function $\rho_Z(z)$ in the population. The subscript u in (7.14) stands for unconditional and is given in order to distinguish partial R-indicators for unconditional representative response P_u from those for conditional representative response, denoted by P_c. The indicator can be computed for any Z, meaning both elements in X that are used to estimate the response propensities as well as variables not in X. In general, it holds that $P_u(Z) \in [0, 1]$. Furthermore $P_u(Z) \in [0, (1-R(\rho_X))/2]$ when Z is one of the elements of the vector X. Contrary to the R-indicator, small values of P_u correspond to representative response with respect to Z. The more dispersed the response for Z, the larger is the unconditional partial R-indicator.

For categorical variables the unconditional partial R-indicator is defined for each category. Let Z be a categorical variable with categories $k = 1, 2, \ldots, K$, and let Z_k be the 0 to 1 variable that indicates whether or not $Z = k$. For example, Z represents age and Z_k is the indicator for being younger than 35 years. The partial R-indicator for a category k is defined as

$$P_u(Z, k) = \sqrt{\frac{N_k}{N}}(\bar{\rho}_{Z_k} - \bar{\rho}), \tag{7.15}$$

with $N_k = \sum_U Z_k$ the number of population units in category k. $P_u(Z, k)$ originates from dividing $P_u(Z)$ over the categories of Z while maintaining the signs between the response propensity $\bar{\rho}_{Z_k}$ and the overall response rate $\bar{\rho}$. Negative values indicate underrepresentation while positive values indicate overrepresentation. Furthermore, $P_u(Z, k) \in [-1, 1]$ and

$$P_u(Z) = \sqrt{\sum_{k=1}^{K} P_u^2(Z, k)}. \tag{7.16}$$

Note that (7.16), the partial R-indicator at the variable level, is in fact the square root of the "between" variance of the response propensities for the categories of variable Z. As such, it is a component of the total variance of the response propensities, $S^2(\rho_X)$, in (7.4), and hence always smaller than or equal to that variance.

If *conditional representativeness* is measured, the impact of a variable, say Z, is adjusted for the impact of the other variables in X. Contrary to the unconditional partial R-indicator, conditional partial R-indicators only have a meaning for variables in X that are used in the definition of response propensities. The conditional partial R-indicator at the variable level is defined as

$$P_c(Z|X) = \sqrt{\frac{1}{N-1}\sum\nolimits_U (\rho_{X,Z}(x_i, z_i) - \rho_X(x_i))^2}, \tag{7.17}$$

which is the distance between propensities based on X plus Z, and propensities based on X alone. For example, X could be a vector containing household composition, household income, and province of residence, while Z equals the age of the head of the household. Response propensities are then defined for the combination household composition, household income, province of residence, and age of the head of the household, and evaluated for the age of the head of the household.

Again, conditional partial R-indicators are also defined for classes of categorical variables by distributing (7.17) over the classes of Z:

$$P_c(Z,k|X) = \sqrt{\frac{1}{N-1}\sum\nolimits_U Z_k(\rho_{X,Z}(x_i, z_i) - \rho_X(x_i))^2}. \tag{7.18}$$

Other than for the unconditional partial indicators, it is not possible to assign a positive or negative sign to the category level conditional partial indicators in (7.18). The reason is that the sign may be different for each subclass of X. In some subclasses, for instance, a certain age of the head of the household may have a positive effect on response, while in others, it has a negative effect. Consequently a positive or negative sign does not have a clear meaning and may best be avoided.

It can be shown that (7.18) is the square root of the "within" variance of the $\rho_{X,Z}$ propensities for a stratification of the population with X. In other words, it is the variation that is left within the cells defined by X. In the example given earlier, it represents the variation in response behavior due to the age of the head of the household given its household composition and income and the province in which the household lives. As the within variance is again a component of the total variance, the conditional partial indicators too cannot exceed the total variance that makes up the R-indicator in (7.4). Furthermore the conditional partial R-indicator for Z is always smaller than the unconditional partial R-indicator for that variable. This makes sense; the impact on response behavior is to some extent removed by accounting for other characteristics of the population unit. In many survey settings, for instance, the impact of gender on response behavior is completely or considerably removed by accounting for the age of the person.

EXAMPLE 7.6 Partial R-Indicators

Unconditional and conditional partial R-indicators can be computed for Example 7.2. The three tables below present the variable level partial

R-indicators and the category level partial R-indicators for gender and sports watching for the 50% and the 62.5% response rates. The partial indicators are multiplied by 1000 to improve the presentation. As a consequence they cannot be bigger than 1000 or smaller than −1000 for the unconditional partial R-indicators.

	Variable level partial R (×1000)			
	Unconditional		Conditional	
X	50%	62.5%	50%	62.5%
Gender	50	125	50	125
Sports watching	10	25	0	0

The unconditional and conditional partial R-indicators for gender are the same in all cases as there is no relation between response and sport watching conditional on gender. The indicators are larger after two months of data collection, reflecting again that the representativity of the response worsened when the response rate was lifted.

	Category level partial R (×1000)			
	Unconditional		Conditional	
Gender	50%	62.5%	50%	62.5%
Male	−4	−9	−4	−9
Female	4	9	4	9

	Category level partial R (×1000)			
	Unconditional		Conditional	
Sport	50%	62.5%	50%	62.5%
<2 hours	4	10	0	0
>2 hours	−8	−22	0	0

The category level partial R-indicators for gender and sport watching are higher for the 62.5% response rate. However, the sign of the indicators does not change. The conditional partial R-indicators for sports watching are equal to zero as the dependence is completely removed by gender.

EXAMPLE 7.7 Partial R-indicators in the STS Survey

The unconditional and conditional partial R-indicators are given for the retail STS survey in Example 7.5 after 15, 30, 45, and 60 days of data collection. Two variables are considered: SBI and GK. SBI is the standard business taxonomy, also referred to as NACE in EU statistics. GK is the classification of the number of employees of the business. GK ranges from 0 to 9, with GK = 0 being businesses without employees and GK = 9 being businesses with over 500 employees.

The first figure contains the unconditional partial R-indicators for SBI and GK. From the plots it becomes clear that after 30 days the partial R-indicators stabilize and changes are relatively small. This indicates that the composition of SBI and GK in the response does not change. The partial R-indicators for GK show that especially the small businesses, GK = 0, are underrepresented in the response. The partial R-indicator values for GK exceed those of SBI, implying that GK has a stronger impact on response than SBI.

Unconditional partial R-indicators at the category level

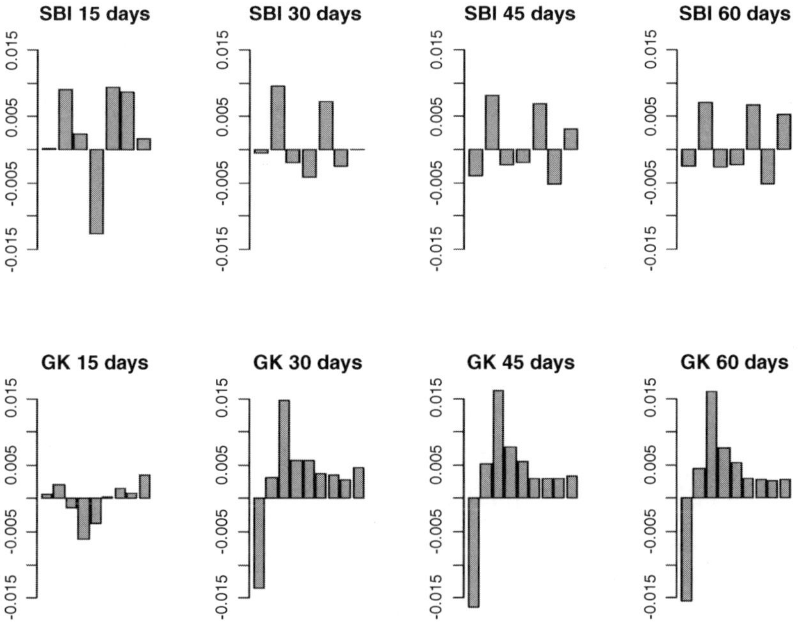

The second figure contains conditional partial R-indicators for SBI and the change in maximal bias due to SBI. The latter is derived as B_m with and without SBI. The conditional partial R-indicators are computed with respect to GK × BTW, where BTW is the turnover reported to the tax board in the previous year.

Conditional partial R-indicators (first row) and the change in maximal bias (second row)

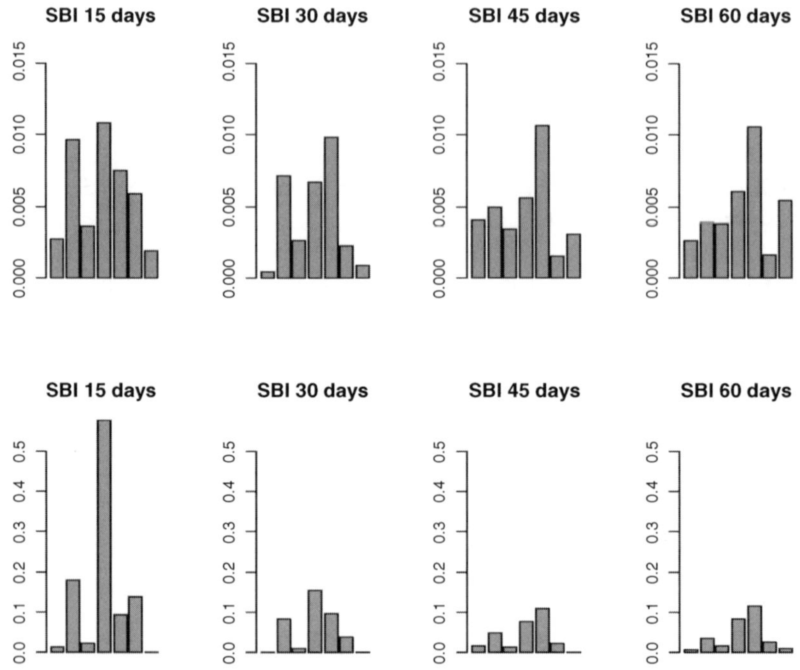

When the conditional partial R-indicators for SBI are compared to the unconditional partial R-indicators, it can be concluded that $X = \text{GK} \times \text{BTW}$ reduces the impact of $Z = \text{SBI}$ only slightly. It is the type 5 businesses (fifth bar in graph) that have the strongest (conditional) impact on the response distribution, while type 6 businesses (sixth bar in graph) have a negligible impact (conditionally).

7.2.5 HOW TO USE R-INDICATORS?

The R-indicators and partial R-indicators can be used in different settings. First, they can be used to compare the response to different surveys that share the same target population, such as households or businesses. Second, the indicators can be used to compare response to the same survey longitudinally, such as monthly, quarterly, or annually. Finally, the indicators can be employed in monitoring the response to a survey during data collection, such as after several days, weeks, or months of fieldwork. The third motive may also be explained as monitoring different reasons for nonresponse like obtaining contact and obtaining participation given contact was established. An important question is how to apply R-indicators with these objectives in mind.

Schouten et al. (2009) and Schouten and Bethlehem (2010) present guidelines for using R-indicators and partial R-indicators in practical settings. They make the following recommendations based on various empirical studies and analyses:

- R-indicators should be evaluated and presented, stating the vector X that was used.
- When comparing different surveys, one should always use the same X.
- R-indicators should be adjoined by a confidence interval.
- When comparing different surveys, fix X beforehand and add all variables in X to the model for the estimation of response propensities. However, stay with the demographic and socioeconomic characteristics that are generally available in many surveys.
- When comparing a survey in time, again fix X beforehand. However, now the vector X may also include variables that correlate to the main survey variables, and variables that relate to the data collection process (paradata). When many variables are available, parsimonious models are favored.
- In comparisons of different surveys, partial R-indicators are supplementary to R-indicators. Models for the estimation of response propensities should be simple and employ general auxiliary variables only.
- In comparisons of a survey over time, partial R-indicators are again supplementary to R-indicators. Models for the estimation of response propensities may be more complex, for example, may define multiple model equations or levels, and may employ paradata additionally to auxiliary variables.
- In the monitoring of data collection, partial R-indicators assist in identifying groups that are underrepresented and may support decisions in responsive or dynamic survey designs or a change in future survey designs. Propensities may be modeled for different response types and data collection steps that lead to missing data.
- In improving the representativity of response, it must always be the objective to increase the response rate and to decrease the variation in response propensities.
- Maximal bias is supplementary to R-indicators when one compares the same survey over time or when one monitors data collection of a survey.

The maximal bias is discussed in Chapter 13 in the context of adaptive survey designs. It will be shown that it can be used as an objective function in the selection and allocation of survey strategies.

It was argued in Section 7.2.1 that the R-indicator $R(\rho_X)$ needs to be estimated because individual response probabilities are unknown. The resulting estimator $\hat{R}(\hat{\rho}_X)$ employs estimated response propensities and sample variances instead of population variances. As a consequence the R-indicator may be biased and have a variation due to the sampling design.

The bias of R-indicators stems from plugging in estimated response propensities in the sample variance $\hat{S}^2(\hat{\rho}_X)$. Estimated propensities have a certain precision or standard error. The quadratic terms in the variance raise the standard error, which results in a small bias. Hence, even if the true response probabilities turn out to be exactly equal to zero and the responses representative, the sample variance will still have a positive value, with the size of the bias inversely proportional to the sample size.

The variance of R-indicators is a result of sampling variation in the estimation of response propensities and sampling variation in the estimation of the population variance. Again, the variance is sample size dependent and diminishes as the sample size grows.

Shlomo et al. (2009) show how to adjust for the bias and how to estimate confidence intervals for the R-indicators. Bias adjustments are usually small given realistic sample sizes. Confidence intervals, however, may be quite wide for sample sizes up to 5000 units.

EXAMPLE 7.8 R-Indicator and Maximal Bias for Different Survey Strategies

In an experimental study for the Dutch Labor Force Survey (LFS) samples of nonrespondents received either a callback or a basic question follow-up, as described in Chapter 10.

The combined composite, weighted response rates (LFS plus callback response and LFS plus basic question response) were 77%, while for the LFS alone it was 62%. Hence both approaches resulted in an additional response of 15%. What is unknown whether these approaches succeeded in improving the representativity of the response.

Response probabilities are estimated using a fixed set of demographic and socioeconomic auxiliary variables, including job status. The values of R and B_m are given for the various response subsets in the table below:

Response	Response rate	R	B_m
LFS	62%	0.79	0.084
LFS + basic question	76%	0.77	0.076
LFS + callback	77%	0.85	0.049

The R-indicator for the LFS with callback response increased, while for the LFS with basic question response it decreased. The example shows that it does not always pay (in terms of representativity) to put extra effort into increasing response rates.

EXAMPLE 7.9 R-Indicator and Partial R-indicator for Contact and Refusal

The Dutch Consumer Sentiments Survey of 2005 was a telephone survey. Nonresponse can, among other reasons, be the result of noncontact or refusal. Those two reasons are nested, meaning participation is possible only when contact is established. The variable level unconditional and conditional partial R-indicators can be computed for contact, for participation given contact, and for response. The R-indicator, maximal bias and response rate values are presented in the following table.

Response	Response rate	R	B_m
Contact	95%	0.92	0.042
Participation (given contact)	71%	0.83	0.124
Response	66%	0.82	0.137

The variable level unconditional and conditional partial R-indicators are computed for six variables: household type, age, ethnicity, degree of urbanization of the area of residence, gender, and house value. The conditional partial R-indicators are computed conditionally on all six variables except the selected variable itself. All indicators are multiplied by 1000.

The table shows that age and gender have the strongest impact on contact without conditioning on other variables. However, conditionally the impact of gender is considerably smaller. The impact of age on contact is only slightly smaller when conditioned on the other variables. One may therefore conclude that contact representativity should be improved first with respect to age.

	Contact		Participation		Response	
(×1000)	$P_u(Z)$	$P_c(Z)$	$P_u(Z)$	$P_c(Z)$	$P_u(Z)$	$P_c(Z)$
Household type	19	9	56	13	67	17
Age	22	19	71	51	58	35
Ethnicity	20	11	38	27	50	34
Urbanization	15	10	15	11	23	9
Gender	22	8	51	7	62	12
House value	15	9	29	8	32	9

For participation the strongest variable is age, both unconditionally and conditionally. Hence, again, it is age that deserves more effort and attention.

The overall impact is strongest for household type. However, the impact is considerably smaller when conditioned on all other variables. The table also shows that the impact of ethnicity is strengthened when noncontact and nonparticipation are combined. It can be concluded that the two response types have a similar impact. For age this is not the case. The combined impact is smaller than for participation alone.

To gain more insight, the category level conditional partial R-indicators are given for variable age. The contact indicators are largest for persons in the age groups 25 to 39 and persons of 69 years and older. The participation indicators are largest for persons between 30 and 44 years of age, persons in age group 55 to 59 years, and again the oldest group.

	$P_c(Z,k)$		
Age	Contact	Participation	Response
< 25	1	3	2
25–29	8	12	2
30–34	10	18	3
35–39	9	14	3
40–44	3	15	7
45–49	1	7	3
50–54	2	8	3
55–59	2	17	14
60–64	1	11	8
65–69	1	7	6
>69	15	15	7

7.3 Application

In this section the theory of R-indicators and partial R-indicators is applied to the General Population Survey (GPS). Table 7.1 lists the auxiliary variables that are available in the GPS. From this list the variables *Urban*, *Ethnic*, *HHType*, *Hasjob*, *Age3*, and *Gender* are selected for the exploration of the representativity of response. The other variables of Table 7.1 are left as exercises.

7.3.1 R-INDICATORS

Three reasons for nonresponse are investigated: not processed, not contacted, and nonparticipation. Unprocessed cases are sampling units that are allocated to interviewers but that are omitted because of time constraints and work loads. Processing comes before contacting, and participation is only possible after a sampling unit is contacted. Hence the three nonresponse types are nested.

Table 7.2 gives relative and cumulative response rates. The overall response rate equals 58.7%. Most of the nonresponse is the result of refusal. The R-indicator is

TABLE 7.1 Available sociodemographic and socioeconomic variables

Variable	Description
Region	Region of the country
Urban	Degree of urbanization
Phone	Has listed phone number
Pnonnat1	Percentage nonnatives in neighborhood
Pnonnat2	Percentage nonwestern nonnatives in neighborhood
Houseval	Average house value in neighborhood
Ethnic	Type of nonnative
HHType	Type of household
HHSize	Size of the household
Marstat	Marital status
Married	Is married
Nonnativ	Is nonnative
Socall	Has social allowance
Age13	Age in 13 age groups
Allowan	Has an allowance
Children	Children in household
Hasjob	Has a job
Age3	Age in 3 age groups
Disaball	Has disability allowance
Gender	Gender
Unempall	Has unemployment allowance

computed for two choices of X:

$$X_1 = \text{Urban} + \text{Age3} + \text{Gender}$$

$$X_2 = \text{Urban} + \text{Ethnic} + \text{HHType} + \text{Hasjob} + \text{Age3} + \text{Gender} + \text{Houseval}$$

Table 7.3 gives the R-indicators for both sets of auxiliary variables and for the different response types. The R-indicators given are based on cumulative contact and response propensities, which are relative to the sample.

Notice that in Table 7.3 the R-indicators for the larger model X_2 are smaller than for the model X_1. This is no coincidence as X_1 is a subset of X_2. R-indicators are

TABLE 7.2 Relative and cumulative rates for response types

Type of nonresponse	Relative rate	Cumulative rate
Processed	92.3%	92.3%
Contacted	93.8%	86.5%
Responded	67.8%	58.7%

TABLE 7.3 Cumulative R-indicators for response types

Type of nonresponse	X_1	X_2
Processed	0.932	0.926
Contacted	0.927	0.900
Participated	0.847	0.807

strictly decreasing when variables are added. From the table it can also be observed that the R-indicators decrease when response types are added. One could therefore conclude that the response types do not cancel each other but have a combined effect on the representativity of response. The strongest drop in R-indicator comes from the nonparticipation. However, given the relatively small size of nonprocessed cases, this type of nonresponse also has a serious impact. The difference between the two models is small for nonprocessed cases. Most of the selectiveness of processing stems from age, gender, and degree of urbanization. The addition of the four variables in X_2 does lead to a stronger drop in R-indicator for contacted cases and nonparticipating cases. Apparently ethnicity, household type, job status, and house value have an additional impact. Partial R-indicators in section 7.3.2 will provide more insight.

Table 7.4 presents the maximal biases for the two models and the three response types. The results are not surprising. Since both R-indicators and response rates are dropping, the maximal biases increase as response types are added.

The data collection of the GPS covered two months. The response rate after one month was 46.4%. Hence, in the second month an extra 12.3% response was obtained. It would be interesting to learn whether the second month of data collection also led to a more representative response. Table 7.5 gives the

TABLE 7.4 Cumulative maximal bias for response types

Type of nonresponse	X_1	X_2
Processed	0.037	0.040
Contacted	0.042	0.058
Participated	0.010	0.164

TABLE 7.5 Cumulative R-indicators and maximal bias after one and two months

		X_1	X_2
R-indicator	Month 1	0.859	0.835
	Month 2	0.847	0.807
Maximal bias	Month 1	0.152	0.178
	Month 2	0.130	0.164

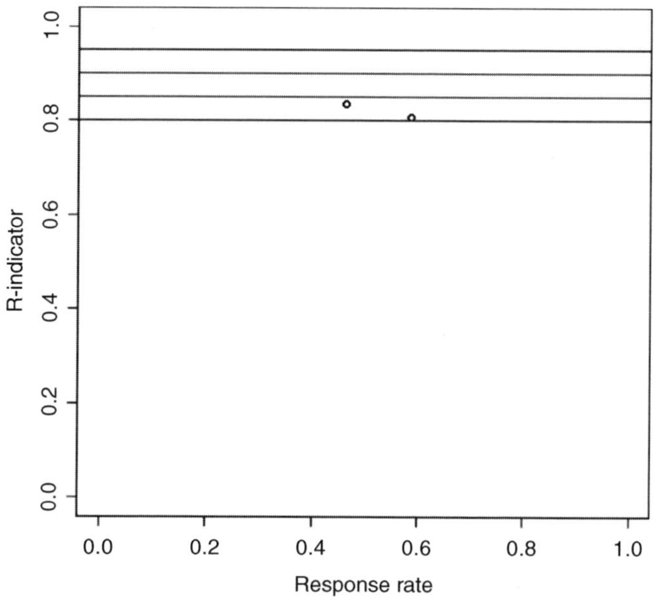

FIGURE 7.1 RR_1 **functions for** $\gamma = 5\%, 10\%, 15\%, 20\%$ **(top to bottom), and** $\alpha = 0.05$.

R-indicators and maximal bias for the two models of the final response, namely without differentiation with respect to response type.

In Table 7.5 we can see that the representativity of response is smaller after two months. Both the R-indicator and the maximal bias are smaller for the two models. Although the data collection was extended, it did not result in a better response.

Figures 7.1 and 7.2 depict the representativity functions RR_1 and RR_2 after one and two months of data collection for model X_2. In Figure 7.1, the R-indicator after one month is well above the threshold $\gamma = 5\%$ but below $\gamma = 10\%$. After two months the indicator is only slightly above $\gamma = 5\%$. In Figure 7.2, the R-indicator after two months is closer to the $\gamma = 15\%$ level than after one month, although it dropped. The decrease in the R-indicator is mediated by the increase in the response rate.

7.3.2 PARTIAL R-INDICATORS

In this section the unconditional and conditional partial R-indicators are applied to the GPS. As in Section 7.3.1, two models are employed:

$$X_1 = \text{Urban} + \text{Age3} + \text{Gender},$$

$$X_2 = \text{Urban} + \text{Ethnic} + \text{HHType} + \text{Hasjob} + \text{Age3} + \text{Gender} + \text{Houseval}$$

Conditional representative response only has a meaning, however, when the representativeness of a variable Z is investigated against a set of (conditional)

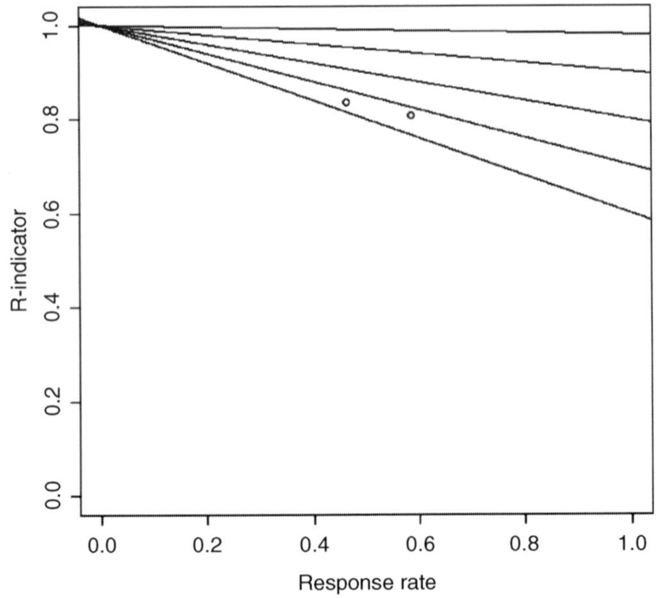

FIGURE 7.2 RR_2 **functions for thresholds** $\gamma = 1\%, 5\%, 10\%, 15\%, 20\%$ **(top to bottom).**

variables X. Here

$$Z_1 = \text{Urban}, Z_2 = \text{Age3}$$

are analyzed against X_1 and against X_2.

Table 7.6 contains the variable level unconditional and conditional partial R-indicators for the two selected variables and the two models. From the table it follows that degree of urbanization has the strongest impact of the two variables. Remarkably, the impact of both variables on the representativity of response is not or little affected by conditioning on the other variables. Hence the two variables have a separate impact on response that is not accounted for by the ethnicity, gender, house value, job status, or household type The unconditional and conditional partial R-indicators are similar in size.

Another observation from Table 7.6 is that the impact of both variables increases considerably when the larger model is selected. This holds only for

TABLE 7.6 Variable level partial R-indicators for Z_1 and Z_2 against X_1 and X_2

	X_1		X_2	
	$P_u(Z)$	$P_c(Z)$	$P_u(Z)$	$P_c(Z)$
Urban (Z_1)	75	75	75	48
Age3 (Z_2)	15	12	15	6

Note: Values are multiplied by 1000.

conditional partial R-indicators. This result makes sense as the model contains four additional variables that may interact with age and degree or urbanization when it comes to response behavior. However, it still pays off to attempt to improve response representativity with respect to degree of urbanization.

Tables 7.7 and 7.8 show the category level partial R-indicators for Z_1 and Z_2. The values for degree of urbanization have a clear pattern; more urbanized areas have a stronger underrepresentation. When conditioning on the other variables, this pattern becomes blurred. The conditional category indicators are closer to each other in size. The unconditional partial R-indicators for age show that younger and older persons are underrepresented. When conditioning on other variables, the impacts of the categories are equal.

For model X_1 the unconditional and conditional partial R-indicators are computed after one month and after two months of data collection. The results are presented in Tables 7.9, 7.10, and 7.11 contain the category level indicator values.

In all cases the values for the different R-indicators are smaller after two months of data collection, the only exception being the category indicator for age group 18 to 34 years. This indicates that the representativity of the response deteriorated in the second month; the additional response sharpened the contrast between respondents and nonrespondents.

TABLE 7.7 Category level partial R-indicators for Z_1 against X_1 and X_2

	X_1		X_2	
Urban	$P_u(Z,k)$	$P_c(Z,k)$	$P_u(Z,k)$	$P_c(Z,k)$
Very strong	−63	62	−63	38
Strong	−8	8	−8	11
Moderate	13	13	13	11
Little	23	23	23	13
Not	32	32	32	21

Note: Values are multiplied by 1000.

TABLE 7.8 Category level partial R-indicators for Z_2 against X_1 and X_2

	X_1		X_2	
Age	$P_u(Z,k)$	$P_c(Z,k)$	$P_u(Z,k)$	$P_c(Z,k)$
18–34	−8	4	−8	4
35–54	11	9	11	3
55 and older	−5	6	−5	3

Note: Values are multiplied by 1000.

TABLE 7.9 Variable level partial R-indicators for Z_1 and Z_2 against X_1 after one and two months of data collection

	Month 1		Month 2	
	$P_u(Z)$	$P_c(Z)$	$P_u(Z)$	$P_c(Z)$
Urban	70	70	75	75
Age3	12	8	15	12

Note: Values are multiplied by 1000.

TABLE 7.10 Category level partial R-indicators for Z_1 against X_1 after one and two months of data collection

	Month 1		Month 2	
Urban	$P_u(Z,k)$	$P_c(Z,k)$	$P_u(Z,k)$	$P_c(Z,k)$
Very strong	−55	54	−63	62
Strong	−14	14	−8	8
Moderate	14	14	13	13
Little	19	19	23	23
Not	33	33	32	32

Note: Values are multiplied by 1000.

TABLE 7.11 Category level partial R-indicators for Z_2 against X_1 after one and two months of data collection

	Month 1		Month 2	
Age3	$P_u(Z,k)$	$P_c(Z,k)$	$P_u(Z,k)$	$P_c(Z,k)$
18–34	−10	7	−8	4
35–54	6	3	11	9
55 and older	4	3	−5	6

Note: Values are multiplied by 1000.

7.4 Summary

For decades the response rate was used as the single quality indicator for survey responses. At best the response rate would be evaluated for subgroups in the population of interest. There is a vast literature on methodology to reduce nonresponse and to adjust for nonresponse. Indicators that measure the representativity of survey response have become more and more popular over recent years.

Definitions for representative response and conditionally representative response are given in this chapter. These definitions provide a mathematical basis for the analysis, evaluation, and monitoring of response quality. Both definitions exploit the random response model and the concept of individual response propensities. Zero variation in response propensities is considered the ideal situation, as then response can be viewed as an additional, random phase in the sampling design. Conditionally representative response refer to the variation of response propensities within classes or subpopulations formed by a predefined set of background characteristics. If no variation is left when conditioned on other variables, then the variable is less interesting from the perspective of nonresponse reduction.

R-indicators and partial R-indicators measure deviations from representative response and conditionally representative response. R-indicators range from 0 to 1 with the value 1 being optimal and equal to representative response. Partial R-indicators come in two types, unconditional and conditional, and can be assessed at the variable and at the category level. Unconditional partial R-indicators reflect the impact of single variables on the representativity of response. Conditional partial R-indicators correspond to the impact of single variables conditional on the other selected variables.

R-indicators and partial R-indicators have three purposes: comparing the quality of response for different surveys or the same survey at different times, evaluating the quality of response to a single survey in time, and monitoring response quality during data collection. For the third purpose response-representativity functions are presented. These functions describe the relationships among response rates, R-indicators, and quality thresholds. From an assumed threshold it can be decided whether a response is acceptable in terms of quality.

It is important to note that R-indicators and partial R-indicators cannot be viewed separately from the background variables used in an analysis. For every selection of variables their values are different. In the chapter simple guidelines are given for the selection of background variables.

Another important aspect of the indicators is the estimation of response propensities. Since these propensities are unknown, they need to be estimated. The estimation results in a small bias for the indicators and sampling variation. The bias can be approximated analytically and adjusted for.

7.5 KEY TERMS

Conditionally representative response Response to a survey is called conditionally representative with respect to some characteristic *X* when individual response propensities are equal within the strata formed by *X*.

Representative response Response to a survey is called representative if (under the random response model) all individual response propensities are equal.

Partial R-indicator A quality indicator expressing the contribution of a single variable to deviations from the representative response. Two types are distinguished: unconditional and conditional. Unconditional partial R-indicators reflect the univariate impact of single variables, while conditional partial R-indicators reflect their remaining impact conditional on the other available variables.

Maximal bias In the worst-case scenario, the absolute nonresponse bias of response means. Maximal bias arises when response and survey variable are fully correlated.

Response-representativity function A function relating representativity and response rate to acceptable quality levels.

Response monitoring Activities employed by the data collection staff in order to assess the rate and composition of response. Indicators are essential tools in the monitoring.

Response propensity The estimated probability of response from an individual population unit. The propensity depends on the selected set of auxiliary variables using the random response model.

R-indicator A quality indicator measuring the extent to which a response deviates from the representative response.

7.6 REFERENCES

Curtin, R., Presser, S., and Singer, E. (2000), The Effects of Response Rate Changes on the Index of Consumer Sentiment. *Public Opinion Quarterly* 64, pp. 413–428.

Groves, R.M. (2006), Nonresponse Rates and Nonresponse Bias in Household Surveys. *Public Opinion Quarterly* 70, pp. 646–675.

Groves, R.M., and Peytcheva, E. (2006), *The Impact of Nonresponse Rates on Nonresponse Bias: A Meta-analysis.* Paper presented at 17th International Workshop on Household Survey Nonresponse, August 28-30, Omaha, NE.

Groves, R. M., Presser, S., and Dipko, S. (2004), The Role of Topic Interest in Survey Participation Decisions. *Public Opinion Quarterly* 68, pp. 2–31.

Heerwegh, D., Abts, K., and Loosveldt, G. (2007), Minimizing Survey Refusal and Noncontact Rates: Do Our Efforts Pay Off? *Survey Research Methods* 1, pp. 3–10.

Keeter, S., Miller, C., Kohut, A., Groves, R.M., and Presser, S. (2000), Consequences of Reducing Nonresponse in a National Telephone Survey. *Public Opinion Quarterly* 64, 125–148.

Kruskal, W., and Mosteller, F. (1979a), Representative Sampling I: Non-scientific Literature, *International Statistical Review* 47, pp. 13–24.

Kruskal, W., and Mosteller, F. (1979b) Representative Sampling II: Scientific Literature Excluding Statistics. *International Statistical Review* 47, pp. 111–123.

Kruskal, W., and Mosteller, F. (1979c) Representative Sampling III: Current Statistical Literature. *International Statistical Review* 47, pp. 245–265.

Little, R.J.A., and Rubin, D.B. (2002), *Statistical Analysis with Missing Data.* Wiley, New York.

Merkle, D.M., and Edelman, M. (2002), Nonresponse in Exit Polls: A Comprehensive Analysis. In: Groves, R.M., Dillman, D.A., Eltinge, J.L., and Little, R.J.A. (eds.), *Survey Nonresponse.* Wiley, New York, pp. 243–258.

Särndal, C., and Lundström, S. (2005), *Estimation in Surveys with Nonresponse.* Wiley, Chichester, UK.

Särndal, C., Swensson, B., and Wretman, J. (2003), *Model-Assisted Survey Sampling.* Springer, New York.

Schouten, B., and Bethlehem, J.G. (2007), *Increasing Response Rates and Representativeness Using Follow-up Surveys with Basic Questions.* Paper presented at ISI 2007, August 23-29, Lisbon.

Schouten, B., Cobben, F., and Bethlehem, J.G. (2009), Indicators for the Representativeness of Survey Response. *Survey Methodology* 35, pp. 101–113.

Schouten, B., and Bethlehem, J.G. (2010), *Representativeness Indicators for Measuring and Enhancing the Composition of Survey Response.* Deliverable EU FP7 project RISQ, available at www.ris-project.eu.

Schouten, B., Morren, M., Bethlehem, J.G., Shlomo, N., and Skinner, C. (2009), *How to Use R-indicators?* Deliverable EU FP7 project RISQ, available at www.risq-project.eu.

Shlomo, N., Skinner, C., Schouten, B., Bethlehem, J.G., and Zhang, L.C. (2009), *Statistical Properties of Representativity Indicators.* Deliverable EU FP7 project RISQ, available at www.risq-project.eu.

Stoop, I. (2005), Surveying Nonrespondents, *Field Methods* 16, pp. 23–54.

Voogt, R. (2004), *I Am Not Interested: Nonresponse Bias, Response Bias and Stimulus Effects in Election Research.* PhD thesis. University of Amsterdam, Amsterdam.

7.7 EXERCISES

Exercise 7.1 Are the R-indicator and maximal bias dependent on the model *X*?

(a) No.

(b) Yes, since they employ response propensities that are a function of *X*.

(c) Yes, since if X is larger, there are more subpopulations.

(d) Yes, since the nonresponse missing-data mechanism may depend on X.

Exercise 7.2 When is the conditional partial R-indicator smaller than the unconditional partial R-indicator?

(a) They are always the same.

(b) If *X* and *Z* are different.

(c) If variables in model *X* relate to each other.

(d) If the relation with response is partially removed by other variables in *X*.

Exercise 7.3 Can we decompose the maximal bias to category level maximal bias like that proposed for the partial R-indicators?

(a) Yes, by replacing the R-indicator term in the maximal bias by the unconditional partial R-indicator.

(b) Yes, by replacing the R-indicator term in the maximal bias by the conditional partial R-indicator.

(c) Yes, but not by using partial R-indicators.

(d) No, that makes no sense.

Exercise 7.4 What variables in X should we select for comparing different surveys?

(a) Any X that is available.

(b) Multiple sets of X.

(c) X that are available in all selected surveys.

(d) That depends on the objective of the comparison.

Exercise 7.5 For which of the following purposes are R-indicators suited?

(a) Monitoring data collection in order to re-assign or re-approach sample cases.

(b) Comparing the quality of survey response in each month for a longitudinal survey.

(c) Selecting variables for nonresponse adjustment techniques.

(d) Assessing whether response has an acceptable quality.

Exercise 7.6 For the General Population Survey investigate (a) which of the available background variables leads to the largest unconditional partial R-indicator at the variable level, and (b) which category of that variable is the largest component in that partial R-indicator;

Exercise 7.7 For the General Population Survey select X = Houseval + Hasjob + Socall + Disaball + Unempall. Investigate which of the variables in X has the largest conditional partial R-indicator at the variable level.

Exercise 7.8 For the General Population Survey produce response-representativity functions for the three response types processed: contact, participation, and selecting X = Age13 + Phone + HHType. When does the R-indicator drop below the 10% threshold in RR_1 and RR_2?

CHAPTER EIGHT

Weighting Adjustment Techniques

8.1 Introduction

Unit nonresponse is the phenomenon that the questionnaire remains empty for some elements in the sample. No answers are obtained from the nonresponding individuals. This reduces the number of available observations, making estimates less precise. If observations are missing completely at random (MCAR), it is still possible to make a valid inference. The confidence intervals will just be somewhat wider.

Unfortunately, nonresponse is often selective. This means that certain groups in the target population of the survey are over- or underrepresented. Hence the responses to the survey cannot be considered representative for the target population. If there is a relationship between the lack of representativity and the survey variables, the estimators will be biased. They will systematically over- or underestimate population characteristics.

Handbook of Nonresponse in Household Surveys, By Jelke Bethlehem, Fannie Cobben, and Barry Schouten

EXAMPLE 8.1 Selective Nonresponse in the Dutch Housing Demand Survey

Statistics Netherlands conducted a Dutch Housing Demand Survey in 1981. The sample size was 82,849. The number of respondents was 58,972, which comes down to a response rate of 71.2%.

The population characteristic to be estimated was the percentage of people with the intention to move within two years. The column "Response" in the table below gives the distribution of the answers received to this question. Apparently the percentage of people wanting to move within two years was $100 \times 17515/58972 = 29.8\%$.

Do you intend to move within 2 years?	Response	Nonresponse	Total
Yes	17,515	3,056	20,571
No	41,457	20,821	62,278
Total	58,972	23,877	82,849

To obtain insight on the nonresponses, a follow-up survey was carried out among the nonrespondents. They were also asked whether they intended to move within two years. The results are summarized in the table above under the column "Nonresponse." The percentage of nonrespondents with the intention to move was $100 \times 3056/23877 = 12.8\%$.

The intention to move turned out to be much higher among respondents than among nonrespondents. Apparently there is a relationship between response behavior and the intention to move.

There is ample evidence that nonresponse can be selective. Bethlehem and Kersten (1985) give some examples:

- A follow-up study of the Dutch Victimization Survey showed that people who are afraid to be home alone at night are less inclined to participate in a survey.
- A Dutch Housing Demand Survey showed that people who refuse to participate have fewer housing demands than people who respond.
- The Survey of Mobility of the Dutch Population showed that more mobile people are underrepresented among respondents.

Selective nonresponses will often cause estimators to be biased, so something has to be done to correct for a possible bias. A frequently used family of techniques is *weighting adjustment*. Weighting adjustment is typically applied in cases of unit nonresponse. Different correction techniques are available for item nonresponses. Chapter 14 fully covers correction for item nonresponse.

Bailar et al. (1978) describe weighting as a frequently used adjustment method to correct for a potential bias due to nonresponse. Platek and Gray (1983) and Lindström et al. (1979) also see weighting as an important method to correct for such a bias. Little (1982) discusses the uses of model, including weighting, for correcting nonresponse. Holt and Smith (1979) describe poststratification as a special weighting method that may also help to improve precision of estimators in case of full response. Kalton and Flores-Cervantes (2003) give an overview of weighting adjustment techniques.

Adjustment weighting is based on the use of *auxiliary information*. Auxiliary information is defined here as a set of variables that have been measured in the survey, and for which the distribution in the population is available. Sometimes the complete sample distribution is available, and not the population distribution. Then this sample distribution can be used as an unbiased estimator of the population distribution. Auxiliary information may also come from another source than the survey. For example, it may be possible to link data from an administrative register to both respondents and nonrespondents.

By comparing the population distribution (or complete sample distribution) of an auxiliary variable with its response distribution, it can be assessed whether or not the response is representative for the population (with respect to this variable). If these distributions differ considerably, one must conclude that nonresponse has resulted in a selective sample. More on the use of auxiliary variables for analysis and correction of nonresponse can be found in Chapters 5 and 9.

EXAMPLE 8.2 Comparing the Population and the Response Distribution

The sample of the Survey on Living Conditions of Statistics Netherlands is selected from the population register. Therefore the degree of urbanization can be established for the area in which the sample persons live. In 2002 the sample consisted of a little over 34,000 persons of age 18 and over. The response rate was 54.9%. The table below compares the distribution of urbanization among respondents and the complete sample.

Urbanization	Response	Sample	Difference
Very strong	12.5%	16.0%	−3.5%
Strong	26.5%	26.6%	−0.1%
Average	21.8%	21.0%	0.8%
Little	23.6%	22.2%	1.4%
Not	15.5%	14.3%	1.2%

People in highly urbanized areas are substantially underrepresented. And people in rural areas are overrepresented. Clearly, the response is not representative with respect to the variable degree of urbanization.

The next step is to use this auxiliary information to compute *adjustment weights*. Weights are assigned to all observed records. Estimates of population characteristics can now be obtained by taking weighted values instead of the unweighted values. The weights are defined in such a way that population characteristics for the auxiliary variables can be computed without error. Then the weighted sample is said to be *representative* with respect to the auxiliary variables used.

Whatever sampling design is used, it is always possible to construct an unbiased estimator. This is called the *Horvitz–Thompson estimator*. To do so, we assume that the sampling is without replacement and denote the first-order inclusion probabilities of the sample elements by π_1, π_2, ..., $\pi_{n,}$. The Horvitz–Thompson estimator is defined by

$$\bar{y}_{HT} = \frac{1}{N}\sum_{i=1}^{n}\frac{y_i}{\pi_i}. \tag{8.1}$$

The *inclusion weight* $d_i = 1/\pi_i$ is introduced as one over the first-order inclusion probability of sample element *i*. Consequently the Horvitz–Thompson estimator is rewritten as

$$\bar{y}_{HT} = \frac{1}{N}\sum_{i=1}^{n} c_i y_i. \tag{8.2}$$

Adjustment weighting replaces this estimator by a new estimator:

$$\bar{y}_W = \frac{1}{N}\sum_{i=1}^{n} w_i y_i, \tag{8.3}$$

where the weight w_i is equal to

$$w_i = g_i \times d_i \tag{8.4}$$

and g_i is a *correction weight* produced by a weighting adjustment technique.

The weighting adjustment techniques impose the condition of representativity to one or more selected auxiliary variables. Suppose that X is such an auxiliary variable. Representativity with respect to X implies that the weights w_i have to be constructed such that

$$\frac{1}{N}\sum_{i=1}^{n} w_i x_i = \bar{X}. \tag{8.5}$$

In words, this expression says that if the weights are used to estimate the population mean of the auxiliary variable, the estimate is exactly equal to the population mean.

If the response can be made representative with respect to several auxiliary variables, and if all these variables have a strong relationship with the phenomena to be investigated, then the weighted sample will also be (approximately) representative with respect to these phenomena. Hence estimates of population characteristics will be more accurate.

Several weighting techniques will be described in this chapter, startings with the simplest and most commonly used one: *poststratification*. Next *linear weighting* is described, which is more general than poststratification. This technique can be applied in situations where the auxiliary information is inadequate for poststratification. Then *multiplicative weighting* is discussed as an alternative for linear weighting. Furthermore an introduction into *calibration* is provided. This can be seen as an even more general theoretical framework for adjustment weighting that includes linear weighting and multiplication as special cases.

Propensity weighting is a weighting technique that does not fit in the framework of this chapter. It is treated separately in Chapter 11.

8.2 Poststratification

8.2.1 THEORY

Stratification has a long history in survey methodology. It comes down to dividing the target population of the survey into a number of exclusive, nonoverlapping subpopulations, also called *strata*. A sample is selected from each subpopulation so that estimates can be computed for each subpopulation separately. As a next step, the subpopulation estimates can be combined into an estimate for the whole population.

Stratification was one of the key ideas when survey sampling was first considered during the meeting of the ISI (International Statistical Institute) in Bern in 1895. It was Anders Kiaer, director of the Norwegian Statistical Institute who proposed using sampling instead of complete enumeration. He argued that good results could be obtained with his *representative method*. His idea was to select a sample that should reflect all aspects of the population as much as possible. One way to realize such a sample was the "balanced sample." The population was divided in subpopulations using variables like gender, age, and region. These subpopulations were called *strata*. The sizes of the strata were supposed to be known. The same percentage of persons was taken in each stratum. Selection of samples took place in some haphazard way (probability sampling had not yet been invented). As a result the sample distribution of variables like gender, age, and region was similar to the distribution in the population. The sample was representative with respect to these variables.

In the 1970s Holt and Smith (1979) note the wide use of stratification. This is not surprising as this technique has two attractive properties: (1) it leads to representative samples, and (2) it improves the precision of estimators. There is, however, also a drawback. To be able to draw a stratified sample, a sampling frame is required for each subpopulation separately. This cannot always be done. There are many situations where membership of a subpopulation can be established only after inspection of the sampled data. For example, to obtain a sample that is representative with respect to age groups, one would need to draw a sample from each age group separately. Usually there is no sampling frame for each age group.

Poststratification is an estimation technique by which the representative sample is constructed after the data have been collected. It comes down to assigning stratum weights whereby respondents in underrepresented groups get a weight larger than 1 and respondents in overrepresented groups get a weight smaller than 1. The weighted values are then substituted into expression (8.3) to obtain an improved estimator.

Strata are obtained by crossing auxiliary variables. Poststratification is particularly effective if the strata are *homogeneous*, meaning that people within strata resemble each other. When this is the case, poststratification will not only improve the precision (as measured by the variance or the standard error of estimators) but also reduce any bias due to nonresponse.

Poststratification is described in two steps. First, the situation is described in which every sample element responds. Then, the properties of this estimator are shown if nonresponse occurs.

To be able to carry out the poststratification, one or more qualitative auxiliary variables are needed. The theory is described for one such variable, but the case of more variables is not essentially different. Suppose that an auxiliary variable X has L categories and the population U is divided into L strata $U_1, U_2, \ldots, U_L$. The number of population elements in stratum U_h is denoted by N_h, for $h = 1, 2, \ldots, L$. So $N = N_1 + N_2 + \cdots + N_L$. These stratum sizes are assumed to be known.

Suppose that a simple random sample of size n is selected from the population without replacement. If n_h denotes the number of sample elements in stratum U_h (for $h = 1, 2, \ldots, L$), then $n = n_1 + n_2 + \cdots + n_L$. Note that the values of the n_h are the result of a random selection process. So they are random variables.

For a sample to be representative with respect to the variable X, the proportion of elements in stratum h should be equal to N_h/N, for $h = 1, 2, \ldots, L$. However, the proportion of sample elements in stratum h is equal to n_h/n. To correct for this, each observed element i in stratum U_h is assigned a correction weight

$$g_i = \frac{N_h/N}{n_h/n}. \tag{8.6}$$

If the values of the inclusion probabilities ($d_i = n/N$) and correction weights (8.6) are substituted into expression (8.3), the result is the *poststratification estimator*

$$\bar{y}_{PS} = \frac{1}{N}\sum_{h=1}^{L} N_h \bar{y}^{(h)}, \tag{8.7}$$

where $\bar{y}^{(h)}$ is the mean of the observed elements in stratum h. So the poststratification estimator is equal to a weighted sum of sample stratum means.

EXAMPLE 8.3 Computing Weights by Means of Poststratification

A sample of size 100 is selected from a population of a town of size 1000. There are two auxiliary variables: Gender (with two categories Male and Female), and AgeClass (with three categories Young, Middle, and Old). The tables below show the population and sample distribution of these variables.

Population

	Male	Female	Total
Young	226	209	435
Middle	152	144	296
Elderly	133	136	269
Total	511	480	1000

Sample

	Male	Female	Total
Young	23	15	38
Middle	16	17	33
Elderly	13	16	29
Total	52	48	100

Weights

	Male	Female
Young	0.983	1.393
Middle	0.950	0.847
Elderly	1.023	0.850

The sample is not representative for the population. For example, the percentage of young females in the population is 20.9%, whereas the corresponding sample percentage is 15.0%. The sample contains too few young females.

The correction weights in the table have been computed using expression (8.6). For example, the weight for a young female is equal to $(209/1000)/(15\,/\,100) = 1.393$. Young females are underrepresented in the sample, and therefore get a weight larger than 1. People in overrepresented strata get a weight less than 1.

The adjustment weights w_i are obtained by multiplying the correction weights g_i by the inclusion weights d_i. Here all inclusion weights are equal to $N/n = 10$. Suppose that these weights are used to estimate the number of young females in the population. The weighted estimate would be

$15 \times 10 \times 1.393 = 209$, and this is exactly the population total. Thus, application of weights to the auxiliary variables results in perfect estimates. If there is a strong relationship between the auxiliary variable and the target variable, also estimates for the target variable will be improved if these weights are used.

Now consider the situation where the sample is distorted by nonresponse. The poststratification estimator takes the form

$$\bar{y}_{R,PS} = \frac{1}{N}\sum_{h=1}^{L} N_h \bar{y}_R^{(h)}, \tag{8.8}$$

where $\bar{y}_R^{(h)}$ denotes the mean of the responding elements in stratum h. It can be shown that the bias of this estimator is equal to

$$B(\bar{y}_{R,PS}) = \frac{1}{N}\sum_{h=1}^{L} N_h B(\bar{y}_R^{(h)}). \tag{8.9}$$

Apparently the bias of this estimator is the weighted sum of the biases of the stratum estimators. We will assume that the random response model applies, whereby each population member has a certain unknown probability of participating in the survey, if asked. The response probability of population element k in stratum h is denoted by $\rho_k^{(h)}$. The bias of the poststratification estimator can be written as

$$B(\bar{y}_{R,PS}) = \frac{1}{N}\sum_{h=1}^{L} N_h (\tilde{Y}^{(h)} - \bar{Y}^{(h)}), \tag{8.10}$$

where $\bar{Y}^{(h)}$ is the mean of the target variable in stratum h, and

$$\tilde{Y}^{(h)} = \frac{1}{N_h}\sum_{k=1}^{N_h} \frac{\rho_k^{(h)}}{\bar{\rho}^{(h)}} Y_k^{(h)}. \tag{8.11}$$

Here $Y_k^{(h)}$ denotes the value of the target variable of element k in stratum h, and

$$\bar{\rho}^{(h)} = \frac{1}{N_h}\sum_{k=1}^{N_h} \rho_k^{(h)} \tag{8.12}$$

is the mean of the response probabilities in stratum h. The bias in expression (8.10) can be rewritten as

$$B(\bar{y}_{PS,R}) = \frac{1}{N}\sum_{h=1}^{L} N_h \frac{R_{\rho Y}^{(h)} S_\rho^{(h)} S_Y^{(h)}}{\bar{\rho}^{(h)}}, \tag{8.13}$$

where $R_{\rho Y}^{(h)}$ is the correlation between Y and ρ in stratum h. $S_\rho^{(h)}$ and $S_Y^{(h)}$ are the standard errors of ρ and Y in stratum h, respectively.

The bias of the poststratification estimator is small if the biases within the strata are small. A stratum bias is small in the following circumstances:

- There is little or no relationship between the target variable and response behavior within all strata. So their correlations are small.
- All response probabilities within a stratum are more or less equal. So their standard errors are small.
- All values of the target variable within a stratum are more or less equal. So their standard errors are small.

These conclusions give some guidance as to the construction of strata. Preferably strata should be used that are homogeneous with respect to the target variable, response probabilities, or both. The more elements that resemble each other within the strata, the smaller the bias will be.

Two variables were used for weighting in Example 8.3: Gender and AgeClass. Strata were formed by crossing these two variables. Therefore this weighting model is denoted by

$$AgeClass \times Gender.$$

The idea of crossing variables can be extended to more than two variables. As long as the table with population frequencies is available, and all response frequencies are greater than 0, weights can be computed. However, if there are no observations in a stratum, the corresponding weight cannot be computed. This leads to incorrect estimates. If the sample frequencies in the strata are very small, say less than 5, weights can be computed, but the estimates will be unstable.

As more variables are used in a weighting model, there will be more strata. Therefore the risk of empty strata or strata with too few observations will be larger. There are two solutions to this problem. One is to use fewer auxiliary variables, but then a lot of auxiliary information is not used. Another is to *collapse strata*. This means merging a stratum with too few observations with another stratum. It is important to combine strata that resemble each other as much as possible. Collapsing strata is not a simple job, particularly if the number of auxiliary variables and strata are large. It is often a manual job.

Another problem with the use of several auxiliary variables is the lack of sufficient population information. It may happen that the population distribution of the complete crossing of all auxiliary variables is simply not available. A possible cause could be that the population distributions of these variables come from different sources. Without the complete population distribution, no weights can be computed.

One way to solve this distribution problem is to use fewer auxiliary variables, but that would mean ignoring all information with respect to the other variables. What is needed is a weighting technique capable of using partial population information. There are two weighting techniques that can do that: linear weighting and multiplicative weighting. These two techniques are described in Sections 8.3 and 8.4.

EXAMPLE 8.4 Incomplete Population Information

The population distributions of the two variables AgeClass and Gender are known separately, but the distribution in the cross-classification is not known. In this case the poststratification *AgeClass* × *Gender* cannot be carried out because the weights cannot be computed for the strata in the cross-classification.

Population

	Male	Female	Total
Young	?	?	435
Middle	?	?	296
Elderly	?	?	269
Total	511	480	1000

Sample

	Male	Female	Total
Young	23	15	38
Middle	16	17	33
Elderly	13	16	29
Total	52	48	100

Weights

	Male	Female
Young	?	?
Middle	?	?
Elderly	?	?

8.2.2 APPLICATION

Poststratification is applied in this section to the General Population Survey (GPS) survey. First, the complete survey data set is used to select a number of auxiliary variables that can potentially be used for the weighting adjustment. Next, various subsets of these auxiliary variables are included in the weighting models. We want to learn how well these weighting models are capable of reducing a nonresponse bias. The selection of weighting variables for the GPS is considered in much more detail in Chapter 9.

We saw in Section 8.2.1 that under the random response model the bias of the poststratification estimator is small if the variation of the response probabilities within strata is small. Candidates for a weighting model are therefore those auxiliary variables that show variation between strata and not within strata. To say it differently, the aim is to select auxiliary variables that show a relationship with response behavior.

Table 8.1 lists the auxiliary variables that are available in the General Population Survey. The survey data file has been linked to the Social Statistics Database (SSD) of Statistics Netherlands. Therefore many variables are available. To explore a possible relationship between an auxiliary variable and response behavior, a table is created by crossing the auxiliary variable with the response variables (with categories Yes and No). A chi-square test for independence is conducted for this table. The hypothesis of independence is rejected if the *p*-value is less than the significance level of the test. The significance level is taken equal to 0.05, which is the standard value.

The chi-square column in Table 8.1 contains the *p*-values for all auxiliary variables. Note that almost all *p*-values are 0. This means that independence is rejected in all these cases. There is a relationship between these auxiliary variables and response behavior. There is only one variable for which independence is not rejected. This is the final variable in the table (Unempall).

TABLE 8.1 Available poststratification variables

Variable	Description	Chi-square	Cramér's *V*
Region	Region of the country	0.000	0.163
Urban	Degree of urbanization	0.000	0.153
Phone	Has listed phone number	0.000	0.150
Pnonnat1	Percentage nonnatives in neighborhood	0.000	0.138
Pnonnat2	Percentage nonwestern nonnatives in neighborhood	0.000	0.133
Houseval	Average house value in neighborhood	0.000	0.115
Ethnic	Type of nonnative	0.000	0.112
HHType	Type of household	0.000	0.106
HHSize	Size of the household	0.040	0.099
Marstat	Marital status	0.000	0.097
Married	Is married	0.000	0.097
Nonnativ	Is nonnative	0.000	0.087
Socall	Has social allowance	0.000	0.077
Age13	Age in 13 age groups	0.000	0.061
Allowan	Has an allowance	0.000	0.061
Children	Children in household	0.000	0.056
Hasjob	Has a job	0.000	0.037
Age3	Age in 3 age groups	0.000	0.030
Disaball	Has disability allowance	0.000	0.021
Gender	Gender	0.040	0.011
Unempall	Has unemployment allowance	0.944	0.000

The *p*-value of the chi-square test is only of limited value in selecting auxiliary variables for a weighting model. The test only detects whether or not there is a relationship between a variable and response behavior. It does not provide an indication of how strong the relationship is. This is also clear in Table 8.1. Since all *p*-values are equal to 0, they do not help us to discriminate. A more meaningful indicator is Cramér's *V*. This quantity is defined by

$$V = \sqrt{\frac{G}{\min(r-1,\ c-1)}},$$

in which *r* and *c* denote the numbers of rows and columns and *G* is the chi-square test statistic. *V* always has a value in the interval [0,1]. A value of 0 indicates independence, and a value of 1 means complete dependence. The variables in Table 8.1 have been ordered by their values of *V*.

Only the most important variables are used in this example. These are the variables with the highest *V*-value. Moreover, if two variables more or less measure the same thing, only one of them was used. Table 8.2 contains the remaining auxiliary variables.

The variables in Table 8.2 have been used in various ways to construct poststratification weighting models. The objective was to estimate the percentage of people receiving a social allowance. Since the population percentage (4.21%) was known, estimates could be compared with this value. The results are summarized in Table 8.3. The first row of this table (*No weighting*) contains the unweighted percentage of persons with a social allowance (2.91%). Apparently the estimate has a substantial bias.

If only one auxiliary variable is used for poststratification, *Pnonnat*1 (percentage of nonnatives) performs best. It causes a shift in the estimate from 2.91% to 3.38%. Still the bias is not completely removed. Table 8.3 also contains all poststratifications based on crossing two auxiliary variables. Crossing the variables *Pnonnat*1 and *Phone* (has listed phone number) gives the best results. The estimate for the percentage of people with a social allowance increases to 3.56%. This weighting model performs better than just using *Pnonnat*1 or *Phone* separately.

A next step could be to include three auxiliary variables in the weighting model. It might be a good idea to extend the model *Phone* × *Pnonnat*1 with either *Urban* or *HHType*. These models are, however, not possible. *Phone* × *Pnonnat1* × *Urban* has empty cells in the population. The model *Phone* × *Pnonnat1* × *HHType* has empty

TABLE 8.2 Variables used for poststratification

Variable	Description	Chi-square	Cramér's *V*
Urban	Degree of urbanization	0.000	0.153
Phone	Has listed phone number	0.000	0.150
Pnonnat1	Percentage non-natives in neighborhood	0.000	0.138
HHType	Type of household	0.000	0.106

TABLE 8.3 Percentages of persons with a social allowance estimated using different weighting models

Weighting model	Number of strata	Estimate
No weighting		2.91%
Urban	5	3.14%
HHType	5	3.16%
Phone	2	3.28%
Pnonnat1	10	3.38%
Urban × HHType	25	3.33%
Urban × Pnonnat1	50	3.40%
Phone × HHType	10	3.44%
Urban × Phone	10	3.45%
Pnonnat1 × HHType	50	3.52%
Phone × Pnonnat1	20	3.62%
Urban × Phone × HHType	50	3.56%
Population value		4.21%

cells for the response. So it is not possible to compute adjustment weights for these complete three-way cross-classifications. Furthermore the model *Pnonnat1* × *HHType* × *Urban* has several empty cells in the population. The only remaining weighting model with three variables is *Urban* × *Phone* × *HHType*. This model does not perform as well as *Phone* × *Pnonnat1*, probably because the variable *Pnonnat*1 is not included.

It may be concluded that *Phone* × *Pnonnat*1 is the best possible weighting model, if weighting is restricted to poststratification and strata are not collapsed. A large part of the bias is removed. Unfortunately, still some bias remains.

8.3 Linear Weighting

8.3.1 THEORY

The technique of linear weighting is based on the theory of *generalized regression estimation*. The general regression estimator is based on a linear model predicting the target variable of the survey from a set of auxiliary variables. This estimator is not only capable of producing precise estimates, but it also can reduce a bias due to nonresponse. So regression estimation is in fact a form of weighting. Correction weights are obtained by adding weight coefficients that are contributed by the various auxiliary variables in the linear model. Therefore it is called *linear weighting*.

After linear weighting has been applied, the sample has become representative with respect to the auxiliary variables in the model. It is shown that poststratification is a special case of linear weighting.

In principle, the auxiliary variables in the linear model have to be quantitative (continuing) variables; that is, they measure a size or value. However, it is also possible to use qualitative (categorical) variables. The trick is to replace a qualitative variable by a number of dummy variables, where each dummy variable indicates whether or not a person belongs to a specific category.

The theory of linear weighting is described for the case of simple random sampling without replacement. The theory can easily by generalized to general sampling designs. For more on this, see Bethlehem (1988).

First, let us consider the case of full response. Suppose that there are p (continuous) auxiliary variables available. The p-vector of values of these variables for element k is denoted by

$$X_k = (X_{k1}, X_{k2}, \ldots, X_{kp})'. \tag{8.14}$$

The prime (′) symbol denotes transposition of a matrix or vector. Let Y be the N-vector of all values of the target variable, and let X be the $N \times p$-matrix of all values of the auxiliary variables. The vector of population means of the p auxiliary variables is defined by

$$\bar{X} = (\bar{X}_1, \bar{X}_2, \ldots, \bar{X}_p)'. \tag{8.15}$$

This vector represents the population information assumed to be available. If the auxiliary variables are correlated with the target variable, then for a suitably chosen vector $B = (B_1, B_2, \ldots, B_p)'$ of regression coefficients for a best fit of Y on X, the residuals $E = (E_1, E_2, \ldots, E_N)'$, defined by

$$E = Y - XB, \tag{8.16}$$

vary less than the values of the target variable. Application of ordinary least squares results in

$$B = (X'X)^{-1}XY' = \left(\sum_{k=1}^{N} X_k X_k'\right)^{-1} \left(\sum_{k=1}^{N} X_k Y_k\right). \tag{8.17}$$

For a simple random sample without replacement, the vector B can be estimated by

$$b = \left(\sum_{k=1}^{N} a_k X_k X_k'\right)^{-1} \left(\sum_{k=1}^{N} a_k X_k Y_k\right) = \left(\sum_{i=1}^{n} x_i x_i'\right)^{-1} \left(\sum_{i=1}^{n} x_i y_i\right), \tag{8.18}$$

where $x_i = (x_{i1}, x_{i2}, \ldots, x_{ip})'$ denotes the p-vector of values of the p auxiliary variables for sample element i (for $i = 1, 2, \ldots, n$). The quantity a_k indicates whether or not element k is selected in the sample. The estimator b is an asymptotically design unbiased (ADU) estimator of B. This means that the bias vanishes for large samples. The *generalized regression estimator* is now defined by

$$\bar{y}_{GR} = \bar{y} + (\bar{X} - \bar{x})'b, \tag{8.19}$$

where $\bar{x}$ is the vector of sample means of the auxiliary variables.

The generalized regression estimator is an ADU estimator of the population mean of the target variable. If there exists a p-vector c of fixed numbers such that $Xc = J$, where J is a p-vector consisting of 1's, the generalized regression estimator can also be written as

$$\bar{y}_{GR} = \bar{X}'b. \tag{8.20}$$

This condition holds if there is a constant term in the regression model. The model also holds if the model contains a set of dummy variables corresponding to all the categories of a categorical variable. It can be shown that the variance of the generalized regression estimator can be approximated by

$$V(\bar{y}_{GR}) = \frac{1-f}{n} S_E^2, \tag{8.21}$$

where S_E^2 is the population variance of the residuals $E_1, E_2, \ldots, E_N$.

Expression (8.21) is identical to the variance of the simple sample mean, but with the values Y_k replaced by the residuals E_k. This variance will be small if the residual values E_k are small. Hence use of auxiliary variables, which can explain the behavior of the target variable, will result in a precise estimator.

EXAMPLE 8.5 Comparing a Simple Estimator with the Regression Estimator

There are 200 dairy farms in the rural part of Samplonia. Surveys are regularly conducted to estimate the average daily milk production per farm. There is a register containing, among other variables, the number of cows per farm. The table below summarizes these variables for the population of 200 farms.

	Mean	Standard deviation	Minimum	Maximum
Milk production (liters per day)	723.5	251.9	10.0	1875.0
Number of cows	28.9	9.0	8.0	67.0

Suppose that a simple random sample without replacement of size 10 is selected to estimate the mean milk production per farm. Then the variance of the sample mean would be equal to

$$V(\bar{y}) = \frac{1-f}{n} S^2 = \frac{1-(10/200)}{10} 251.9^2 = 6028.1.$$

A linear relationship is likely to exist between milk production per farm (Y) and number of cows per farm (X). If the linear model $Y = B_0 + B_1X$ is applied, there are two auxiliary variables involved. The first is always equal to 1, representing the constant term in the model, and the second is equal to X.

There are also two regression coefficients: B_0 and B_1. Application of formula (8.18) would result in values $B_0 = -31.4$ and $B_1 = 26.1$. The variance of the regression estimator is equal to

$$V(\bar{y}_{GR}) = \frac{1-f}{n} S_E^2 = \frac{1-(10/200)}{10} 732.8 = 69.6.$$

Because of the strong relationship between milk production and number of cows, the variance of the regression estimator is substantially smaller.

In the case of nonresponse the following modified version of the general regression estimator is introduced:

$$\bar{y}_{GR,R} = \bar{y}_R + (\bar{X} - \bar{x}_R)' b_R = \bar{X}' b_R, \tag{8.22}$$

where b_R is defined by

$$b_R = \left(\sum_{k=1}^{N} a_k R_k X_k X_k' \right)^{-1} \left(\sum_{k=1}^{N} a_k R_k X_k Y_k \right). \tag{8.23}$$

The quantity R_k indicates whether or not sample element k did respond. So b_R is the analogue of b, but just based on the response data. Bethlehem (1988) shows that the bias of estimator (8.22) is approximately equal to

$$B(\bar{y}_{GR,R}) = \bar{X} B_R - \bar{Y}, \tag{8.24}$$

where B_R is defined by

$$B_R = \left(\sum_{k=1}^{N} \rho_k X_k X_k' \right)^{-1} \left(\sum_{k=1}^{N} \rho_k X_k Y_k \right). \tag{8.25}$$

The bias of estimator (8.24) vanishes if $B_R = B$. Thus the regression estimator will be unbiased if nonresponse does not affect the regression coefficients. Practical experience shows that nonresponse often seriously affects estimators such as means and totals but less often causes estimates of relationships to be biased. Particularly, if relationships are strong (the regression line fits the data well), the risk of finding wrong relationships is small. This assumption is used in Chapter 9 to derive a selection criterion for weighting variables.

By writing

$$B_R = B + \left(\sum_{k=1}^{N} \rho_k X_k X_k' \right)^{-1} \left(\sum_{k=1}^{N} \rho_k X_k E_k \right), \tag{8.26}$$

the conclusion can be drawn that the bias will be small if the residuals are small.

Theoretically use of the generalized regression estimator has the potential of improving precision and reducing bias in cases of ignorable nonresponse. Therefore it forms the basis for linear weighting adjustment techniques.

Bethlehem and Keller (1987) have shown that the generalized regression estimator (8.19) can be rewritten in the form of the weighted estimator (8.3). The adjustment weight w_i for observed element i is equal to $w_i = v'X_i$, and v is a vector of weight coefficients, which is equal to

$$v = n\left(\sum_{i=1}^{n} x_i x_i'\right)^{-1} \bar{X}. \tag{8.27}$$

Poststratification is a special case of linear weighting, where the auxiliary variables are qualitative variables. To show this, qualitative auxiliary variables are replaced by sets of dummy variables. Suppose that there is one auxiliary variable with L categories. Then L dummy variables $X_1, X_2, \ldots, X_L$ can be introduced. For an observation in a certain stratum h, the corresponding dummy variable X_h is assigned the value 1, and all other dummy variables are set to 0. Consequently the vector of population means of these dummy variables is equal to

$$\bar{X} = \left(\frac{N_1}{N}, \frac{N_2}{N}, \ldots, \frac{N_L}{N}\right), \tag{8.28}$$

and v is equal to

$$v = \frac{n}{N}\left(\frac{N_1}{n_1}, \frac{N_2}{n_2}, \ldots, \frac{N_L}{n_L}\right)'. \tag{8.29}$$

If this form of v is used to compute $w_i = v'X_i$ and the result is substituted in expression (8.3) of the weighted estimator, the poststratification estimator is obtained.

EXAMPLE 8.6 Poststratification as a Special Case of Linear Weighting

A sample has been selected from the imaginary country of Samplonia. The population size is 1000. The realized sample size is 100. Suppose that there are two discrete auxiliary variables: *Gender* and *AgeClass* (in 3 categories). Crossing these two variables produces a table with $2 \times 3 = 6$ cells. A dummy variable is introduced for each cell. So there are six dummy variables, $X_1, X_2, \ldots, X_6$.

The possible values of these dummy variables are shown in the table below. For example, X_4 is the dummy variable for the stratum consisting of young females. Note that always one dummy variable has the value 1, while all other 5 dummy variables have the value 0.

The table also contains the vector of population means of the auxiliary variables. These values are equal to the population fractions in the cells of the population table. So the fraction of young females in the population is equal to 0.209.

Sex	AgeClass	X_1	X_2	X_3	X_4	X_5	X_6
Male	Young	1	0	0	0	0	0
Male	Middle	0	1	0	0	0	0
Male	Elderly	0	0	1	0	0	0
Female	Young	0	0	0	1	0	0
Female	Middle	0	0	0	0	1	0
Female	Elderly	0	0	0	0	0	1
Population means		0.226	0.152	0.133	0.209	0.144	0.136
Weight coefficients		0.983	0.950	1.023	1.393	0.847	0.850

The weight coefficients in the vector v are given in the bottom row of the table. They have been computed using expression (8.29).

These weight coefficients are used to compute the adjustment weights for the observed elements. The weight of each person is obtained by summing the relevant weight coefficients. In poststratification there is always only one relevant weight coefficient. So the weight is equal to this weight coefficient. For example, the weight for a young male is equal to 0.983.

Section 8.2.1 described how the application of poststratification can be complicated by the presence of empty cells. Linear weighting can address the problem of insufficient population information. It offers the possibility to include variables in the weighting scheme without having to know the population frequencies in the cells obtained by cross-tabulating all variables. The trick is to use a different set of dummy variables. Instead of defining one set of dummy variables for the complete crossing of all auxiliary variables, a set of dummy variables is defined for each variable separately or for each crossing of subsets of variables separately.

Consider, for example, three auxiliary variables, X_1, X_2, and X_3. Poststratification would come down to crossing the three variables, using one set of dummy variables. If only the marginal population distributions of the three variables can be used, there are three sets of dummy variables, each corresponding to the categories of one auxiliary variable. And if the population distribution of the crossing of X_1 and X_2 is available and only the marginal distribution of X_3, there are two sets of dummy variables: one for $X_1 \times X_2$ and one for X_3. Of course, other combinations and subsets are possible, depending on the available auxiliary information and the number of observations in each cell of each cross-classification.

In poststratification the weight is equal to one of the weight coefficients. If the weighting model contains more than one set of dummy variables, there will also be more weight coefficients contributing to the weight. In fact each set contributes a weight coefficient, and these weights are added to obtain the weight.

It should be noted that a weighting model containing more than one set of dummy variables will use less information than the model for the complete crossing of all auxiliary variables. Nevertheless, such a weighting model uses more information than a poststratification corresponding to one of the subsets.

EXAMPLE 8.7 Linear Weighting Using Only Marginal Distributions

Continuing Example 8.6, we now learn how to use only the marginal distributions of *Gender* and *AgeClass*. Two sets of dummy variables are introduced: one set of two dummy variables for the categories of *Gender*, and another set of three dummy variables for the categories of *AgeClass*. Then there are 2 + 3 = 5 dummy variables. In each set, always one dummy has the value 1, whereas all other dummies are 0. The possible values of the dummy variables are shown in the table below.

Sex	AgeClass	X_1	X_2	X_3	X_4	X_5	X_6
Male	Young	1	1	0	1	0	0
Male	Middle	1	1	0	0	1	0
Male	Elderly	1	1	0	0	0	1
Female	Young	1	0	1	1	0	0
Female	Middle	1	0	1	0	1	0
Female	Elderly	1	0	1	0	0	1
Population means		1.000	0.511	0.489	0.435	0.296	0.269
Weight coefficients		0.991	−0.033	0.033	0.161	−0.095	−0.066

The first dummy variable X_1 represents the constant term in the regression model. It always has the value 1. The second and third dummy variable relate to the two gender categories, and the last three dummies represent the three age categories. The vector of population means is equal to the fractions for all dummy variables separately. Note that in this weighting model always three dummies in a row have the value 1.

The weight for an observed element is now obtained by summing the appropriate elements of this vector. The first value corresponds to the dummy X_1, which always has the value 1. So there is always a contribution 0.991 to the weight. The next two values correspond to the categories of *Gender*. Note that their sum equals zero. For males, an amount 0.033 is subtracted, and for females, the same amount is added. The final three values correspond to the categories of *AgeClass*. Depending on the age category a contribution is added or subtracted. For example, the weight for a young male is equal to $0.991 - 0.033 + 0.161 = 1.119$.

No information is used for the crossing of *Gender* by *AgeClass* in Example 8.7. Only the marginal distributions are included in the computation of the weights. Therefore a different notation is introduced. This weighting model is denoted by

$$Gender + AgeClass.$$

Because of the special structure of the auxiliary variables, the computation of the weight coefficients v cannot be carried out without imposing extra conditions. Here for every qualitative variable the condition is imposed that the sum of the weight coefficients for the corresponding dummy variables must equal zero.

Examples 8.6 and 8.7 use only two auxiliary variables. More variables can be included in a weighting model. This makes it possible to define various weighting models with these variables. Suppose that the three auxiliary variables are *Gender*, *AgeClass*, and *MarStat* (marital status). If the complete population distribution on the crossing of all three variables is available, then the weighting model

$$Gender \times AgeClass \times MarStat$$

can be applied. If only the bivariate population distributions of every crossing of two variables are available, the following weighting scheme can be applied:

$$(Gender \times AgeClass) + (AgeClass \times MarStat) + (Gender \times MarStat).$$

Note that this scheme comes down to doing three poststratifications simultaneously. If only marginal frequency distributions are available, the model

$$Gender + AgeClass + MarStat$$

could be considered. More details about the theory of linear weighting can be found in Bethlehem and Keller (1987).

Up to now only linear weighting with qualitative auxiliary variables was described. It is also possible, however, to apply linear weighting with quantitative auxiliary variables, or a combination of qualitative and quantitative variables.

If there is only one quantitative variable, linear weighting comes down to applying the simple regression as discussed, for example, by Cochran (1977). If there are more quantitative auxiliary variables, the generalized regression estimator as defined in (8.19) can be used. The population information used is the vector of population means of the auxiliary variables.

It is also possible to combine qualitative and quantitative auxiliary variables, and even crossings of those variables. For example the discrete variable *Gender* can be crossed with continuous variable *Age*. Application of the generalized regression estimator requires the mean ages for males and females separately to be known.

EXAMPLE 8.8 Linear Weighting with a Discrete and Continuous Variable

Continuing Example 8.7, we want to combine the qualitative variable *Gender* and the quantitative variable *Age* in a weighting model. Three different weighting models are described. The table below shows the first group of cases for each model.

In the first weighting model only the variable *Age* is used. If continuous variables are used in a weighting model, then always a column of constants must be included in the model to ensure that the resulting estimator is unbiased. The matrix X for this model consists of the two columns X_1 (the constant) and X_2 (age) in the table.

The table also contains the population means. The value 34.369 denotes the mean age in the population. The row "Weight coefficients 1" contains the weight coefficients for this model. There are only two coefficients: one corresponding to the constant term, and one for *Age*. The second weight coefficient is negative (−0.003). This implies that the weights decrease with age: younger people are given higher weight than older people. Apparently young people are underrepresented in the survey, while older people are overrepresented.

Age	Sex	X_1	X_2	X_3	X_4	X_5	X_6
65	Male	1	65	1	0	65	0
36	Male	1	36	1	0	36	0
73	Female	1	73	0	1	0	73
6	Male	1	6	1	0	6	0
33	Female	1	33	0	1	0	33
82	Female	1	82	0	1	0	82
2	Male	1	2	1	0	2	0
32	Male	1	32	1	0	32	0
66	Female	1	66	0	1	0	66
2	Female	1	2	0	1	0	2
Population means		1.000	34.369	0.511	0.489	33.509	35.268
Weight coefficients 1		1.101	−0.003				
Weight coefficients 2		1.101	−0.003	−0.032	0.032		
Weight coefficients 3		1.087				−0.001	−0.004

The second weighting model uses both variables *Age* and *Gender*. The weighting model is denoted by *Age* + *Gender*. This model uses columns X_1 (constant term), X_2 (age), X_3 (dummy for male), and X_4 (dummy for female). The row "Weight coefficients 2" contains the weight coefficients for this model. There are four coefficients: one for the constant term, one for age, one for males, and one for females. The second weight coefficient is negative. This implies, again, that weights decreases with age: for males, an extra amount is subtracted from the weight, and for females, the same amount is added. A closer look at the adjustment weights would reveal that, for example, young females are underrepresented, and that old males are overrepresented. The model *Age* + *Gender* should be a better model than the one just containing Age, since more population information is used.

The third example of a weighting model is a model where the qualitative variable *Gender* and the quantitative variable *Age* are crossed. This weighting model is denoted by *Age* × *Gender*. Theory allows only one quantitative variable to be crossed with a number of qualitative variables in each term of the model. Crossing a quantitative variable with discrete variables means that instead of the relative sizes of the strata in the population, the population means of the quantitative variables in the strata has to be known. Hence, for the model *Age* × *Gender*, mean ages of males and females are required. For this model, columns X_1, X_5, and X_6 are used. Note that the age for males is set to zero in the column for females, and vice versa, the ages of females are set to zero in the column for males. The resulting weights can be found in the row "Weight coefficients 3." There are three coefficients: one for the constant term, one for the age of males, and one for the age of females. The weight coefficients for both strata are negative. This means that both for males and females weights decrease with age. The weight coefficient for females is more negative than that for males. So females get a lower weight than males.

8.3.2 APPLICATION

Linear weighting is now applied to the General Population Survey (GPS) survey. The objective is to estimate the percentage of people receiving a social allowance. Many auxiliary variables are available. These variables are listed in Table 8.1 of Section 8.2.1. Candidates for a weighting model are those auxiliary variables that show a relationship with response behavior. Based on Cramér's *V* a number of promising variables were selected. These variables are listed in Table 8.4.

A number of poststratification weighting models have already been explored in Section 8.2.2. The best turned out to be *Phone* × *Pnonnat1*. Because of the problems with empty strata, it is often not possible to use three or more variables in a poststratification model. Linear weighting overcomes these problems. Some new weighting models are included in Table 8.5. The program Bascula was used for the computations of the weights and the estimates. For more information about this software, see Bethlehem (1996).

If no weighting is applied, the estimate for the percentage of people receiving an allowance is 2.91%. This is far from the true value of 4.21%. The best possible stratification model produces an estimate of 3.62%. Linear weighting allows us to include all four weighting variables where only the marginal distributions are used

TABLE 8.4 Variables used for linear weighting

Variable	Description	Cramér's *V*
Urban	Degree of urbanization	0.153
Phone	Has listed phone number	0.150
Pnonnat1	Percentage nonnatives in neighborhood	0.138
HHType	Type of household	0.106

TABLE 8.5 Percentages of persons with a social allowance estimated with different linear weighting models

Weighting model	Estimate
No weighting	2.91%
Phone × Pnonnat1 (*Post-stratification*)	3.62%
Urban + HHType + Phone + Pnonnat1	3.68%
(Urban × HHType) + (Urban × Phone) + (Urban × Pnonnat1) + (HHType × Phone) + (HHType × Pnonnat1) + (Phone × Pnonnat1)	3.69%
(Urban × Phone × HHType) + (Phone × Pnonnat1)	3.69%
Population value	4.21%

to compute weights and not the cross-classifications. This leads to an improved estimate of 3.68%. Also included in the table is a weighting model containing all bivariate cross-classifications of the auxiliary variables. This results in a small increase to 3.69%, However, when, the two best poststratifications were combined in one linear model, this did not produce a further improvement.

It is interesting that the simple linear model containing only the marginal distributions of the four auxiliary variables performs almost as well as more complex models. The interactions between auxiliary variables are often not important.

Linear weighting was able to produce better estimates in this example than poststratification. Nevertheless, again weighting was not able to remove the bias completely.

8.4 Multiplicative Weighting

8.4.1 THEORY

If linear weighting is applied, the correction weights obtained are computed as the sum of a number of *weight coefficients*. There is an alternate way to compute correction weights, namely as the product of a number of *weight factors*. This weighting technique is called *raking, raking ratio estimation,* or *iterative proportional fitting*. Here it is denoted by *multiplicative weighting* because weights are obtained as the product of a number of factors contributed by the various auxiliary variables.

Multiplicative weighting can be applied in the same situations as linear weighting as long as only qualitative auxiliary variables are used. The correction weights are computed by an iterative procedure. The resulting weights are the product of factors contributed by all cross-classifications.

The iterative proportional fitting technique was described by Deming and Stephan (1940). Skinner (1991) discusses the application of this technique in multiple frame surveys. Little and Wu (1991) describe the theoretical framework, and show that this technique comes down to fitting a loglinear model for the probabilities of getting observations in strata of the complete cross-classification, given the probabilities for marginal distributions. To compute the weight factors, the following scheme must be carried out:

Step 1. Introduce a weight factor for each stratum in each cross-classification term. Set the initial values of all factors to 1.

Step 2. Adjust the weight factors for the first cross-classification term so that the weighted sample becomes representative with respect to the auxiliary variables included in this cross-classification.

Step 3. Adjust the weight factors for the next cross-classification term so that the weighted sample is representative for the variables involved. Generally, this will disturb representativeness with respect to the other cross-classification terms in the model.

Step 4. Repeat this adjustment process until all cross-classification terms are dealt with.

Step 5. Repeat steps 2, 3, and 4 until the weight factors do not change any more.

EXAMPLE 8.9 Multiplicative Weighting

Multiplicative weighting is illustrated using the same data as in the Examples 8.6 to 8.8. Two variables are used in the weighting model, *Gender* and *AgeClass*.

Suppose that only the marginal population distributions of *Gender* (2 categories) and *AgeClass* (3 categories) are available, and not the cross-classification. The table below contains the starting situation. The upper-left part of the table contains the weighted relative frequencies in the sample for each combination of *AgeClass* and *Gender*.

Starting situation

	Male	Female	Weight factor	Weighted sum	Population distribution
Young	0.230	0.150	1.000	0.380	0.435
Middle	0.160	0.170	1.000	0.330	0.296
Elderly	0.130	0.160	1.000	0.290	0.269
Weight factor	1.000	1.000			
Weighted sum	0.520	0.480		1.000	
Popul. Distr.	0.511	0.489			1.000

The row and column denoted by "Weight factor" contain the initial values of the weight factors (1.000). The values in the row and column denoted by "Weighted sum" are obtained by first computing the weight for each sample cell (by multiplying the relevant row and column factor), and then summing the weighted cell fractions. Since the initial values of all factors are equal to 1, the weighted sums in the table are equal to the unweighted sample sums. The row and column denoted by "Population distribution" contain the fractions for *AgeClass* and *Gender* in the population.

The iterative process must result in row and column factors with such values that the weighted sums match the population distribution. This is clearly not the case in the starting situation. First, the weight factors for the rows are adjusted. This leads to weight factors 1.145, 0.897, and 0.925 for the categories *Young*, *Middle*, and *Elderly*; see the table below. The weighted sums for the rows are now correct, but the weighted sums for the columns are 0.527 and 0.473, and thus still show a discrepancy.

Situation after adjusting for AgeClass

	Male	Female	Weight factor	Weighted sum	Population distribution
Young	0.230	0.150	1.145	0.435	0.435
Middle	0.160	0.170	0.897	0.296	0.296
Elderly	0.130	0.160	0.928	0.269	0.269
Weight factor	1.000	1.000			
Weighted sum	0.527	0.473		1.000	
Popul. distr.	0.511	0.489			1.000

The next step will be to adjust the weight factors for the columns such that the weighted column sums match the corresponding population frequencies. Note that this adjustment for *Gender* will disturb the adjustment for *AgeClass*. The weighted sums for the age categories no longer match the relative population frequencies. However, the discrepancy is much smaller than in the initial situation.

The process of adjusting for *AgeClass* and *Gender* is repeated until the weight factors do not change any more. The final situation is reached after a few iterations. The table below shows the final results.

Situation after convergence

	Male	Female	Weight factor	Weighted sum	Population distribution
Young	0.230	0.150	1.151	0.435	0.435
Middle	0.160	0.170	0.895	0.296	0.296
Elderly	0.130	0.160	0.923	0.269	0.269
Weight factor	0.968	1.035			
Weighted sum	0.511	0.489		1.000	
Popul. distr.	0.511	0.489			1.000

The adjustment weight for a specific sample element is now obtained by multiplying the relevant weight factors. For example, the weight for a young male is equal to $1.151 \times 0.968 = 1.114$. For this example the adjustment weights differ only slightly from those obtained by linear weighting as described in Section 8.3.

There are many situations in which both linear and multiplicative weighting can be applied. This raises the question of which weighting method should be preferred. A number of observations may help one take a decision.

In the first place, linear weighting is based on a simple linear model that describes the relationship between a target variable and a number of auxiliary variables. If this model fits well, weighting adjustment will be effective. For multiplicative weighting there is no straightforward model allowing a simple interpretation.

In the second place, computations for linear weighting are straightforward. Weights are obtained by application of ordinary least squares. The weights of multiplicative weighting are obtained as the solution of an iterative process. There is no guarantee that this process will always converge.

In the third place, since linear weighting comes down to generalized regression estimation, an analytical expression of the variance of weighted estimates can be derived. No simple expressions are available for estimates based on multiplicative weighting.

In the fourth place, weights produced by linear weighting may sometimes turn out to be negative. This seems counterintuitive, but it is simply a consequence of the linear model applied. Negative weights usually indicate that the linear model does not fit to well. A disadvantage of negative weights is that some statistical analysis packages do not accept negative weights. This may prevent weighted analysis of the survey outcomes.

In the fifth place, it has been shown (see Section 8.5) that in many situations estimators based on linear weights have asymptotically the same properties as those based on multiplicative weights.

From a statistical point of view, it can be concluded that linear weighting should be preferred.

8.4.2 APPLICATION

Multiplicative weighting is now applied to the General Population Survey (GPS). The objective is to estimate the percentage of people receiving a social allowance. Many auxiliary variables are available. These variables are listed in Table 8.1 of Section 8.2.1. Candidates for a weighting model are those auxiliary variables that show a relationship with response behavior. Based on Cramér's V a number of promising variables were selected. These variables are listed in Table 8.6.

Some poststratification weighting models were explored earlier in Section 8.2.2. The best turned out to be *Phone* $\times$ *Pnonnat1*. Because of empty

TABLE 8.6 Variables used for linear weighting

Variable	Description	Cramér's V
Urban	Degree of urbanization	0.153
Phone	Has listed phone number	0.150
Pnonnat1	Percentage nonnatives in neighborhood	0.138
HHType	Type of household	0.106

strata, it is often not possible to use three or more variables in a poststratification model. Like linear weighting, multiplicative weighting overcomes these problems. Some new weighting models are included in Table 8.7. The program Bascula was used for the computations.

If no weighting is applied, the estimate for the percentage of people having an allowance is 2.91%. This is far from the true value of 4.21%. The best possible stratification model produces an estimate of 3.62%. Multiplicative weighting allows us to include all four weighting variables, where only the marginal distributions are used to compute weights and not the cross-classifications. This leads to an improved estimate of 3.69%. Also included in the table is a weighting model containing all bivariate cross-classifications of the auxiliary variables. This results in a small increase to 3.70%, However, when the two best poststratifications were combined in one multiplicative model, this did not produce a further improvement.

It is interesting that the simple multiplicative model containing only the marginal distributions of the four auxiliary variables performs almost as well as more complex models. The interactions between auxiliary variables are apparently not important.

Multiplicative weighting is able to produce better estimates in this example than poststratification. The estimates are only slightly better than those of similar linear weighting models. Also here, however, weighting is not able to remove the bias completely.

TABLE 8.7 Percentages of persons with a social allowance estimated using different multiplicative weighting models

Weighting model	Estimate
No weighting	2.91%
Phone × Pnonnat1 (*Post-stratification*)	3,62%
Urban + HHType + Phone + Pnonnat1	3.69%
Urban × HHType + Urban × Phone + Urban × Pnonnat1 + HHType × Phone + HHType × Pnonnat1 + Phone × Pnonnat1	3.70%
Urban × Phone × HHType + Phone × Pnonnat1	3.70%
Population value	4.21%

8.5 Other Weighting Issues

8.5.1 CALIBRATION ESTIMATION

Deville and Särndal (1992) and Deville et al. (1993) have created a general framework for weighting within which linear and multiplicative weighting are special cases. Assuming simple random sampling, their starting point is that correction weights g_i in $w_i = g_i \times d_i$ have to satisfy two conditions:

- The correction weights g_i have to be as close as possible to 1.
- The weighted sample distribution of the auxiliary variables has to match the population distribution,

$$\bar{x}_W = \frac{1}{N}\sum_{i=1}^{n} w_i x_i = \bar{X}. \tag{8.30}$$

The first condition sees to it that resulting estimators are unbiased, or almost unbiased, and the second condition guarantees that the weighted sample is representative with respect to the auxiliary variables used.

Deville and Särndal (1992) start from a general distance measure $D(g_i, 1)$ for the difference between g_i and 1. The problem is then to minimize

$$\sum_{i=1}^{n} D(g_i, 1) \tag{8.31}$$

under condition (8.30). This problem can be solved by the method of Lagrange. By choosing the proper distance function, they show that linear and multiplicative weighting can be obtained as special cases of this general approach. For linear weighting the distance function is defined by

$$D(g_i, 1) = (g_i - 1)^2, \tag{8.32}$$

which is the Euclidean distance, and for multiplicative weighting, the distance

$$D(g_i, 1) = g_i \log(g_i) - g_i + 1 \tag{8.33}$$

must be used.

Deville and Särndal (1992) and Deville et al. (1993) only consider the full-response situation. They show that estimators based on weights computed within their framework, have asymptotically the same properties. This means that for large samples it does not matter whether linear or multiplicative weighting is applied. Estimators based on both weighting techniques will behave approximately the same. Note that although the estimators behave the same way, the individual weights computed by linear or multiplicative weighting may differ substantially.

Under nonresponse the situation is different. Generally, the asymptotic properties of linear and multiplicative weighting will not be equal under nonresponse. The extent to which the chosen weighting technique is able to reduce the nonresponse bias depends on how well the corresponding underlying model can be estimated using the observed data. Linear weighting assumes a linear model to hold, with the target variable as dependent variable and the auxiliary variables as explanatory variables. Multiplicative weighting assumes a loglinear model for the cell frequencies. An attempt to use a correction technique for which the underlying model does not hold will not help to reduce the bias.

8.5.2 CONSTRAINING THE VALUES OF WEIGHTS

There are several reasons why survey researchers may want to have some control over the values of the adjustment weights. One reason is that extremely large weights are generally considered undesirable. Large weights usually correspond to population elements with rare characteristics. Use of such weights may lead to unstable estimates of population parameters. To reduce the impact of large weights on estimators, a weighting method is required that is able to keep adjustment weights within prespecified boundaries, and that at the same time enables valid inference.

Another reason to have some control over the values of the adjustment weights is that application of linear weighting may produce negative weights. Although theory does not require weights to be positive, negative weights should be avoided, since they are counterintuitive. Negative weights cause problems in subsequent analyses, and they are an indication that the regression model does not fit the data well.

Negative weights can be avoided by using a better regression model. However, it is not always possible to find such models. Another solution is to use the current model, and force weights within certain limits. Several techniques have been proposed for this. A technique developed by Deville et al. (1993) comes down to repeating the (linear) weighting process a number of times. First, a lower bound L and an upper bound U are specified. After the first run, weights smaller than L are set to L and weights larger than U are set to U. Then, the weighting process is repeated, but records from the strata with the fixed weights L and U are excluded. Again, weights not satisfying the conditions may be produced. These weights are also set to either the value L or U. The weighting process is repeated until all computed weights fall within the specified limits. Convergence of this iterative process is not guaranteed. Particularly if the lower bound L and upper bound U are not far apart, the algorithm may not converge.

Huang and Fuller (1978) use a different approach. Their algorithm produces weights that are a smooth, continuous, monotone increasing function of the original weights computed from the linear model. The algorithm is iterative. At each step the weights are checked against a user-supplied criterion value M.

This value M is the maximum fraction of the mean weight by which any weight may deviate from the mean weight. For example, if M is set to 0.75, then all weights are forced into the interval with lower bound equal to 0.25 times the mean weight and upper bound equal to 1.75 times the mean weight. Setting the value to 1 implies that all weights are forced to be positive. Huang and Fuller (1978) prove that the asymptotic properties of the regression estimator constructed with their algorithm are asymptotically the same as those of the generalized regression estimator. So restricting the weights has (at least asymptotically) no effect on the properties of population estimates computed with these weights.

8.5.3 CONSISTENT PERSON AND HOUSEHOLD WEIGHTS

Another issue is the computation of weights that are consistent for persons and households. Some statistical surveys have complex sample designs. One example of such a complex design is cluster sampling. Many household surveys are based on cluster samples. First, a sample of households is selected. Next, all persons in the selected households are interviewed. The collected information can be used to make estimates for two populations: the population consisting of all households, and the population consisting of all individual persons. In both situations, weighting can be carried out to correct for nonresponse. This approach results in two weights assigned to each record: one for the household and one for the individual. Having two weights in each record complicates further analysis.

If the aim of the survey is to make inference on the population of all individual persons, the process is fairly straightforward. The unit of measurement is the individual person. The data file must be approached as a file of records with data on persons. Available population information on the distribution of personal characteristics can be used to compute adjustment weights, and these weights are assigned to the individual records.

For making inference on the population of households, the same approach can be used. However, this approach poses a problem if there is no or limited information available on the population distribution of household variables. Even information on simple variables like size of the household and household composition is usually lacking. This makes it impossible to carry out an efficient weighting procedure.

Since it is possible to compute weights for the members of the household, one may wonder whether it is possible to use the individual weights to compute household weights. Possible approaches could be to take (1) the weight of the head of the household, (2) the weight of a randomly selected household member, or (3) to compute some kind of average weight of the household members. Whatever approach is used, there are always complications. If household weights are applied to members of the households, weighted estimates of individual characteristics will not match known population frequencies. This discrepancy will not occur if the individual weights are used. Furthermore inconsistencies may turn up. For example, an estimate of the total income through the households will not be equal to an estimate based on the individual persons.

Generalized regression estimation offers a solution to this problem. The trick is to sum the dummy variables corresponding to the discrete auxiliary variables for the individuals over the household. Thus continuous auxiliary variables are created at the household level.

The resulting weights are assigned to the households. Furthermore all elements within a household are assigned the same weight, and this weight is equal to the household weight. This approach forces estimates computed using element weights to be consistent with estimates based on cluster weights. For an application of consistent weighting, see Nieuwenbroek (1993).

EXAMPLE 8.10 Individual Weights and Household Weights

Suppose a household survey is conducted in which all household members of age 18 and over have to complete a questionnaire. There are two auxiliary variables: *Gender* and *AgeClass* (in 3 categories). The table below contains the values of these variables for a few sample persons. Note that the auxiliary variables have been transformed into dummy variables.

Person	0–29	30–59	60 +	Male	Female
1	0	1	0	1	0
2	0	1	0	0	1
3	1	0	0	0	1
4	0	0	1	1	0
...	...	...	...	...	...

To compute household weights, a matrix H is required indicating to which household the persons belong. An example is given in the table below. The household are identified by *H1*, *H2*, *H3*, ...

Person	*H1*	*H2*	*H3*	...
1	1	0	0	...
2	1	0	0	...
3	1	0	0	...
4	0	1	0	...
...	...	...	...	...

A value of 1 means that the particular person is a member of the particular household. So persons 1, 2, and 3 belong the household H1, and person 4 belongs to household H2. Note that each row of the table always contains one value 1. All other values are 0.

The table with auxiliary information at the household level is now obtained by computing the matrix $Z = H'X$. The result is shown in the table below.

Household	0–29	30–59	60 +	Male	Female
H1	1	2	0	1	2
H2	0	0	1	1	0
...	...	...	...	...	...

The rows contain the number of people in the respective categories. So the first household has one person in the age category 0 to 29, and two in the age category 30 to 50. There are two females and one male. A weighting method can now be applied in which this sample information is confronted with the population distribution of these variables.

8.6 Summary

Nonresponse can cause estimators of population characteristics to be biased. This happens when specific groups are over- or underrepresented in the survey response and these groups behave differently with respect to the survey variables.

Weighting adjustment techniques can help reduce such a bias. These techniques assign weights to the observed elements, where underrepresented elements get a weight larger than 1 and overrepresented elements get a weight smaller than 1.

To be able to compute adjustment weights, auxiliary variables are required. Such variables have to be measured in the survey, and also the population distribution (or complete sample distribution) must be available. Sometimes it is also possible to retrieve the values of auxiliary variables for the sample elements from a different source, like an administrative register.

A weighting adjustment will only help reduce a nonresponse bias if there is a strong relationship between the survey variables and the auxiliary variables and/or response behavior and auxiliary variables.

Poststratification is the most frequently used weighting adjustment technique. Using auxiliary variables, the population is divided into a number of strata (subpopulations). All observed elements in a stratum are assigned the same weight. Poststratification works well in reducing a bias if the strata are homogeneous, meaning all elements within a stratum resemble each other.

Practical limitations may hinder application of poststratification. If many auxiliary variables are used to form strata, there may be strata without observations.

Consequently it is impossible to compute weights for such strata. It can also happen that an insufficient population is available with respect to the distribution of the auxiliary variables.

If it is not possible to carry out poststratification, there are two alternative weighting methods. One is linear weighting. It is based on a linear regression model that predicts the values of the target variable of the survey from a set of auxiliary variables. Such regression models offer more flexibility with respect to the way auxiliary information is used to compute adjustment weights. Another possibility is to use multiplicative weighting. This weighting model is based on iterative proportional fitting. It can be shown that weighted estimators based on linear and multiplicative weighting behave approximately the same in many situations.

Calibration is a theoretical framework for adjustment weighting. Poststratification, linear weighting, and multiplicative weighting are special cases for this framework. It has more possibilities, such as imposing constraints on the values of the weights.

8.7 KEY TERMS

Auxiliary variable A variable that has been measured in the survey and for which also the distribution in the population (or the complete sample) is available. Sometimes it is also possible to retrieve the values of auxiliary variables for the sample elements from a different source, like an administrative register.

Homogeneous A stratum (subpopulation) is called homogeneous if all its elements resemble each other with respect to the target variables of the survey.

Linear weighting A weighting technique where weights are computed from a generalized linear regression model predicting the target variable of the survey from a set of auxiliary variables.

Multiplicative weighting A weighting technique based on iterative fitting procedure that adjusts the weight so that the weighted sample distributions of the auxiliary variables fit their population distributions. This technique is also known as iterative proportional fitting or raking ratio estimation.

Poststratification A weighting method that divides the population in strata and subsequently assigns the same weight to all observed elements within a stratum.

Representative The (weighted) survey response to a variable if the (weighted) response distribution is equal to its population distribution.

Stratification A division of the population into a number of subpopulations (strata) by cross-classifying a number of auxiliary variables.

8.8 REFERENCES

Bailar, B.A., Bailey, L., and Corby, C. (1978), A Cmparison of some Ajustment and Weighting Procedures for Survey Data. In: Krishnan Namboodiri, N. (ed.), *Survey Sampling and Measurement*. Academic Press, Cambridge, UK.

Bethlehem, J.G. (1988), Reduction of Nonresponse Bias through Regression Estimation. *Journal of Official Statistics* 4, pp. 251–260.

Bethlehem, J.G. (1996), *Bascula for Weighting Sample Survey Data, Reference Manual.* Statistics Netherlands, Statistical Informatics Department, Voorburg/Heerlen.

Bethlehem, J.G., and Keller, W.J. (1987), Linear Weighting of Sample Survey Data. *Journal of Official Statistics* 3, pp. 141–154.

Bethlehem, J.G., and Kersten, H.M.P. (1985), On the Treatment of Non-response in Sample Surveys. *Journal of Official Statistics* 1, pp. 287–300.

Cochran, W.G. (1977), *Sampling Techniques*. Wiley, New York.

Deming, W.E., and Stephan, F.F., (1940). On a Least Squares of Adjustment of a Sampled Frequency Table When the Expected Totals Are Known. *Annals of Mathematical Statistics* 11, pp. 427–444.

Deville, J.C.and Särndal, C.E. (1992), Calibration Estimation in Survey Sampling. *Journal of the American Statistical Association* 87, pp. 376–382.

Deville, J.C., Särndal, C.E., and Sautory, O. (1993), Generalized Raking Procedures in Survey Sampling. *Journal of the American Statistical Association*, 88, pp. 1013–1020.

Holt, D., and Smith, T.M.F. (1979), Post Stratification. *Journal of the Royal Statistical Society, Series A*, 142, pp. 33–46.

Huang, E.T., and Fuller, W.A. (1978), Nonnegative Regression Estimation for Survey Data. *Pr*oceedings of the Social Statistics Section of the American Statistical Association,* Washington, DC pp. 300–303.

Kalton, G., and Flores-Cervantes, I. (2003), Weighting Methods. *Journal of Official Statistics* 19, pp. 81–97.

Kiaer, A.N. (1895), Observations et Expériences Concernant des Dénombrement Réprésentatifs. *Bulletin of the International Statistical Institute* 9 (Book 2), pp. 176–183.

Lindström, H., Wretman, J., Forsman, G., and Cassel, C. (1979). *Standard Methods for Non-response Treatment in Statistical Estimation.* Statistics Sweden, Stockholm.

Little, R.J.A. (1982), Models for Nonresponse in Sample Surveys. *Journal of the American Statistical Association* 77, pp. 237–250.

Little, R.J.A., and Wu, M.M. (1991), Models for Contingency Tables with Known Margins When Target and Sampled Populations Differ. *Journal of the American Statistical Association*, 86, pp. 87–95.

Nieuwenbroek, N.J. (1993), *An Integrated Method for Weighting Characteristics of Persons and Households Using the Linear Regression Estimator.* Research Paper. Department of Statistical Methods, Statistics Netherlands, Heerlen.

Platek, R., and Gray, G.B. (1983), Imputation Methodology: Total Survey Error Introduction. In: Nisselson, H. (ed.), *Incomplete Data in Sample Surveys*, Academic Press, New York, pp. 249–268.

Skinner, C.J. (1991), On the Efficiency of Raking Ratio Estimates for Multiple Frame Surveys. *Journal of the American Statistical Association* 86, pp. 779–784.

8.9 EXERCISES

Exercise 8.1 Which property of an auxiliary variable makes it useful for including in a weighting adjustment model?

(a) The response distribution of the variable is approximately equal to its population distribution.

(b) The sample distribution of the variable is approximately equal to its population distribution.

(c) The response distribution of the variable differs considerably from its sample distribution.

(d) The response distribution of the variable is approximately equal to its sample distribution.

Exercise 8.2 A large company has 2500 employees. The management has installed coffee machines everywhere in the building. After a while, the management wants to know whether or not the employees are satisfied with the coffee machines.

(a) Determine the sample size under the condition that the margin of the 95% confidence interval is not allowed to exceed 4%.

It is decided to draw a simple random sample without replacement of 500 employees. It turns out that 380 employees complete the questionnaire form. Of those, 310 are satisfied with the coffee machines.

(b) Compute the 95% confidence interval of the percentage of employees in the company who are satisfied with the coffee machines.

Only 380 of the 500 selected employees responded. So there is a nonresponse problem.

(c) Compute a lower bound and an upper bound for the percentage of employees in the sample who are satisfied with the coffee machines.

Previous research has shown that employees with a higher level of education are less satisfied with the coffee facilities. The management knows the level of education of each employee in the company: 21% has a high education and 79% has a low education. The table below shows the relationship between coffee machine satisfaction and level of education for the 380 respondents:

	Low education	High education	Total
Satisfied	306	4	310
Not satisfied	40	30	70
Total	346	34	380

A weighting adjustment procedure is carried out to reduce the nonresponse bias.

(d) Compute weights for low and high educated employees.

(e) Compute the weighted estimate of the percentage of employees in the company satisfied with the coffee facilities.

Exercise 8.3 There are plans in The Netherlands to introduce a system of road pricing. It means car drivers are charged for the roads they use. Such a system could lead to better use of the available road capacity, and therefore could reduce traffic congestion. An Automobile Association wants to know what the attitude of the Dutch is toward road pricing. It conducts a survey in which a simple random sample of 1000 people is selected. Selected people are asked two questions:

- Are you in favor of road pricing?
- Do you have a car?

Unfortunately, not everybody wants to participate in the survey. Due to nonresponse only a part of the selected persons answers the questions. The results are summarized below:

	In favor of road pricing?	
Has a car?	Yes	No
Yes	128	512
No	60	40

(a) Compute the response percentage.

(b) Using the available data, compute the percentage in favor of road pricing.

(c) Using the available data, compute a lower bound and upper bound for the percentage in the complete sample in favor of road pricing.

(d) From another source it is known that 80% of the target population owns a car, and 20% does not have one. Use this additional information to apply a weighting adjustment. Compute a weight for car owners and a weight for those without a car.

(e) Make a table like the one above, but with weighted frequencies.

(f) Compute a weighted estimate for the percentage in favor of road pricing.

(g) Explain the difference between the weighted and unweighted estimate.

Exercise 8.4 A transport company carries out a survey to determine how healthy its truck drivers are. From the population of all its drivers a simple random sample has been selected. Of course, there is nonresponse. Therefore data on only 21 drivers become available. Each respondent was asked whether he has visited a doctor because of medical problems. Also experience of the driver (little, much) and age (young, middle, old) were recorded. The results are in the table below

Driver	Age	Experience	Doctor visits
1	Young	Much	2
2	Young	Much	3
3	Young	Much	4
4	Young	Little	3
5	Young	Little	4
6	Young	Little	4
7	Young	Little	5
8	Middle	Much	5
9	Middle	Much	6
10	Middle	Much	7
11	Middle	Little	5
12	Middle	Little	6
13	Middle	Little	6
14	Middle	Little	7
15	Old	Much	8
16	Old	Much	10
17	Old	Much	10
18	Old	Much	8
19	Old	Little	8
20	Old	Little	9
21	Old	Little	10

(a) Estimate the average number of doctor visits, assuming that the response can be seen as a simple random sample.

(b) Assume that the population distributions of experience and age are available for the population of all drivers of the company:

Experience	Percentage
Much	48%
Little	52%

Age	Percentage
Young	22%
Middle	30%
Old	48%

Establish whether or not the response is selective. Explain which of these two auxiliary variables should be preferred for computing adjustment weights.

(c) For each auxiliary variable, separately carry out the weighting adjustment. Compute weights for each of the categories of the auxiliary variable.

(d) Compute for both weighting adjustments a weighted estimate of the average number of doctor visits.

(e) Compare the outcomes under Exercises 8.3a and 8.3d. Explain differences and/or similarities.

CHAPTER NINE

Selection of Auxiliary Variables

9.1 Introduction

This chapter is complementary to Chapters 5 and 8. In Chapter 5 it is illustrated what methods are available to analyze the impact of nonresponse on estimators for population parameters of interest. It will be shown in this chapter that such analysis techniques are also important tools in the selection of auxiliary variables; they provide means to differentiate the potential of these variables to adjust for nonresponse. In Chapter 8 a broad spectrum of methods is presented that adjust or correct standard estimators for the impact of nonresponse on estimators. Additional methods are presented in Chapters 11 and 12, Vital components of these adjustment methods are the auxiliary variables; without auxiliary variables there can be no adjustment. However, even when auxiliary variables are available, nonresponse adjustment will only be effective if the variables are informative on nonresponse behavior and on the key survey items. The selection of auxiliary variables is therefore at the core of a nonresponse adjustment.

In the context of this chapter, auxiliary variables are variables that are external to the survey; they are (also) measured outside the survey. Three options are possible: (1) the values of the auxiliary variable are available for all population units including respondents and nonrespondents, (2) the values of the auxiliary variable are available for all sample units, or (3) the values of the auxiliary variable are available for respondents and on an aggregated population level. The third option

Handbook of Nonresponse in Household Surveys, By Jelke Bethlehem, Fannie Cobben, and Barry Schouten

implies that population means or population counts are available. In the following discussion the three options are referred to as *population level*, *sample level*, and *aggregated population level* auxiliary variables. Population level and sample level auxiliary variables are intrinsically the same when it comes to the selection of auxiliary variables. However, the variables are included in a different way in nonresponse adjustments. Auxiliary variables at the sample level arise from interviewer or data collection observations. They are often termed process data or *paradata*. Examples are observations on the neighborhood or dwelling or the contactability or response propensity of sample units. Auxiliary variables at the aggregated population level are usually produced by national statistical institutes.

The auxiliary variables used for nonresponse adjustment are often termed weighting variables, so adhering to the fact that they form the input to adjustment weights. Since it is possible to adjust for nonresponse without the use of weights, the term weighting variables is avoided in this chapter.

Most nonresponse adjustment methods do not have strict guidelines or algorithms for the selection of variables in the models underlying the methods. Usually variables are selected in multiple steps based on experience and best practices. A simple good rule to follow is that adjustment weights in calibration estimators should not exceed a value of three and should preferably be positive (see Chapter 8). Large weights or negative weights are considered to be symptoms of an ill-conceived weighting model. Also useful is the rule of thumb that five propensity score adjustment classes suffice to adjust for nonresponse (Cochran, 1968; Little, 1986). Use of heuristics and experience comes from the great variety of surveys and survey topics in the literature.

Recently a number of attempts have been made to formalize and methodize the selection of auxiliary variables. Särndal and Lundström (2009) and Schouten (2007) suggest some criteria to use in judging and assessing the effectiveness of auxiliary variables in reducing nonresponse bias. These methods are not yet available in standard statistical software, but the criteria can be implemented and computed easily.

In Section 9.2 the various variable selection approaches are described. In Section 9.3 the methods are applied.

9.2 Theory

9.2.1 THE AUXILIARY VARIABLE SELECTION PROBLEM

The auxiliary variable selection problem can be summarized as the search for variables that replace the unknown relation between target variables and response behavior. Figure 9.1 shows how an auxiliary variable X may bridge the gap between response behavior R and target variable Y. By definition, the relation between a target variable or the list of target variables and response behavior is unknown because of the nonresponses. However, the relations between auxiliary variables and response behavior and between auxiliary variables and target variables(s) may

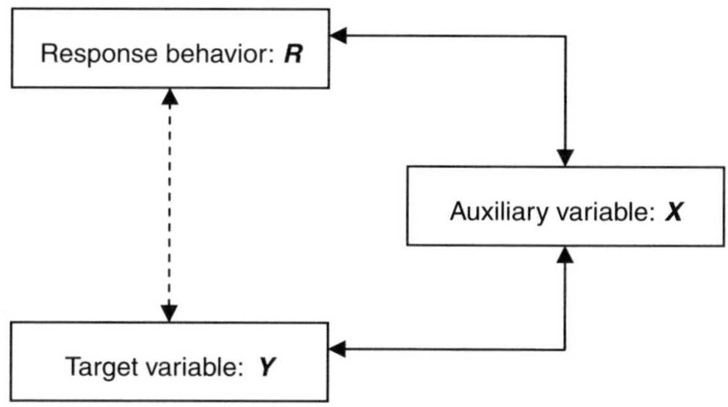

FIGURE 9.1 Relationships between target variable, nonresponse and auxiliary variable(s).

be known and may form the input to adjustment methods. In Figure 9.1 both Y and X may consist of vectors of variables rather than single variables.

Figure 9.1 is particularly helpful in illustrating the selection problem. Suppose that a set of auxiliary variables $\aleph$ is available for nonresponse adjustment. Variables in $\aleph$ that relate strongly to both the target variables in the survey and to response behavior, or in other words to the missing-data mechanism, are the candidates for nonresponse adjustment. Variables in $\aleph$ that relate strongly to target variables only are still interesting from a variance point of view. These variables do not help in understanding or explaining nonresponse. However, the variables allow for small prediction errors for the nonobserved answers of both nonrespondents and nonsampled population units. Variables in $\aleph$ that relate strongly only to response behavior are no candidates for nonresponse adjustment. These variables are useful, in general, in understanding motives and reasons behind nonresponse, so they may lead to clues for the reduction of nonresponse. Nonetheless, they do not contribute to adjustment of nonresponse bias.

Särndal and Lundström (2005) conclude that an ideal auxiliary variable has three features: (1) it explains well the response behavior, (2) it explains well the survey variables, and (3) it identifies the most important domains for publication of the survey statistics. By important domains, they mean subpopulations that appear in publications of statistics based on the survey.

The selection problem concerns the difference between nonresponse analysis and nonresponse adjustment. Nonresponse analysis is about causes of nonresponse and response behavior, regardless of the survey statistics, whereas nonresponse adjustment is about the impact of nonresponse on the survey statistics.

Figure 9.1 can also be used to understand the size of the bias of response means under the random response model (for a discussion, see Chapter 2). Let Y_k be a survey target variable for population unit k, X_k be an auxiliary variable for unit k, R_k the 0–1 indicator or dummy variable for response of unit k, and ρ_k the response probability of unit k. The population consists of N units, and thus $1 \leq k \leq N$. The bias of the response mean $\bar{y}_R$ equals

$$B(\bar{y}_R) = \frac{R_{\rho Y} S_\rho S_Y}{\bar{\rho}}, \tag{9.1}$$

with $\bar{\rho}$ the population average response probability, $R_{\rho Y}$ the population correlation between response probabilities and target variable, and S_Y and S_ρ the standard deviations of the target variable and response probabilities, respectively. Schouten (2007) showed that the correlation $R_{\rho Y}$ can be bounded from below and above by

$$R_{\rho X}R_{XY}-\sqrt{1-R_{XY}^2}\sqrt{1-R_{\rho X}^2} \leq R_{\rho Y} \leq R_{\rho X}R_{XY}+\sqrt{1-R_{XY}^2}\sqrt{1-R_{\rho X}^2}. \quad (9.2)$$

The lower and upper level in (9.2) hold for any choice of X and correspond to worst-case scenarios; the lower level represents a maximal remaining negative relation between target variable and response conditional on X and the upper level a maximal remaining positive relation conditional on X. Hence (9.2) leads to an interval where the lowest and highest values are the consequence of extreme response behavior. If either $|R_{\rho X}| = 1$ or $|R_{XY}| = 1$, meaning X is a perfect predictor of response behavior or of the target variable, then (9.2) reduces to a single point

$$R_{\rho X}R_{XY}, \quad (9.3)$$

and the correlation between target variable and response behavior can be identified perfectly. In other words, X bridges the gap in Figure 9.1. Generally, neither one of the correlations will be equal to one. However, when the relation of X with both Y and ρ becomes stronger, then the interval becomes smaller, and the less room is left for unobserved dependence between target variable and response behavior. The width of interval (9.2) equals

$$2\sqrt{1-R_{XY}^2}\sqrt{1-R_{\rho X}^2}. \quad (9.3)$$

Note that the correlation $R_{\rho Y}$ appears in the bias of the response mean. The nonresponse bias is directly proportional to the correlation. Auxiliary variables for which $|R_{\rho X}|$ and $|R_{XY}|$ are closer to 1 enable a stronger grip on that component of the bias and more effective nonresponse adjustments. Section 9.2.5.2 explores the use of interval (9.2) further as a criterion in the selection of auxiliary variables.

Little and Vartivarian (2005) discuss the impact of the selection of weighting variables on standard errors of survey statistics, namely the precision and confidence intervals attached to the adjusted estimators. They show that nonresponse adjustment methods that include auxiliary variables that are noninformative about the target variables can lead to an increase in standard errors and, as a consequence, to larger confidence intervals. This result is not surprising, since more, and thus smaller, adjustment cells or strata are created. Smaller strata allow more sampling variation that is not counteracted by smaller prediction errors. Auxiliary variables that correlate to target variables but not to response behavior, however, may lessen the number of errors. It is therefore important to keep in mind that while the selection of auxiliary variables in nonresponse adjustment should be primarily directed toward bias reduction, precision must be considered as well. This holds especially true for small samples or for surveys that have many detailed statistics.

EXAMPLE 9.1 Correlation Intervals

From the population of the imaginary country Samplonia a survey sample of 1000 persons is drawn from persons between 18 and 25 years of age. For each person the age and marital status is known. Marital status has two categories: not married and married. In the survey it is asked whether the person commutes to work or school and the years of work experience. The sampling distribution, response probabilities, and commuting and work experience probabilities for respondents are given in the tables.

Sample composition

	Not married	Married	Total
18 years	120	8	128
19 years	125	10	135
20 years	106	8	114
21 years	105	15	130
22 years	103	27	130
23 years	98	25	123
24 years	80	40	120
25 years	65	65	130
Total	802	198	1000

Response probabilities

	Not married	Married	Total
18 years	0.40	0.50	0.41
19 years	0.40	0.55	0.41
20 years	0.42	0.56	0.43
21 years	0.44	0.62	0.46
22 years	0.48	0.65	0.52
23 years	0.47	0.68	0.51
24 years	0.50	0.72	0.57
25 years	0.55	0.75	0.65
Total	0.45	0.68	0.50

The overall response rate is 50%. The response rate for married persons is higher than for unmarried persons, and the response rate is also slowly increasing with age.

Commuting probabilities

	Not married	Married	Total
18 years	0.52	1.00	0.56
19 years	0.60	0.91	0.63
20 years	0.70	0.45	0.67
21 years	0.74	0.86	0.76
22 years	0.73	0.85	0.76
23 years	0.85	0.82	0.84
24 years	0.90	0.87	0.89
25 years	0.84	0.92	0.89
Total	0.73	0.87	0.77

Of the respondents, 77% is commuting to work or school. For married persons this rate is 87%, 14% higher than for not married persons. The commuting rate is increasing with age and ranges from 56% to 89%

More than 2 *years' work experience*

	Not married	Married	Total
18 years	0.04	0.00	0.04
19 years	0.06	0.00	0.05
20 years	0.22	0.00	0.20
21 years	0.26	0.11	0.23
22 years	0.36	0.34	0.36
23 years	0.46	0.35	0.43
24 years	0.88	0.80	0.84
25 years	0.92	0.90	0.91
Total	0.37	0.59	0.43

Overall 43% of the respondents has more than two years of work experience. This proportion increases rapidly with growing age from 4% to 91%. Married people more often tend to have more than two years of work experience, 59% opposed to 37%.

The sample correlations between response probabilities and the two auxiliary variables age and marital status are 0.73 and 0.87, respectively. Hence the correlation is strong for both variables.

The response correlations with the target variable commuting are 0.26 for age and 0.15 for marital status. Both variables thus have a relatively weaker relation to this target variable than to the response behavior. The central point of the correlation interval (9.2), $R_{\rho X} R_{XY}$, equals 0.19 for age and 0.13 for marital status. The bias interval width (9.4), $2\sqrt{1-R_{XY}^2}\sqrt{1-R_{\rho X}^2}$, for age is 1.33 and for marital status is 0.96. Marital status provides a better explanation of response behavior than age, but the reverse is true with respect to commuting. The net impact of marital status is, however, largest and leads to the biggest reduction in bias interval width.

The response correlations with the target variable work experience are 0.62 for age and 0.20 for marital status. The correlation to age is strong. $R_{\rho X}R_{XY}$ equals 0.45 for age and 0.17 for marital status and $2\sqrt{1-R_{XY}^2}\sqrt{1-R_{\rho X}^2}$ for age is 1.07 and for marital status is 0.95. The net impact of marital status still leads to the biggest reduction in bias interval width.

The correlation intervals for commuting to work or school based on age and marital status are, respectively, $[-0.48, 0.85]$ and $[-0.35, 0.61]$. Hence there is still a lot of room left for the correlation between response behavior and commuting, although correlations between the two auxiliary variables and response behavior are large.

For work experience the correlation intervals are $[-0.09, 0.99]$ for age and $[-0.30, 0.64]$ for marital status. Since obviously the correlation between response behavior and work experience is fixed, the intersection of the two intervals can be taken. Hence the combined correlation interval is $[-0.09, 0.64]$, which has a width of 0.73.

It can be concluded from this example that it pays to search for variables that supplement each other in the explanation of response behavior and target variables.

9.2.2 THE CONSTRUCTION OF AUXILIARY VARIABLES

So far the type and measurement level of auxiliary variables were ignored. Both do, however, play an important role in the adjustment for nonresponse. A number of decisions need to be made before a selection of variables is conducted:

1. What categories to use for qualitative auxiliary variables.
2. Whether quantitative auxiliary variables are to be transformed into qualitative variables. If so, how are their categories chosen? If not, should quadratic and higher order terms be included?
3. Which interactions between auxiliary variables should be included

The first decision concerns the level of detail of the qualitative variables entered in models for response behavior and for survey target variables. For some auxiliary variables like gender, there are only two categories and the classification is obvious. For the majority of qualitative variables, however, this is not true, and employing the qualitative variables without any recoding or aggregation may often not be practical or feasible. Good examples are the nationality of persons, the classification of businesses according to activities, and the type of household. The full details of such variables are too large for practical purposes. There are, generally, two approaches to reduce the detail of auxiliary variables.

In many practical settings, the classifications used are the same as in publications, since they are widely accepted and can produce consistent statistics over different surveys and registrations. Standard classifications ensure consistency in calibration methods such as generalized regression weighting or multiplicative weighting. Different estimates from different surveys are calibrated to the same population totals.

Another approach is to employ a selection method that aggregates categories based on statistical tests. Examples of such selection methods are classification and regression trees (Breiman et al. 1984; Hastie et al. 2009). A well-known and widely used classification tree method is CHAID (chi-square automated interaction detector; Kass 1980). Tree methods start with the full population and make repeated splits of the population based on some test statistic that is applied to all possible splits with the available auxiliary variables. The split with the lowest p-value, meaning the split that is statistically speaking farthest away from independence, is selected. The resulting subpopulations are then new candidates for further splitting. The process is repeated until no splits are statistically significant or until some stopping rule is satisfied. The tree methods, clearly, have to follow the categories of variables but do not necessarily follow the classifications of variables as a split may consist of sets of categories.

EXAMPLE 9.2 The Classification of Qualitative Variables

In Example 9.1 two auxiliary variables are used: age and marital status. Suppose that users of the survey are interested in the age classifications 18 to 23 years and 24 to 25 years. Hence the eight categories are merged into two age intervals. The tables with response probabilities and work experience probabilities are as listed below.

Response probabilities

	Not married	Married	Total
18–23 years	0.43	0.62	0.46
24–25 years	0.52	0.74	0.61
Total	0.45	0.68	0.50

More than 2 years' work experience

	Not married	Married	Total
18–23 years	0.23	0.22	0.23
24–25 years	0.90	0.86	0.88
Total	0.37	0.59	0.43

Again, 43% of the respondents has more than two years of work experience. The two age groups differ considerably in the proportion with three or more years of work experience.

The variance of the response probabilities $S^2(\rho)$ did not change, but the sample correlations with age did as a result of the recoding. The sample correlation between response probabilities and age dropped from 0.73 to 0.64. Hence the correlation is still considerable although a lot of detail was lost in the recoding of age. The response correlation between age and work experience also dropped, but only very little, from 0.62 to 0.61. Hence most of the explanatory power of age for work experience was preserved.

$R_{\rho X}R_{XY}$ now equals 0.39 and the width of the correlation interval $2\sqrt{1-R_{XY}^2}\sqrt{1-R_{\rho X}^2}$ is equal to 1.23. For work experience the correlation interval based on age has become wider; it is [−0.22, 0.99] instead of [−0.09, 0.99].

The less detailed classification of age had little impact on the strength of the relation to work experience but affected the relation to response behavior. Consequently the potential of age for nonresponse adjustment decreased slightly.

The second decision that needs to be made is how quantitative auxiliary variables, if present in the data, are treated. Examples of quantitative auxiliary variables are income, age, house value, number of months employed or unemployed, and degree of urbanization. There are two options: (1) the variables are made qualitative by clustering values. or (2) the variables are treated as continuous.

In general, the decision may have far-reaching consequences as the number of parameters in models for response behavior and target variables may vary greatly. A simple example may help explain the difference. When income is entered in a linear regression as a quantitative independent variable, there are only two parameters: an intercept and a slope parameter.

When income values are transformed to 10 income classes, the number of parameters is equal to 10. Hence, by a simple recoding, the number of parameters increased considerably and along the potential to adjust for nonresponse.

Alternatively, higher order polynomial terms for quantitative variables may be entered, such as quadratic or cubic terms. For example, if next to income also the squared income is added to the model, the number of parameters becomes three. In most practical settings, quantitative variables are made qualitative unless there is an intrinsic (polynomial) relation between a quantitative variable and the target variables of interest. When a quantitative variable is categorized, the choice again needs to be made on how to do this.

The same approaches as for qualitative auxiliary variables apply; either category follows explicitly from publication or is implicitly created by the selection methods. For the second approach it should be noted that most tree methods allow for the definition of interval or ordinal variables. The order of such variables is then respected in splitting the population. Hence a regression tree will not make the split "income smaller than 1,000 or bigger than 10,000" versus "income between 1,000 and 10,000." A split "income smaller than 1,000" versus "income bigger than 1,000" is allowed.

EXAMPLE 9.3 Age as a Quantitative Variable

In Example 9.1 age is implicitly treated as a quantitative variable in the calculation of sample correlations with the two target variables and response behavior. This makes sense as differences between the age categories do have a meaning. The recoding of age to a two-category variable in Example 9.2 removes this interpretation.

The labeling is arbitrary and the differences do not have a meaning. If age is treated as quantitative, then it becomes interesting to investigate higher order terms. In other words, instead of the labels 18, 19, 20,. . ., 25, polynomials of these labels may be used. For instance, if age squared is added as a candidate variable, then the labels become 324, 361, 400,. . ., 625.

Along with the polynomial labels, the sample correlations will change. The table below shows the sample correlations of polynomials of age with commuting and with response behavior. Also given are the correlation intervals.

Polynomial	R_{XY}	$R_{\rho X}$	Interval
Age	0.26	0.73	[−0.48, 0.85]
Age^2	0.26	0.73	[−0.47, 0.85]
Age^3	0.26	0.74	[−0.46, 0.84]
Age^4	0.25	0.74	[−0.46, 0.84]
Age^5	0.25	0.74	[−0.46, 0.83]
Age^6	0.25	0.75	[−0.46, 0.83]

Taking higher polynomials of age affects sample correlations only very mildly. An inspection of the response probabilities and commuting probabilities reveals indeed a linear pattern with respect to age. In this case adding higher order terms of age will not be helpful in the nonresponse adjustment.

The last decision concerns the interactions among auxiliary variables that are accounted for in the modeling of response behavior and survey target variables. In fact interactions among variables imply that new variables are created from the existing set of auxiliary variables. In regression models it means that apart from main effects, cross terms are also included.

Interaction terms can be added to the selection both for quantitative and qualitative auxiliary variables, although usually interactions are restricted to qualitative variables. However, the number of variables and parameters in models may increase very rapidly, leading to insufficient information in the survey sample. The question is whether adding interaction terms really improves the nonresponse adjustment.

There are four motivations to include some or all of the interactions. First, publications may again request that certain variables be crossed. For instance,

Statistics Netherlands generally disseminates tables where age and gender are crossed. Second, the auxiliary variables may be strongly collinear. This means that there is a strong correlation between the variables or between some categories of the variables. Good examples are age and marital status or province or state of residence and degree of urbanization of the residence area. Third, response behavior may be different for interactions, and fourth, target variables may be different for interactions.

Literature shows that it is rarely the case that strong interactions appear in models for response behavior. For some target variables, however, interactions can play an important role. An example is the number of children that women ideally would like to have in the future. The interactions between age and the number of children already in the household obviously are strongly predictive.

EXAMPLE 9.4 Interactions

In Examples 9.1 to 9.3 no interactions were considered between age and marital status. Both variables are, however, collinear. Only 4.1% of persons is married and younger than 22 years. In practical settings age and marital status will be combined into a new variable where some of the categories are merged. A classification that accounts for the collinearity is given below. The values represent the numbers of persons in the new strata.

Sample composition

	Not married	Married
18 years	120	
19 years	125	
20 years	106	
21 years	105	93
22 years	103	
23 years	98	
24 years	80	105
25 years	65	

9.2.3 LINKED DATA AND POPULATION TOTALS

Auxiliary information can be available at different levels. Three options are possible: (1) the auxiliary variable is available for all population units, including respondents and nonrespondents, (2) the auxiliary variable is available for all sample units, and (3) the auxiliary variable is available for respondents and on an aggregated population level. The third option corresponds to the availability of population distributions. The three options are referred to as *population level*, *sample level*, and *aggregated population level* auxiliary variables.

When a survey suffers from nonresponse, two random processes are in play. First, there is the sampling distribution. The sampling design that underlies this distribution is usually under control of the survey researcher. Second, there is the response distribution or response behavior. The response distribution is at best partially under control of the survey researcher.

Auxiliary information can be used for two purposes: variance reduction and bias adjustment. Variance reduction is the traditional purpose of weighting in sampling theory. The auxiliary variables function as predictors of the survey target variables.

Many of the traditional weighting methods can be carried out with aggregated population level auxiliary information (see Chapter 8). Population means, population counts, and population contingency tables are, in statistical terms, sufficient statistics for the computation of estimators. Hence most traditional methods employ aggregated population level information only.

Aggregated population level auxiliary variables are often produced by national statistical institutes and are most commonly available to all survey researchers. They are sometimes treated as "golden standards" by survey organizations that do not have access to data sources other than the actual survey. Whereas in the traditional weighting aggregated population level information is used for variance reduction, it can also be deployed for nonresponse adjustment. It must, however, be accepted that the composition of response is the result of sampling and response distribution. As a consequence the predictive power of the auxiliary variables for nonresponse is reduced by the sampling variation.

Aggregated population level variables must be survey items. It is therefore important to carefully evaluate whether they are defined the same as in the population tables. For this reason variables that are prone to measurement errors should be avoided.

Population level auxiliary variables are also called *linked data* because they arise from a linkage to registered or administrative data. The survey sample is drawn from a sampling frame that carries unique identification keys connected to other registrations. Variables in the registers can be selected for nonresponse adjustment purposes.

For example, at Statistics Netherlands the unique identification key is the social security number. This number is used throughout all the government organizations. At Statistics Netherlands the number is scrambled for internal use so that it is not recognizable. All copies of registrations that are used for statistical purposes carry the scrambled social security numbers. Additionally all registrations are accessible only at closed computer networks and after a confidentiality pledge is signed. This scheme offers a wide range of auxiliary variables that are potentially informative on response behavior and survey target variables. Other types of registrations are the municipal population registers, tax and income records from the tax authorities, land registry records, and hospital records.

Population level auxiliary variables are available for all population units, and thus for both respondents and nonrespondents. Data available at the population level can, of course, be used for nonresponse adjustment directly. Contrary to the aggregated population level variables, population level variables allow for a direct comparison of response to sample. As such the information carried by the variables is not reduced by the sampling variation. For large survey samples, however, the aggregated and nonaggregated population level variables provide similar adjustments.

Sample level auxiliary variables are different. They are not available on the population level, so no corresponding population tables exist. Because auxiliary variables at the sample level arise from interviewer or data collection observations, sample level auxiliary variables are suited for nonresponse adjustment but not for variance reduction. These auxiliary variables are often termed process data or *paradata*. In other words, the auxiliary variables become available during data collection for both respondents and nonrespondents. Examples are the age and gender of sample persons as observed by interviewers, observations on the neighborhood, the causes for nonresponse, interviewer assessments of the willingness to participate, at home patterns, and the characteristics of interviewers themselves.

EXAMPLE 9.5 **Auxiliary Information Levels**

In the Dutch Health Survey of 2009 it was asked whether a person voted in the last National Elections of 2006. This variable was used as an auxiliary variable for weighting. The survey sample can also be linked to the Social Statistical Database (SSD), which is comprised of all government registrations. The linked data provide information on, for example, employment status, social allowance status, age, ethnicity, and household composition. During the survey data collection, interviewers process data about the at home pattern of the sampled persons, the presence of physical impediments, and the causes for nonresponse.

Although it is recorded whether or not, persons voted the resulting voting register is one of the few registrations that cannot be linked to any other government registration. Consequently population tables for voting behavior cannot be crossed against sociodemographics or other population characteristics, other than region of residence. The table below shows the population and response distribution for voting in the 2006 National Elections.

	Population distribution	Response distribution
Voted	57.6%	59.0%
Entitled to vote, but did not vote	14.3%	11.2%
Not entitled to vote, native	3.6%	1.7%
Not entitled to vote, <18 years	24.5%	28.1%

> The 2009 Health survey contains auxiliary variables on the aggregated population level, voting behavior, on the population level (e.g. age and gender), and on the sample level (e.g. presence of impediments).

Population level and sample level auxiliary variables are intrinsically the same when it comes to the selection of auxiliary variables for nonresponse bias adjustment. Both types of variables allow for a direct modeling of response behavior. If one is not interested in variance reduction, then there is no conceptual difference between the two.

Because two random processes are at play, sampling and nonresponse, two estimation steps are distinguished. The first step is from response to sample, and the second step is from sample to population. Different auxiliary variables may be used for both steps. However, sample level auxiliary variables can only be included in the first step from response to sample. In practice, very rarely two estimation steps are performed, and nonresponse adjustment and weighting coincide, and so they serve both variance reduction and bias adjustment. This consideration is important to keep in mind in the selection of auxiliary variables, as they may relate only to survey target variables and still be useful.

The differences in the selection of auxiliary variables based on the level of information are important. First, incorporating auxiliary variables as survey items in a questionnaire or as observations in a data collection process calls for a preselection step. Since the questionnaire must be limited in length, and interviewers spend a finite time at the doorstep, it must be carefully decided what interviewers are asked to record and what not. This means that aggregated population level and sample level auxiliary variables need to be selected before the survey is started. It is still possible to select a subset of auxiliary variables and disregard some of the variables, but this requires an exploratory analysis and literature study. Subsets cannot be used for linked data and population level auxiliary variables; otherwise, there is no limit to the number of auxiliary variables or contents of available registrations. Another difference between the various levels lays in the nonresponse adjustment methods. A different level of information leads to a different range of possible models and estimators. For instance, propensity weighting methods (as discussed in Chapter 11) are seldom applied in cases where aggregated population level variables are the only source of information. Since ultimately the selection of variables takes place within the chosen modeling frame by statistical testing, different levels of auxiliary information can lead to different selection strategies.

Särndal and Lundström (2005) show how the general calibration framework proposed by Deville and Särndal (1992) can be made fit for all levels of auxiliary information. In the notation of Chapter 8, let the calibration weight $w_i = g_i d_i$ be the product of the design weight d_i and the correction weight g_i, then the calibration amounts to minimizing

$$\sum\nolimits_{i \in r} D(g_i, 1), \tag{9.4}$$

with D an arbitrary distance function and r the response, subject to

$$\bar{x}_W = \frac{1}{N}\sum_{i\in r} w_i(x_{1i}, x_{2i})' = (\bar{x}_1, \bar{X}_2)'. \tag{9.5}$$

In (9.5) x_1 is a vector of sample level variables and x_2 is a vector of the aggregated population level variables. Population level variables may be part of both vectors as they can be translated to both the sample and the aggregated population level. The right-hand side of (9.5) consists of sample means for x_1 and population means for x_2.

9.2.4 VARIABLE SELECTION STRATEGIES

In the previous sections the selection problem and its basic ingredients—the auxiliary variables—are introduced. This section moves to the strategies that underlie the selection of auxiliary variables. First a general scheme is presented. Then, some parts of the scheme are discussed in more detail in subsequent sections.

9.2.4.1 Selection Strategies in General In practice the selection of auxiliary variables for nonresponse adjustment is often based on simple heuristics and experience. Nevertheless, a general set of steps can be distinguished. These are listed below.

1. Preselect the auxiliary variables from the literature on similar surveys that relate to response behavior and to the main survey variables.
2. Link to population level variables from the available registers.
3. Include any additional auxiliary variables in the survey questionnaire.
4. Identify any additional process data observed by interviewers and data collection staff.
5. Model the missing-data mechanism of nonresponse.
6. Model the main survey variables.
7. Combine the auxiliary variable results from the computations of steps 5 and 6.
8. Check the weights and if necessary return to step 7.

The variables used in the literature may help in identifying the key survey target variables and the nonresponse missing-data mechanism, as explained in Section 9.2.1. Auxiliary variables that relate only to survey variables could be added as a variance reduction measure. Recall that these variables lead to a reduction of standard errors, and hence to more precision and smaller confidence intervals.

There is a vast literature on correlates of nonresponse. It is paramount, however, that the literature considered be about similar surveys with similar data collection strategies. For instance, because the mode of interview and the mode of contact are critical to obtaining a response, it is advisable that the literature consulted be on surveys with the same mode or mix of modes. Still the general literature has identified correlates of nonresponse that have relevance for almost all surveys such as age, type of household, and urbanization of residence area. A description of general correlates of nonresponse can be found in Chapter 5 and in Groves and Couper (1998), Groves et al. (2002), Voogt (2204), Stoop (2005), and Cobben (2009).

Recall also our earlier discussion about auxiliary information obtained from registrations, and from questionnaires or from observations related to the process data. Since linked data, if available, do not lengthen the questionnaire or the data collection, this form of data is much preferred. Moreover linked data usually do not contain measurement errors due to the questionnaire, context, or interviewer bias.The two alternatives, inclusion of questions to obtain auxiliary variables and collection of process data by interviewers and supervisors, involve additional effort.

The more common of the two is to include the auxiliary variables as questions in the survey. Usually these questions are asked at the beginning or end of the questionnaire if they are not related directly to the topic of the survey. Surveys at Statistics Netherlands, for instance, start with a base module that consists of the same questions for all surveys. Survey organizations that do not have access to registration records mostly depend on auxiliary variables being added to their questionnaires.

The second way of obtaining auxiliary variables from process data or paradata has been little mentioned in the literature. Only recently the literature (e.g., Kreuter and Kohler, 2009; Kreuter et al., 2010 has explored the potential of this type of data. Clearly, since process data consist only of observations, the range of auxiliary variables from process data is limited. The literature does, however, give some examples. Beaumont (2005) describes the use of the number of contact attempts. At Statistics Netherlands, the mode of interview is used in the nonresponse adjustment of the Dutch Safety Monitor.

EXAMPLE 9.6 Number of Contacts as Auxiliary Variable

An interesting example is the number of contact attempts needed to reach a person or household. Contactability relates to important survey variables like employment or unemployment, social activities and free time. In some surveys the sample level auxiliary variable "yes or no more than k contact attempts were needed" is used in weighting models.

Let A_i^k be the 0–1 indicator for sample unit i. Let $\bar{A}^k$ be the number of persons contacted before attempt k. A_i^k may be a candidate in the nonresponse adjustment and $\bar{A}^k$ may be a sample parameter that is calibrated.

It is, however, important to recognize that contact alone is a consequence of the survey design. If a number of at most 10 attempts is fixed, then all contacts score a one on the variable A_i^{10}. Hence there are no respondents to compare nonrespondents to. Variable A_i^1 will have zeros and ones for contacts, but the sample units contacted after the first attempt are more heterogeneous and less similar to noncontacts after 10 attempts. This means a balance needs to be found between homogeneity and size of the groups.

Steps 5 to 7 are the actual variable selection steps. Variables are included in models for response and models for a number of target variables. Two sets of auxiliary variables result. From these two sets a final set of auxiliary variables is extracted. Usually the final selection is ad hoc, as it depends to some extent on the methodologist or researcher performing that chose. In surveys with a broad spectrum of questions, step 6 is sometimes omitted. The range of target variables is then considered sufficiently wide to justify the inclusion of all auxiliary variables that relate to response. The consequence is, however, that auxiliary variables are also deployed for survey target variables to which they do not relate or relate only moderately. Little and Vartivarian (2005) show that this can have a negative impact on the standard errors and precision. Therefore it is best to be conservative in adding auxiliary variables to the nonresponse adjustment.

In combining the auxiliary variables from steps 5 and 6 in step 7, the interactions of auxiliary variables that are listed in contingency tables for publication are often adopted. See Särndal and Lundström (2005). They refer to these as the domains of publication. In these tables the crossed variables are treated as marginal data.

Step 5 pertains to modeling response behavior. Sections 5.2.3 and 5.3.2 of Chapter 5 give a full account of how to model the 0–1 response indicator. The common model for nonresponse is the logistic regression model. However, any multivariate model for 0–1 indicators or dummy variables is suitable. Agresti (2002) is a good reference to consult. Model fitting and evaluation are strongly dependent on the model that is used and the estimation strategy that is customary for that model. However, in most settings the selection is done through statistical tests in forward and backward selection algorithms. For example, in regression models the forward and backward searches are used for explanatory or independent variables in likelihood ratio tests or Wald tests.

A different line of models is formed by classification trees. This data-mining technique selects auxiliary variables or more accurately selects interactions among these variables in a top-down fashion. The most important interactions are selected first. During the selection, stopping rules are applied such as at a minimum number of sample units before and after interactions or at a specified significance level. Some of the tree methods also "prune" the resulting tree, thus accounting for overfitting. Well-known methods are CHAID and CART. Details can be found in Kass (1980), Breiman et al. (1984), and Hastie et al (2009).

EXAMPLE 9.7 A Classification Tree For Nonresponse

Statistics Netherlands produces documentation for each survey, describing the nonresponse adjustment techniques that were applied. For the Dutch Family Survey of 2003 the documentation contained the following classification tree

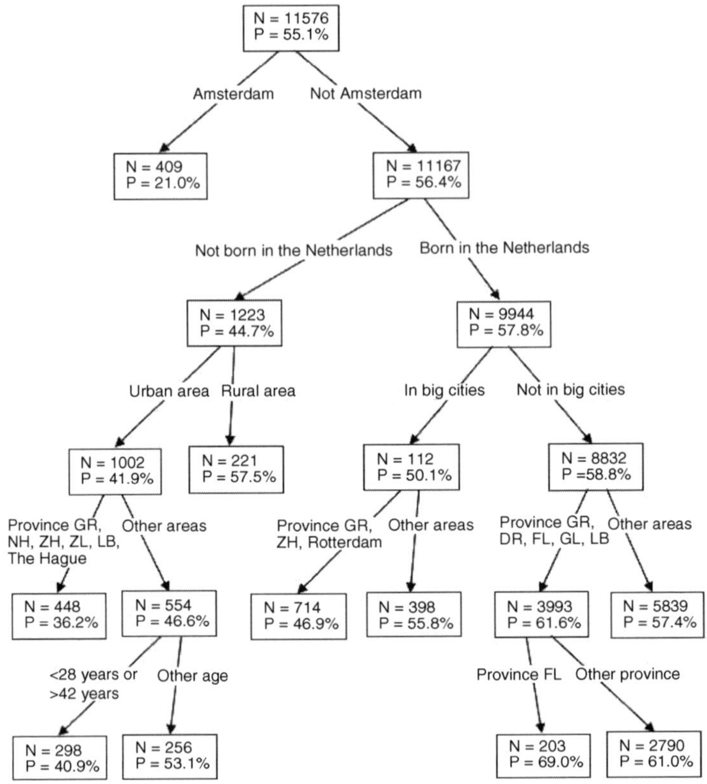

The survey sample consisted of 11,576 persons and the overall response rate was 55.1%. The strongest auxiliary variable was the area of residence. When the area was Amsterdam, the capitol of The Netherlands, the response rate was only 21.0%. The response rate in the remaining areas was 56.4%. Subsequently the selected interactions were based on the ethnicity of the person, the degree of urbanization, the province of residence, and age.

The classification tree resulted in 10 end nodes also called "leaves." Hence the selection suggests 10 strata that are a combination of area of residence ethnicity, urbanization, and age.

In step 6 the key survey items are modeled. The models employed depend on the nature and measurement level of the survey variables. The range of possible survey variables is wide. Although a general account of multivariate models goes beyond the scope of this handbook, in Section 9.3 some examples were provided, and a few brief observations follow.

A frequently applied nonresponse adjustment is the generalized regression estimators (GREG) modified to nonresponse. GREG is based on a linear model for the relationship between the auxiliary variables and the target variable. A special case of this estimator is poststratification, where all auxiliary variables are qualitative and are crossed.

Because the model underlying the regression estimator is a linear model, the objective is to reduce variance in samples with full response. It can be proved that regardless of the survey target variable, the regression estimator leads to the same adjustment weights. The survey target variables can be ignored in the computation of the weights, since a linear model between target variables and auxiliary variables is implicitly assumed. However, the choice of auxiliary variables that enter the model may be based on the relation to both nonresponse and target variables.

Another observation should be made about the adjustment weights. Many survey organizations prefer a single set of adjustment weights for each survey data set. From the point of view of accuracy, this is not optimal. Ideally a nonresponse adjustment is done separately for each survey target variable, since the strength of $X - Y$ relation in Figure 9.1 obviously differs for each Y. In practice, one therefore restricts analysis to a few key statistics from the survey.

The final step, step 8, concerns investigating whether the selected auxiliary variables lead to negative or extreme adjustment weights. Chapter 8 briefly describes the methodology used for nonnegative adjustment weights and to limit the maximum value of weights. Negative and large weights are signs of an ill-conceived model for nonresponse and target variables. This can mean that auxiliary variables are mutually dependent or collinear or that adjustment cells or strata are too small. Based on weight diagnostics like weight histograms and box plots, step 8 may lead to a re-selection of auxiliary variables.

EXAMPLE 9.8 Weight Histogram

For the 2003 Family Survey of Example 9.7 weighting was conducted by way of the generalized regression estimator (GREG). The weighting model was

$$Age + Urbanization + Marital\ status + Ethnicity.$$

The auxiliary variable marital status is added because it relates strongly to the survey target variables. The histogram of correction weights is

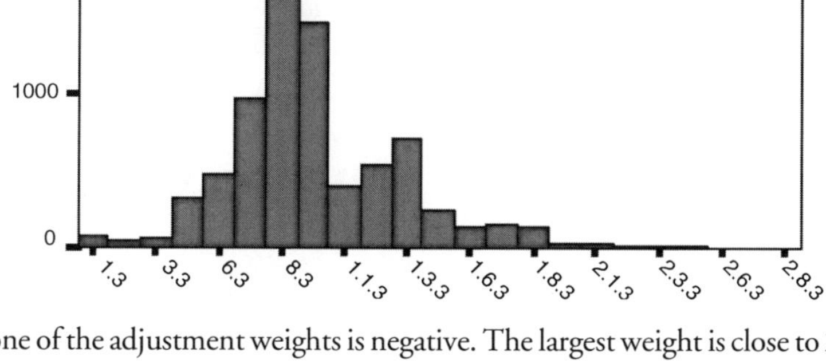

None of the adjustment weights is negative. The largest weight is close to 2.5.

Recent literature suggests some criteria to impact mathematical rigor to steps 5 to 9 (see Schouten, 2007; Särndal and Lundström 2009). These criteria call for the selection of auxiliary variables to be done in a single step. Sections 9.2.4.2 and 9.4.2.3 cover two such criteria, the variance of calibration weights and the maximal worst-case bias.

9.2.4.2 Variance of Calibration Weights Särndal and Lundström (2008) present a measure for the selection of variables in calibration estimators as

$$Q^2 = \left[\sum_{k=1}^{N} \rho_k\right]^{-1} \left[\sum_{k=1}^{N} \rho_k(\phi_k - \bar{\phi}_{\rho U})^2\right], \tag{9.6}$$

with ρ_k the response probability of population unit k, $\phi_k = \rho_k^{-1}$ the inverse of the response probability, and $\bar{\phi}_{\rho U}$ the ρ_k-weighted mean of the ϕ_k,

$$\bar{\phi}_{\rho U} = \left[\sum_{k=1}^{N} \rho_k\right]^{-1} \left[\sum_{k=1}^{N} \rho_k \phi_k\right]. \tag{9.7}$$

This measure is, as Särndal and Lundström argue, the *proportion of the relative near bias* (earlier defined in Särndal and Lundström, 2005) that is independent of survey target variables. Hence (9.6) can be applied independently of the target variables in the survey. The near bias is an approximation to the bias of response means of the survey target variables due to the nonresponse. Measure (9.6) is large whenever an auxiliary variable leads to strongly deviating calibration weights. It is the variance of calibration correction weights. The ϕ_k are termed *response influences*.

Since both the response probabilities and response influences are unknown, Q^2 needs to be estimated from the survey sample. Särndal and Lundström (2008) use the following estimator for the case where population level and sample level auxiliary variables are included:

$$q^2 = \left[\sum_{k=1}^{N} a_k R_k d_k\right]^{-1} \left[\sum_{k=1}^{N} a_k R_k d_k (\hat{\phi}_k - \hat{\bar{\phi}}_r)^2\right], \tag{9.8}$$

where $\hat{\bar{\phi}}_r = \left(\sum_{k=1}^{N} a_k R_k d_k \hat{\phi}_k\right) / \left(\sum_{k=1}^{N} a_k R_k d_k\right)$ is the weighted response mean of estimated response influences, d_k the inclusion weight, a_k the 0–1 sample indicator, R_k the 0–1 response indicator, and the estimated response influences are given by

$$\hat{\phi}_i = X'_i \left[\sum_{k=1}^{N} a_k R_k d_k X_k X_k'\right]^{-1} \left[\sum_{k=1}^{N} a_k d_k X_k\right] \tag{9.9}$$

With X as the candidate vector of auxiliary variables.

EXAMPLE 9.9 Stepwise Forward Selection Using q^2

Särndal and Lundström (2008) provide an example of the selection of auxiliary variables in the Swedish National Crime Victim and Security Study of 2006.

Step	Auxiliary variable	q^2 (× 1000)
0	Empty model	0
1	Country of birth	20.0
2	Income group	27.6
3	Age	31.3
4	Gender	35.1
5	Marital status	38.6
6	Region	40.7
7	Family size	41.4
8	Days unemployed	41.9
9	Urban center dweller	42.3
10	Occupation	42.7

Notice that the strongest variable with respect to q^2 is country of birth, followed by income group. The increments of q^2 get smaller when variables are added, and after 10 steps no more variables are entered.

9.2.4.3 Worst-Case Maximal Bias Schouten (2007) introduces a criterion by which auxiliary variables in generalized regression estimators (GREG) can be selected. The selection criterion is based on minimizing the maximal absolute bias under a weaker assumption than MAR (missing completely at random). Recall that the GREG estimator is based on a linear regression model with a survey target variable as dependent variable and the available auxiliary variables as independent variables. Schouten (2007) assumes that the estimated slope parameter β in the linear regression can be estimated without bias, but that the constant term or intercept $\hat{\alpha}$ in the linear regression is biased. Loosely speaking, it means that the slope of the regression remains unaffected by nonresponse, but the level can change. In a simple poststratification the assumption means that all stratum means have the same bias; that is to say, the relative distances between the stratum means are preserved under nonresponse.

One can even show that in terms of sample size, the bias in the slope parameter is of a higher order than the bias in the intercept. This means that for large sample sizes the nonresponse bias of the GREG estimator will be dominated by the bias in the intercept.

In Section 6.1 it was shown that under the random response model the bias of the response mean can be written as

$$B(\bar{y}_R) = \frac{R_{\rho Y} S_\rho S_Y}{\bar{\rho}}, \tag{9.10}$$

where $R_{\rho Y}$ is the correlation between the target variable and the response probability, S_Y is the standard deviation of the variable Y, and S_ρ is the standard deviation of the response probabilities. Hence the response mean is unbiased if there is no correlation between the target variable and nonresponse. The stronger the relationship is between the target variable and nonresponse, the larger is the bias. The size of the bias is affected by the amount of nonresponse, meaning the response rate. The more people are inclined to cooperate in a survey, the higher the average response probability will be, resulting in a smaller bias.

The bias of the response mean always lies in the interval

$$\frac{S_\rho S_Y}{\bar{\rho}} \left(R_{\rho,Z} R_{Z,Y} \pm \sqrt{1 - R_{\rho,Z}^2} \sqrt{1 - R_{Z,Y}^2} \right), \tag{9.11}$$

for any auxiliary variable Z.

Let X be the vector of auxiliary variables that is used in the GREG estimator and β_{XY} be the true slope parameter in the linear regression. Suppose that X does not contain a constant term, but an intercept α_{XY} is added to the model. The subscripts in α_{XY} and β_{XY} are given in order to distinguish between different auxiliary variables and survey target variables. Also suppose that for qualitative auxiliary variables, X contains all but one dummy variable. This is necessary in order to avoid nonidentifiable parameters due to the inclusion of the intercept. Let $\hat{\alpha}_{XY}$ and $\hat{\beta}_{XY}$ be the GREG estimators for both parameters

$$\hat{\alpha}_{XY} = \bar{y}_R - \hat{\beta}_{XY} \bar{x}_R, \tag{9.12a}$$

$$\hat{\beta}_{XY} = \left(\sum_{i=1}^{n} R_i d_i X_i X_i' \right)^{-1} \sum_{i=1}^{n} R_i d_i X_i Y_i. \tag{9.12b}$$

Schouten (2007) proves that by the unbiased slope assumption, meaning $E(\hat{\beta}_{XY}) = \beta_{XY}$, the bias of the GREG estimator can be bounded approximately by

$$\frac{S_\rho S_Y}{\bar{\rho}} \left(R_{\rho,Z} R_{Z,Y} - R_{\rho,X'\beta_{X,Y}} R_{X'\beta_{X,Y},Y} \pm \sqrt{1 - R_{\rho,Z}^2} \sqrt{1 - R_{Z,Y}^2} \right) \tag{9.13}$$

for any variable Z.

From a comparison of expression (9.13) to (9.11) it follows that the GREG estimator produces a shift of the interval for the bias of size

$$R_{\rho,X'\beta_{X,Y}} R_{X'\beta_{X,Y},Y}. \tag{9.14}$$

This is exactly the center of the bias interval when $Z = X'\beta_{X,Y}$, namely when Z is the slope predictor for sample units that carry score X on the auxiliary variables. This observation implies that the GREG estimator moves the bias interval to zero without affecting the width of the interval itself. If $Z = X'\beta_{X,Y}$, then (9.14) changes to

$$\frac{S_\rho S_Y}{\bar{\rho}}\left(-\sqrt{1-R^2_{\rho,Z}}\sqrt{1-R^2_{Z,Y}}, \sqrt{1-R^2_{\rho,Z}}\sqrt{1-R^2_{Z,Y}}\right). \tag{9.15}$$

The maximal absolute bias of the regression estimator can be derived from (9.15). It equals two times the width of (9.15):

$$2\frac{S_\rho S_Y}{\bar{\rho}}\sqrt{1-R^2_{\rho,Z}}\sqrt{1-R^2_{Z,Y}}. \tag{9.16}$$

Schouten (2007) suggests that expression (9.16) be used as the basis for selecting auxiliary variables. Then a vector X is sought that minimizes (9.16).

Since $2(S_\rho S_Y/\bar{\rho})$ is independent of the choice of auxiliary variables, it suffices to minimize

$$W(X) = \sqrt{1-R^2_{\rho,X^T\beta_{X,Y}}}\sqrt{1-R^2_{X^T\beta_{X,Y},Y}}. \tag{9.17}$$

Criterion (9.17) can, however, not be computed as it contains unknown population parameters, and needs to be estimated using the response to the survey. The correlation between response probabilities and the linear predictor $Z = X'\beta_{X,Y}$ is estimated by the sample correlation between estimated response probabilities and estimated predictor $\hat{Z} = X'\hat{\beta}_{X,Y}$. The correlation between survey target variable Y and the linear predictor $Z = X'\beta_{X,Y}$ is estimated by the response correlation between Y and estimated predictor $\hat{Z} = X'\hat{\beta}_{X,Y}$. This results in

$$\hat{W}(X) = \sqrt{1-\hat{R}^2_{\hat{\rho},X'\hat{\beta}_{X,Y}}}\sqrt{1-\hat{R}^2_{X'\hat{\beta}_{X,Y},Y}}. \tag{9.18}$$

The estimated maximal bias width (9.18) is used to determine the weighting model. The proposed selection strategy starts with a simple model having only one weighting variable, namely the variable that minimizes (9.18). In subsequent steps iteratively variables are added in forward steps and removed in backward steps. Variables are only added or removed when the values of (9.18) change significantly for some prescribed significance level. Usual levels are 1% and 5%. Details are given in Schouten (2007).

EXAMPLE 9.10 Step-Wise Forward-Backward Selection Using $\hat{W}$

The General Population Survey (GPS) contains many auxiliary variables. One of the auxiliary variables is *Social allowance*. The variable is a 0–1 indicator for receiving a form of social allowance in one of the months of the reference year. The selection of auxiliary variables is illustrated by treating *Social allowance* artificially as a survey target variable. The values of the indicator for nonrespondents are set to zero.

The following table shows the forward and backward steps for *Social allowance*. The GREG estimate, the estimated correlations with nonresponse and *Social allowance*, and the selection criterion $\hat{W}$ are given. *Age* and *Marital status* are crossed because of their collinearity.

Model	$\bar{y}_{GREG}$	$\hat{R}_{X'\hat{\beta}_{X,Y},Y}$	$\hat{R}_{\hat{\rho},X'\hat{\beta}_{X,Y}}$	$\hat{W}(X)$
Empty	0.105	—	—	1
Age × Marital status	0.109	0.34	−0.05	0.940
Age × Marital status + House value	0.112	0.36	−0.08	0.930
Age × Marital status + House value + Phone	0.114	0.37	−0.11	0.925

Three variables were selected, *Age* × *Marital status*, the average house value, and the 0–1 indicator for the availability of fixed landline phone number. $\hat{W}$ drops from 1 to 0.925 when the three variables are added. None of the backward steps led to a removal of an auxiliary variable. The estimated correlation with the survey target variable is 0.37, which is reasonable. The estimated correlation with nonresponse is −0.11. The estimate for the proportion of persons with a social allowance moves from 10.5% to 11.4%.

The sample mean is 12.1%. Hence the nonresponse adjustment did lead to a smaller nonresponse bias but was not able to remove all of it.

Schouten and De Nooij (2005) present an alternative selection algorithm based on a tree method. The tree method repeatedly splits the population into subpopulations, by which it adds interactions between auxiliary variables. The interactions selected first lead to the largest drops in $\hat{W}$. The selection of new splits is stopped when none of the interactions lead to a significant decrease in $\hat{W}$ at some specified significance level. The algorithm is tree-shaped as each split of a subpopulation results in two new subpopulations that are candidates for successive splits employing the remaining auxiliary variables.

The tree selection algorithm leads to a stratification of the population based on the end nodes or leaves. Consequently the algorithm always leads to a poststratification estimator, whereas the forward-backward selection algorithm leads to a GREG estimator in which many of the interactions are omitted. However, the stratification that results from the tree selection algorithm may have an exotic form compared to the more traditional poststratification. Two strata may have only one split in common if they share only one variable in their derivation.

EXAMPLE 9.11 A Tree Selection Algorithm Based on $\hat{W}$

The auxiliary variables search using the tree method is again done for the GPS. The target variable is social allowance. The resulting tree is depicted below.

The tree algorithm selects splits based on age, employment status, gender, marital status, house value, and ethnicity. The strongest decrease of the maximal bias criterion is produced by age group 55 to 64 years (node 1 to nodes 2 and 3). The population of persons younger than 55 or older than 64 years is further split based on age. Persons with an age in the range 20 to 54 years lead to the second strongest decrease (node 3 to nodes 4 and 5).

After k splits the tree consists of $2k + 1$ nodes of which k nodes are end modes, also called leaves. From the figure it can be deduced that the age group of 55 to 64 years is split in the sixth step of the algorithm based on employment status. The selection stops after 16 splits, thus leading to 16 nonresponse adjustment strata.

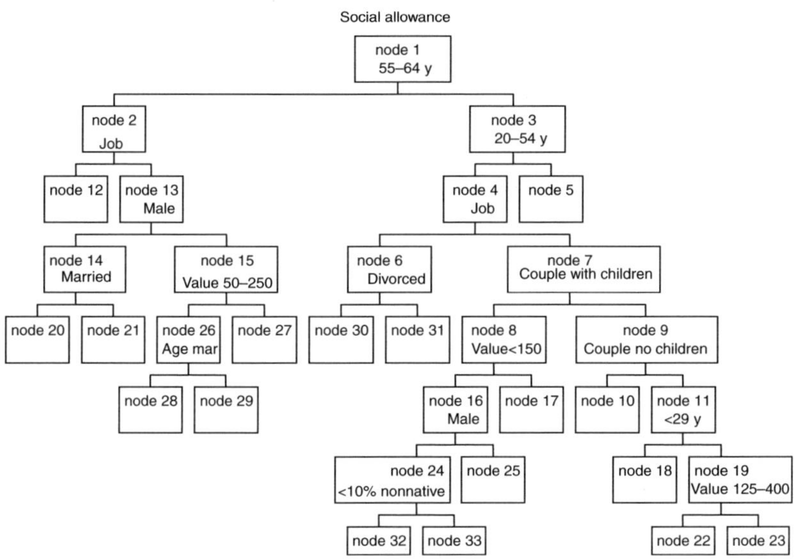

9.3 Application

In this section the variable selection strategies of Section 9.2 are applied to the General Population Survey (GPS). Table 9.1 lists the auxiliary variables that are available in the GPS and that are linked from the Social Statistics Database (SSD) of Statistics Netherlands.

In the GPS no specific auxiliary variables were added to the questionnaire nor were auxiliary variables observed as paradata. This choice was made because of the wide range of survey target variables.

9.3.1 MODELING NONRESPONSE

In Chapter 5 the nonresponse to the GPS was modeled using a logistic regression. Logistic regression models are widely used to explain 0–1 dependent variables, like

TABLE 9.1 Available sociodemographic and socioeconomic variables

Variable	Description
Region	Region of the country
Urban	Degree of urbanization
Phone	Has listed phone number
Pnonnat1	Percentage nonnatives in neighborhood
Pnonnat2	Percentage nonwestern nonnatives in neighborhood
Houseval	Average house value in neighborhood
Ethnic	Type of nonnative
HHType	Type of household
HHSize	Size of the household
Marstat	Marital status
Married	Is married
Nonnativ	Is nonnative
Socall	Has social allowance
Age13	Age in 13 age groups
Allowan	Has an allowance
Children	Children in household
Hasjob	Has a job
Age3	Age in 3 age groups
Disaball	Has disability allowance
Gender	Gender
Unempall	Has unemployment allowance

the 0–1 indicator for response, with both qualitative and quantitative independent variables. For the GPS the list of independent variables is given in Table 9.1.

It is important to observe that some of the auxiliary variables in Table 9.1 are strongly collinear. Examples are the two percentages of nonnatives (*Pnonnat1* and *Pnonnat2*), household size and household type (*HHType* and *HHSize*), age and marital status (*Age13* and *Marstat*), and the various indicators for allowances and employment (*Socall, Disaball, Allowan, Unempall, Hasjob*). Consequently the classifications of the variables and the simultaneous selection of the collinear variables must be carefully done.

Parameter estimates and model fit diagnostics for the logistic regression are fully discussed in Chapter 4, Section 4.3.2.2. This section therefore refrains from the details. The model selection was done by a forward search using Wald's chi-square statistics (Agresti, 2002). The resulting selection is given in Table 9.2. Since the GPS has equal inclusion probabilities, it was not needed to account for the design weights.

In total, 12 variables were selected. For simplicity, interactions were not included in the search for variables. However, the model may be refined by treating a number of the interactions as potential auxiliary variables for the nonresponse

TABLE 9.2 Variable selection for nonresponse

Variable	Wald χ^2											
Region	817	191	159	174	163	160	159	160	163	164	164	165
Urban		90	50	22	20	16	16	15	15	16	16	16
Phone			416	344	275	252	251	244	234	238	242	242
Houseval				96	74	37	34	31	27	25	26	25
Ethnic					70	84	97	108	92	91	96	94
Hhtype						117	31	16	17	15	14	24
Hhsize							53	52	53	53	52	53
Marstat								50	51	65	77	74
Socall									15	8	9	9
Hasjob										29	18	24
Age3											13	11
Gender												15
pseudo R^2	1.9%	2.2%	3.1%	3.3%	3.5%	3.8%	3.9%	4.0%	4.0%	4.1%	4.1%	4.2%
χ^2	843	933	1348	1444	1515	1631	1684	1734	1748	1778	1791	1806
df	4	8	9	20	24	28	32	35	36	37	39	40

adjustment. Note that Table 9.2 contains relatively few collinear variables. Exceptions are household size and household type, age and marital status, and job and social allowance. Variable *Age3* has only three age categories, thus diminishing the collinearity with marital status. When household size is added, the Wald χ^2 for household type drops considerably. The same happens to social allowance when job status is added, though to a smaller extent.

In general, the Wald χ^2 provides a way to investigating the roles of various auxiliary variables. An interesting finding is the contribution of region. As degrees of urbanization are added to the model, they largely displace the regional contributions. So the two variables are related. While some regions are more urbanized than others, the impact is stronger than expected. Apparently the urbanization of the region has a big effect on response behavior. Another interesting finding is the drop in contributions of those having a registered phone number when ethnicity is selected. It is known from the literature that telephone registration is much lower for nonnative households. The impact of having a phone number on nonresponse is to some extent dependent on the ethnicity.

From Table 9.2 it is clear that a forward search using the auxiliary variables as they are may potentially lead to a loss of efficiency. Auxiliary variables that carry similar explanations for nonresponse are selected simultaneously, leading to a relatively large number of parameters and adjustment strata. In Section 9.2 it was argued that more efficient methods exist, the classification and regression trees.

In SPSS 14.0 a CHAID classification tree was made with the auxiliary variables in Table 9.1 as explanatory variables and the 0–1 response indicator as the variable to be classified. The classification was restricted to three levels, so any of the

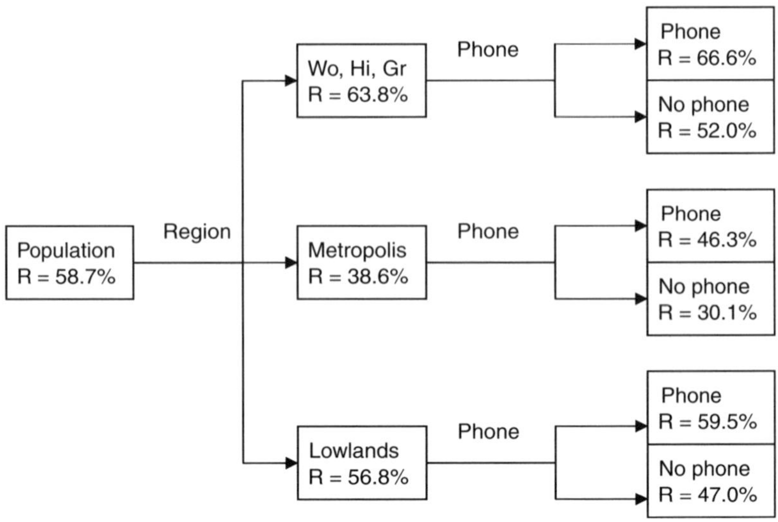

FIGURE 9.2 First two levels of the CHAID classification tree for response. For each node the precentage of response is given as well as the variables responsible for the splits.

resulting classification tree end nodes follow from the interaction of at most three variables. Figure 9.2 depicts the first two levels as they are produced by SPSS. In total 15 end nodes or leaves are created. The first split of the population was again made based on region. However, three of the regions—Woodlands, Highlands, and Greenfields—were aggregated into one node, while two—Metropolis and Lowlands—were separated. The second level splits of the tree were all made with the 0–1 indicator for a registered phone number. On the same level, splits can be made with different variables and within variables based on different clusters of categories, but in all of the clusters having a phone is the strongest auxiliary variable. Thus, as for the logistic regression model, this indicator turns out to be a strong predictor of response. It can also be concluded that the impact of the degree of urbanization was implicitly incorporated in the model through the aggregation of the three region categories. The variables that were selected in the third level are household size, the degree of urbanization, marital status, allowance, and ethnicity. The value of the house is the only variable that appears in the forward selection of the logistic regression model and that is not used a split variable.

In the subsequent sections the stratification induced by the CHAID classification tree is added as a potential auxiliary variable for nonresponse adjustment along with the Table 9.1 variables. This variable is qualitative and has 15 categories corresponding to each of the end nodes. It is referred to as CHAID.

9.3.2 MODELING SURVEY TARGET VARIABLES

Two target variables are chosen from the GPS survey questionnaire: owner of a house (*OwnHouse*) and owner of a PC (or laptop). Both variables are 0–1 indicators where a value one reflects that the person owns a house or PC, respectively. Since the two target variables are 0–1 indicators, again the obvious model is a logistic regression. Tables 9.3 and 9.4 contain the selections of auxiliary variable based on a forward search using the Wald χ^2. As for the logistic regression model for nonresponse, no interactions were included in the selection.

A look at both selections in Tables 9.3 and 9.4 immediately reveals that the number of selected auxiliary variables is considerable. In fact there are only few variables that were not selected. However, only a small number of variables contribute strongly to the explanation of the target variables. Nagelkerke's pseudo R^2 reflects this impact. The increase in R^2 is by far the strongest in the first step and R^2 then slowly converges. The implication is that already with a few auxiliary variables a reasonable fit can be obtained. Both final values of R^2 of 38.1% and 43.5%, are high for qualitative variables and indicate a good fitting model, respectively. For ownership of a PC, the age of the sampled person, the average house value in the neighborhood and the type of household provide good explanations. Age alone leads to a R^2 of 28.1%. The selected variables are as expected. Households with computers are especially found under younger people, households with children and households with a higher income. For ownership of a house it is the average house value of the neighborhood, the type of household, the percentage of nonnative people in the neighborhood, and the age that contribute most. Not surprisingly, the biggest impact follows from the house value. In general, rental houses are cheaper than owner-occupied houses. Furthermore older households and households with children more often own houses. Districts with a relatively high percentage of nonnative persons more often have more rental houses.

The two target variables were also predicted using the CHAID classification tree method (figures not shown). A maximum of three levels was allowed. As for the logistic regression models, the number of selected variables is large. In total, respectively, 87 and 84 end nodes were selected for ownership of a PC and ownership of a house. The trees thus imply a stratification of the population with more than 80 strata or subpopulations. In the classification tree for PC, the first level follows from a split on age categories. Hence the same variable is selected in the logistic regression and in CHAID. In the second level, the variables allowance, ethnicity, average value of the house, degree of urbanization, and marital status are the split variables. Contrary to the classification tree for nonresponse, the second level splits are based on a variety of auxiliary variables. In the third level (not shown) the auxiliary variables age, household size, number of children, job status, and ethnicity are used. In the classification tree for ownership of a house, the first split is made with the average house value. It can be concluded that, again, the two methods agree. In the second level the percentage of nonnative persons in the neighborhood, the type of household, and the 0–1 indicators for married and phone are the inputs to further splits of the average house value strata. In the third level job status, allowance, the number of children, and the region are selected.

From the findings of the two multivariate methods, logistic regression, and CHAID, it can be concluded that there is a consensus which variables provide the strongest explanation for the two target variables.

TABLE 9.3 Variable selection for ownership of a PC or laptop

Variable	Wald χ^2														
Age13	2962	2936	1970	1997	2021	2002	1890	1894	1921	1923	1316	1314	1319	1121	1121
Houseval		698	460	376	323	382	368	367	342	303	302	308	303	302	281
Hhtype			346	364	314	360	40	35	31	27	27	26	21	16	15
Ethnic				176	153	183	198	199	165	118	115	114	109	106	97
Allowan					142	142	139	143	132	127	93	95	55	55	55
Urban						116	125	128	140	138	135	85	85	84	85
Hhsize							81	83	82	84	89	87	88	90	91
Gender								58	57	58	40	41	39	33	33
Phone									37	34	34	35	33	32	31
Pnonnat1										42	42	40	39	39	20
Hasjob											21	21	19	19	19
Region												22	22	21	23
Socall													11	10	10
Marstat														13	13
Pnonnat2															13
pseudo R^2	28.3%	32.4%	34.2%	35.2%	36.0%	36.6%	37.0%	37.3%	37.5%	37.7%	37.8%	37.9%	38.0%	38.1%	38.1%
χ^2	4444	5182	5527	5713	5854	5973	6054	6112	6148	6194	6216	6238	6249	6262	6275
df	12	23	27	31	32	36	40	41	42	49	50	54	55	58	64

TABLE 9.4 Variable selection for ownership of a house

Variable	Wald χ^2													
Houseval	3412	2762	1929	1879	1780	1880	1822	1750	1732	1719	1715	1716	1668	1659
Hhtype		894	880	507	415	383	360	340	329	124	123	122	122	46
Pnonnat2			680	722	670	487	450	127	126	128	106	108	106	105
Age13				326	355	355	379	391	413	401	402	263	262	251
Socall					149	156	147	145	94	87	81	77	76	77
Region							174	146	150	147	147	151	123	123
Phone							62	54	50	48	40	39	40	38
Pnonnat1								60	59	56	59	58	54	53
Allowan									33	29	28	18	18	17
Marstat										35	39	44	43	42
Ethnic											40	39	39	40
Hasjob												22	22	23
Urban													17	17
Hhsize														14
pseudo R^2	29.2%	34.4%	38.3%	40.0%	41.1%	42.1%	42.4%	42.7%	42.8%	43.0%	43.2%	43.3%	43.4%	43.5%
χ^2	4528	5457	6190	6522	6746	6935	6997	7057	7089	7124	7167	7189	7206	7220
df	11	15	21	33	34	38	39	46	47	50	54	55	59	63

9.3.3 COMBINING MODELS FOR NONRESPONSE AND TARGET VARIABLES

In the previous two sections auxiliary variables were selected for the 0–1 response indicator and for two target variables. In order not to overload the nonresponse adjustment, from each of the models the first five auxiliary variables are taken. This results in the following lists:

1. Nonresponse: *Region, Urban, Phone, Houseval, Ethnic*
2. PC: *Age13, Houseval, HHType, Ethnic, Allowan*
3. House: *Houseval, HHType, Pnonnat2, Age13, Socall*

The only auxiliary variable that appears in each of the selections is the average value of the house in the neighborhood. The variable ethnicity is a predictor of response and of ownership of a PC, and the variables age and type of household only appear in the models for the two target variables. All other variables appear only once in the top five selected variables. The union of the three sets leads to the following selection of auxiliary variables: *Region, Urban, Phone, Houseval, Ethnic, Age13, HHType, Allowan, Pnonnat2* and *Socall.*

Since receiving some form of allowance and receiving a social allowance are strongly collinear, it is deemed best to choose just one. For this reason the 0–1 indicator for social allowance is omitted, leading to a nonresponse adjustment based on nine auxiliary variables. Furthermore it needs to be decided whether the full classifications of region and degree of urbanization are employed. It seems best to combine the two variables into a new aggregated variable. The recoding of the two auxiliary variables may be combined with the final step of the variable selection strategy, the diagnostics for the nonresponse adjustment weights. Large or negative weights indicate that nonresponse adjustment strata or cells need to be collapsed.

Only two target variables were explored here, as the range of target variables in the GPS is wide. If the target variables are ignored and the selection of auxiliary variables is fully based on the explanation of nonresponse, then the selection shows a somewhat different picture. The first, say 10, variables in the model for nonresponse may be selected for nonresponse adjustment. This would lead to *Region, Urban, Phone, Houseval, Ethnic, Hhtype, Hhsize, Marstat, Socall, Hasjob.*

Hence, when the selection is to be completely based on the relation to nonresponse, the auxiliary variables age, and percentage of nonnative persons in the neighborhood may not be selected. Age, however, is the strongest predictor for ownership of a PC. It thus makes sense to include at least a few of the important target variables.

9.3.4 SELECTION BASED ON VARIANCE OF CALIBRATION WEIGHTS

In Section 9.2.2 Särndal and Lundström's variable selection criterion q^2 was described. In this section its use is illustrated for the GPS.

TABLE 9.5 Selection criterion for GPS auxiliary variables

Variable	q^2 (×1000)
CHAID	144
Region	75
Urban	58
Phone	56
Pnonnat1	51
Pnonnat2	49
Houseval	32
HHType	26
HHSize	22
Marstat	20
Married	20
Nonnativ	18
Socall	17
Allowan	8
Age13	8
Children	6
Hasjob	3
Disaball	1
Gender	0
Unempall	0

Table 9.5 lists the values of q^2 for all available auxiliary variables ordered from high to low. A high value is preferred, so the variable with the highest q^2 is selected first. In Table 9.5 the value for the CHAID tree variable of Section 9.3.1 is also evaluated. This variable is a composite auxiliary variable with 15 categories that follow from splits with region, phone, household size, the degree of urbanization, marital status, allowance, and ethnicity.

The CHAID variable, not surprisingly, turns out to have the highest selection criterion. Hence with only 15 categories a value can be obtained that is at least twice as large as any other single auxiliary variable. The single auxiliary variables that score best are region, urbanization, phone, and the percentage of nonnatives. This is the same as in the previous sections.

Subsequent auxiliary variables are selected by concatenation with already selected auxiliary variables. As such, interaction effects are ignored and essentially a regression with main effects is the result. Tables 9.6 and 9.7 show the selected variables when the CHAID tree variable is added and when this variable is left out of the selection. When the CHAID variable is added, ultimately six variables are chosen, and q^2 increases from 144 to 166. The stratification induced by the CHAID tree is mostly based on region and landline phone, so these variables no longer contribute here. However, still some variance in calibration weights can be

TABLE 9.6 Selection criterion for GPS auxiliary variables with variable CHAID

Variable	q^2 ($\times 1000$)
CHAID	144
CHAID + Hhtype	151
CHAID + Hhtype + Pnonnat2	156
CHAID + Hhtype + Pnonnat2 + Age13	160
CHAID + Hhtype + Pnonnat2 + Age13 + Marstat	164
CHAID + Hhtype + Pnonnat2 + Age13 + Marstat + Houseval	166

found in some other variables. When the CHAID variable is not added, more variables are selected. After region eight more variables are chosen, and q^2 grows from 75 to 149. Hence, after the selection is stopped, q^2 is much bigger and comparable to the selection with the CHAID stratification alone. The CHAID stratification is efficient as it allows for interactions among auxiliary variables.

9.3.5 SELECTION BASED ON WORST-CASE NONRESPONSE BIAS

In this section both the forward-backward selection algorithm and the tree selection algorithm of Section 9.2.3 are applied to the GPS. The two target variables selected in Section 9.3.2 are again used for the illustration: owner of a house (*OwnHouse*) and owner of a PC (or laptop). Both variables are 0–1 indicators where a value one reflects that the person owns a house or PC. In the selection all auxiliary variables of Table 9.1 are candidates. Tables 9.8 and 9.9 give the results of the forward-backward selection algorithm for both target variables.

Both selections lead to five forward steps, and no auxiliary variables are removed. For ownership of a house, the strongest auxiliary variable is the average

TABLE 9.7 Selection criterion for GPS auxiliary variables without variable CHAID

Variable	q^2 ($\times 1000$)
Region	75
Region + Phone	114
Region + Phone + Pnonnat2	126
Region + Phone + Pnonnat2 + Hhtype	134
Region + Phone + Pnonnat2 + Hhtype + Age13	139
Region + Phone + Pnonnat2 + Hhtype + Age13 + Marstat	143
Region + Phone + Pnonnat2 + Hhtype + Age13 + Marstat + Nonnativ	146
Region + Phone + Pnonnat2 + Hhtype + Age13 + Marstat + Nonnativ + Houseval	148
Regionv + Phone + Pnonnat2 + Hhtype + Age13 + Marstat + Nonnativ + Houseval + Allowan	149

TABLE 9.8 Results of the selection strategy for ownership of a house

Step	Model	$\bar{y}_{GREG}$	$\hat{R}_{X'\hat{\beta}_{X,Y},Y}$	$\hat{R}_{\hat{\rho},X'\hat{\beta}_{X,Y}}$	$\hat{W}(X)$
0	Empty	0.633	—	—	1
1	Houseval	0.612	0.47	0.11	0.875
2	Houseval + HHtype	0.605	0.52	0.13	0.849
3	Houseval + Hhtype + Pnonnat1	0.598	0.54	0.15	0.829
4	Houseval + Hhtype + Pnonnat1 + Region	0.594	0.56	0.16	0.820
5	Houseval + Hhtype + Pnonnat1 + Region + Age13	0.594	0.57	0.16	0.813

house value in the neighborhood. This selection is not surprising as the auxiliary variable relates strongly to house ownership. The higher the house value, the higher is the probability that a house is owned. The correlation between the linear predictor based on house value and ownership of a house is equal to 0.47, which is high for qualitative variables. Surprisingly, average house value also turns up in the selection for ownership of a PC, but as the second strongest variable. For ownership of a PC, age is by far the strongest auxiliary variable. The correlation between the linear predictor based on age and ownership of a PC is 0.50, which is even higher than the correlation between house value and house ownership. After the inclusion of the other four variables, the correlations between the linear predictor and the two target variables increase to 0.57 and 0.54, respectively. In both selections the resulting correlation with nonresponse is, however, relatively modest, 0.16 and 0.12, respectively.

Apart from ownership of a PC and ownership of a house, the forward-backward selection was applied to three other GPS target variables. The resulting selections are listed in Table 9.10. In all cases the number of selected auxiliary variables is small, between three and five. Remarkable is the small model of only three variables for target variable *Active in sports club*. Another interesting finding is that the models are quite different. For instance, for religion and educational level the two selections of auxiliary variables are disjoint. This implies that when the nonresponse adjustment is focused on single target variables, completely different

TABLE 9.9 Results of the selection strategy for ownership of a PC

Step	Model	$\bar{y}_{GREG}$	$\hat{R}_{X'\hat{\beta}_{X,Y},Y}$	$\hat{R}_{\hat{\rho},X'\hat{\beta}_{X,Y}}$	$\hat{W}(X)$
0	Empty	0.598	—	—	1
1	Age13	0.585	0.50	0.06	0.866
2	Age13 + Houseval	0.579	0.52	0.09	0.850
3	Age13 + Houseval + Ethnic	0.574	0.53	0.11	0.842
4	Age13 + Houseval + Ethnic + HHsize	0.573	0.54	0.11	0.837
5	Age13 + Houseval + Ethnic + Hhsize + Socall	0.572	0.54	0.12	0.835

TABLE 9.10 Variable selections for GPS target variables

Variable	Model
Employment	Age13 + Ethnic + Phone + Houseval + Gender
Owner of a house	Houseval + HHType + Nonnativ + Region + Age13
Active in a sports club	Age13 + Nonnativ + Phone
Owner of a PC	Age13 + Houseval + Ethnic + HHSize + Socall
Educational level	Age13 + Nonnativ + Region + Houseval
Religion	Ethnic + Region + Phone + Hasjob

choices can be made. A single nonresponse adjustment would have to comprise of a large set of auxiliary variables in order to achieve the same effect but may lead to larger standard errors.

Next the selection criterion $\hat{W}$ was optimized for ownership of a PC using the classification tree search of Section 9.2.3. Figure 9.3 shows the resulting tree, and Table 9.11 lists the various iterations in the tree search. In total 17 splits were made, resulting in a stratification of the population using 17 end nodes or leaves. The splits in the tree are based on age, household size, average house value, allowance, ethnicity, and the indicator for a nonnative person. The strongest split is on age; persons older and younger than 59 years are separated. The second split is on household size, dividing persons younger than 59 years in households with four or more members from households with less than four members.

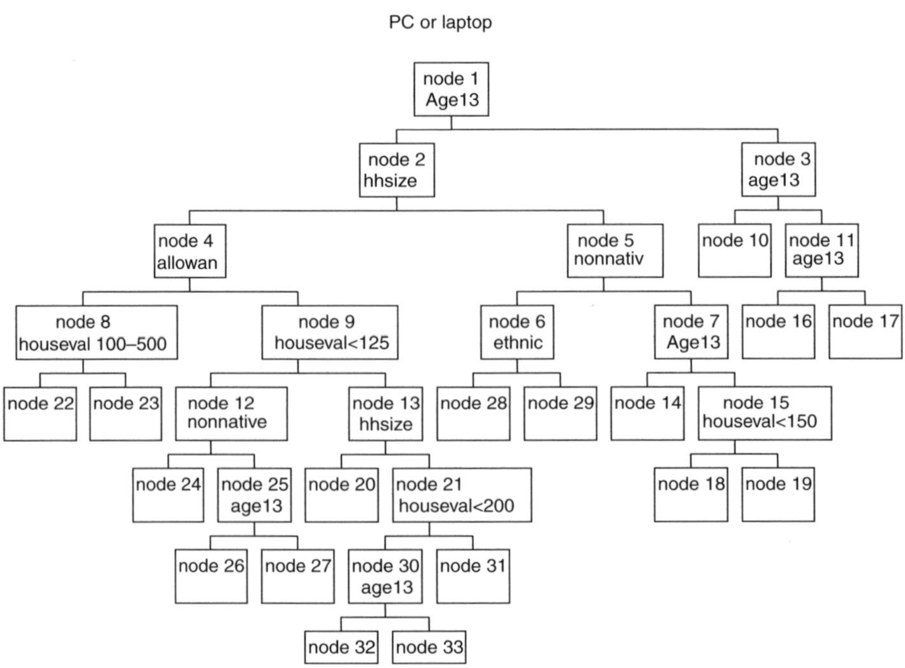

FIGURE 9.3 Classification tree for target variable ownership of a PC.

TABLE 9.11 Selection criterion $\hat{W}$ for the first 17 iterations of the tree selection algorithm applied to target variable PC

Iteration	$\hat{W}$	$\Delta\hat{W}$	Standard error	p-value
1	0.879	0.121	0.0045	26.77
2	0.862	0.017	0.0013	13.43
3	0.854	0.008	0.0010	8.46
4	0.847	0.006	0.0010	6.73
5	0.841	0.006	0.0012	5.17
6	0.838	0.003	0.0008	3.91
7	0.835	0.003	0.0005	5.03
8	0.833	0.002	0.0006	4.23
9	0.831	0.002	0.0006	3.53
10	0.829	0.002	0.0004	4.30
11	0.828	0.001	0.0005	2.77
12	0.826	0.001	0.0003	4.84
13	0.825	0.001	0.0004	3.59
14	0.823	0.001	0.0004	3.07
15	0.822	0.001	0.0005	2.53
16	0.821	0.001	0.0003	2.79
17	0.820	0.001	0.0004	2.13

Note: Also given are the change in $\hat{W}$, $\Delta\hat{W}$, its standard error and the standardized number of standard deviations under a normal approximation.

Table 9.11 gives the selection criterion $\hat{W}$, the difference in $\hat{W}$ when a variable is added, and the corresponding standard error and standardized difference in $\hat{W}$. The standard error is approximated by resampling the sample using a naive Bootstrap. The standardized difference is evaluated assuming a normal distribution; in other words, if the standardize difference exceeds 1.95, then the difference is significant at the 5% level.

From Table 9.11 we learn that it takes only a few splits to have a considerable decrease in the selection criterion $\hat{W}$. After 5 splits, the additional decrease is modest but still significant at the 5% level. After 18 splits, the standardize difference is smaller than 1.96. The value of $\hat{W}$ is 0.820 and is smaller than the value of 0.835 that followed from the forward-backward algorithm. The classification tree allows for more detail and refinement in the selections because interactions between variables are accounted for.

$\hat{W}$ can also be assessed for the CHAID stratification variable of Section 9.3.1. Table 9.12 gives the values for the target variable ownership of a PC The CHAID variable has 15 categories and its construction is purely based on the prediction of nonresponse. For ownership of a PC, the resulting value of $\hat{W}$ is larger than the values based on the selections using either the forward-backward or the classification tree search. The results indicate that a selection based on nonresponse alone leads to a smaller decrease in the selection criterion. This is not surprising as the correlation to the target variable is ignored.

TABLE 9.12 Selection criterion and underlying correlations for the CHAID stratification variable

Variable	$\hat{R}_{X'\hat{\beta}_{X,Y},Y}$	$\hat{R}_{\hat{\rho},X'\hat{\beta}_{X,Y}}$	$\hat{W}(X)$
Ownership of a PC	0.498	0.172	0.855

9.3.6 A COMPARISION OF THE VARIOUS SELECTION STRATEGIES

Table 9.13 provides an overview of the various selections that were made previously. Selections for all target variables occur simultaneously in the first three methods—the logistic regression, the selection criterion q^2, and the combination of logistic regressions. The last two methods give variable-specific selections. In general, the specific selections lead to parsimonious models and to fewer numbers of auxiliary variables in the nonresponse adjustment.

9.4 Summary

This chapter discusses the selection of auxiliary variables for nonresponse adjustment. To have an effective nonresponse bias correction, the variables selected must be informative. A general selection strategy is described. First, auxiliary variables need to be identified from public registers and administrative data, or from the survey questionnaire and paradata. Second, a selection needs to be made either by modeling nonresponse and survey target variables separately or by performing a single selection based on a specified criterion. Then the selection is adapted based on diagnostics of adjustment weights and on classifications of variables that are used in publications.

TABLE 9.13 Summary of selected variables for the various selection methods

Selection method	Ownership of a PC	Ownership of a house
Logistic regression nonresponse	Region, Urban, Phone, Houseval, Ethnic, Hhtype, HHsize, Marstat, Socall, Hasjob	Region, Urban, Phone, Houseval, Ethnic, HHType, HHSize, Marstat, Socall, Hasjob
q^2	Region, Phone, Pnonnat2, HHType, Age13, Marstat, Nonnativ, Houseval, Allowan	Region, Phone, Pnonnat2, HHType, Age13, Marstat, Nonnativ, Houseval, Allowan
Combining logistic regressions—overall	Region, Urban, Phone, Houseval, Ethnic, Age13, HHType, Allowan, Pnonnat2, Socall	Region, Urban, Phone, Houseval, Ethnic, Age13, HHType, Allowan, Pnonnat2, Socall
Combining logistic regressions—vari able specific	Region, Urban, Phone, Houseval, Ethnic, Age13, HHType, Allowan	Region, Urban, Phone, Houseval, Ethnic, HHType, Pnonnat2, Age13, Socall
W forward-backward	Age13, Houseval, Ethnic, HHSize, Socall	Houseval, HHType, Nonnativ, Region, Age13

The variable selection problem includes identification of auxiliary variables that relate both to nonresponse and to key target variables of the survey. The convention is to use weighting methods to reduce variance and to decrease standard errors of survey statistics. For this reason auxiliary variables need to be good predictors of survey target variables only. However, for nonresponse bias reduction, variables need also to be capable of predicting nonresponse.

Auxiliary variables can, in some settings and for some survey organizations, be linked to the survey sample from registered data. While this allows for modeling response directly, the nature of auxiliary variables can be restrictive and more informative variables are needed. Alternatively, the variables could be added either to the questionnaire as questions or to paradata as observations. Paradata consist of observations made by interviewers and the data collection staff. Because use of observations rather than questions limits the applicability of paradata, this remains a topic of interest in research on nonresponse analysis and nonresponse adjustment.

The selection of variables is an empirical process that calls for considerable expertise and simple heuristics are often employed. Recently, two criteria were developed that structure and optimize the selection process. Simple heuristics are useful as are models for nonresponse and key survey variables from major publications that can be combined. The two criteria, the variance of calibration weights and the worst-case bias, allow for overall selection as well as variable-specific selections. Variable-specific selections call for each survey variable to be adjusted separately based on smaller and more effective models. As a consequence there is a reduced risk of large standard errors. Nevertheless, nonresponse adjustment cannot be done with a single set of weights.

9.5 KEY TERMS

Aggregated population level auxiliary variables: Auxiliary variables that are available for respondents and in aggregated form (sums, counts, means) for the population as a whole.

Classification tree: Data-mining techniques that explain qualitative variables following a tree-shaped selection of auxiliary variables.

Forward-backward selection: Multiple selections done according to some criterion. Every iteration consists of two steps. In the first step, the strongest variable is added if significant; in the second step, the weakest variable is removed when nonsignificant.

Paradata: Auxiliary variables, available at the sample level, that describe the data collection process and consist of observations made by interviewers, agents, fieldwork supervisors or data collection management.

Population level auxiliary variables: Auxiliary variables that are available for all population units, including respondents and nonrespondents.

Regression tree: Data-mining techniques that explain quantitative variables following a tree-shaped selection of auxiliary variables.

Sample level auxiliary variables: Auxiliary variables that are available for all sample units.

Selection criterion: An objective function that identifies which auxiliary variables are preferred to others.

Weighting variables: Auxiliary variables used to produce individual adjustment weights.

9.6 REFERENCES

Agresti, A. (2002), *Categorical Data Analysis.* Wiley, New York.

Beaumont, J.F. (2005), On the Use of Data Collection Process Information for the Treatment of Unit Nonresponse through Weight Adjustment. *Survey Methodology* 31, pp. 227–231.

Breiman, L., Friedman, J.H., Olshen, R.A., and Stone, C.J. (1984), *Classification and Regression Trees.* Chapmann and Hall, Boca Raton, FL.

Cobben, F. (2009), *Nonresponse in Sample Surveys. Methods for Analysis and Adjustment.* PhD thesis. University of Amsterdam, Amsterdam.

Cochran, W. (1968), The Effectiveness of Adjustment by Subclassification in Removing Bias in Observational Studies. *Biometrics* 24, pp. 205–213.

Deville, J.C., and Särndal, C.E. (1992), Calibration Estimation in Survey Sampling. *Journal of the American Statistical Association* 87, pp. 376–382.

Groves, R.M., and Couper, M.P. (1998), *Nonresponse in Household Interview Surveys.* Wiley, New York.

Groves, R.M., Dillman, D., Eltinge, J., and Little, R. (2002), *Survey Nonresponse.* Wiley, New York.

Hastie, T., Tibshirani, R., and Friedman, J. (2009), *The Elements of Statistical Learning: Data Mining, Inference and Prediction.* Springer, New York.

Kalton, G., and Flores-Cervantes, I. (2003), Weighting Methods. *Journal of Official Statistics* 19, pp. 81–97.

Kass, G.V. (1980), An Explanatory Technique for Investigating Large Quantities of Categorical Data. *Journal of the Royal Statistical Society C, Applied Statistics* 29, pp. 119–127.

Kreuter, F., and Kohler, U. (2009), Analyzing Contact Sequences in Call Record Data. Potential and Limitations of Sequence Indicators for Nonresponse Adjustments in the European Social Survey. *Journal of Official Statistics* 26, pp. 203–226.

Kreuter, F., Olson, K., Wagner, J., Yan, T., Ezzati-Rice, T., Casas-Cordero, C, Lemay, M., Peytchev, A., Groves, R.M., and Raghunathan, T.E. (2010), Using Proxy Measures and other Correlates of Survey Outcomes to Adjust for Non-response: Examples from Multiple Surveys. *Journal of the Royal Statistical Society A* 173, pp. 389–407.

Little, R.J.A. (1986), Survey Nonresponse Adjustments for Estimates of Means. *International Statistical Review* 54, pp. 139–157.

Little, R.J.A., and Rubin, D.B. (2002), *Statistical Analysis with Missing Data.* Wiley, New York.

Little, R.J.A., and Vartivarian, S. (2005), Does Weighting for Nonresponse Increase the Variance of Survey Means? *Survey Methodology* 31, pp. 161–168.

Särndal, C.E., and Lundström, S. (2005), *Estimation in Surveys with Nonresponse*. Wiley, Chichester, UK.

Särndal, C.E., and Lundström, S. (2008), Assessing Auxiliary Vectors for Control of Nonresponse Bias in the Calibration Estimator. *Journal of Official Statistics* 24, pp. 167–191.

Schouten, J.G. (2006), A Selection Strategy for Weighting Variables under a Not-Missing-at-Random Assumption. *Journal of Official Statistics* 23, pp. 51–68.

Schouten, B., and De Nooij, G. (2005), *Nonresponse Adjustments Using Classification Trees.* Discussion paper 05001. Statistics Netherlands, Voorburg.

Schouten, B., Cobben, F., and Bethlehem, J.G. (2009), Indicators for the Representativeness of Survey Response. *Survey Methodology* 35, pp. 101–113.

Stoop, I., (2005), *The Hunt for the Last Respondent. Nonresponse in Sample Surveys.* PhD thesis. University of Utrecht, Utrecht.

Voogt, R. (2204), *I Am Not Interested: Nonresponse Bias, Response Bias and Stimulus Effects in Election Research*, PhD thesis. University of Amsterdam, Amsterdam.

9.7 EXERCISES

Exercise 9.1 Which of the following statements is false when a variable is used for nonresponse adjustment?

(a) The auxiliary variable must relate to nonresponse.

(b) The auxiliary variable should relate to the at least one key target variable of the survey.

(c) The auxiliary variable may define an important domain for users.

(d) The auxiliary variable must relate both to a target variable and to nonresponse.

Exercise 9.2 Why is variable selection a complex task?

(a) There may be many variables that have an equal impact.

(b) A survey contains many target variables and for each variable the optimal selection may be different.

(c) Nonresponse adjustment affects both nonresponse bias and precision.

(d) There are many nonresponse adjustment methods, and they all lead to different selections.

Exercise 9.3 How could one optimize nonresponse adjustment for each target variable?

(a) Include all available auxiliary variables in the adjustment.

(b) Include all auxiliary variables in the adjustment that relate to at least one target variable.

(c) Optimization is not possible.

(d) Perform separate nonresponse adjustments for each target variable.

Exercise 9.4 Consider the following correlations between two 0–1 auxiliary variables, X_1 and X_2, one target variable Y, and nonresponse. Which statement is true?

	Nonresponse	Y
X_1	0.4	0.6
X_2	0.9	0.2

(a) X_2 is to be selected because it relates to Y while X_1 relates only moderately to Y.

(b) X_2 is to be selected because it relates very strongly to nonresponse and also has sufficient correlation with Y.

(c) X_1 leads to the smallest maximal bias interval while X_2 provides the strongest nonresponse adjustment.

(d) X_1 leads to the smallest maximal bias interval and also provides the strongest nonresponse adjustment.

Exercise 9.5 The GPS data collection takes two months. The GPS survey file contains a 0–1 indicator for response after the first month. Select the auxiliary variables that explain nonresponse after the first month. What variables are selected? What differences are observed when the selection is compared to the selection after a full two months?

Exercise 9.6 Compute the selection criterion q^2 for all auxiliary variables after one month of data collection. Which variable is the strongest? And which variable is selected before q^2 changes by less than 1%?

Exercise 9.7 Compute the variable selections for one of the other target variables in Table 9.11: employment, educational level, religion, or active in a sports club.

CHAPTER TEN

Re-approaching Nonrespondents

10.1 Introduction

Nonresponse can cause estimators to be biased. How to assess whether this is indeed the case, taking into account that nonrespondents do not provide information? There are two ways to obtain information on the nonrespondents:

- *Linking data sets*. Sometimes it is possible to link the survey sample to one or more external administrative sources, for instance, a population register. From these registers, data can be retrieved on both respondents and nonrespondents.
- *Re-approaching nonrespondents*. The literature suggests two methods: the *callback approach* (CBA) and the *basic-question approach* (BQA). In the callback approach, a sample is selected from the nonrespondents, and the persons in this sample are re-approached by specially trained interviewers. In the basic-question approach, attempts are made to persuade nonrespondents to cooperate by offering them a much shorter questionnaire containing just a few basic questions.

The *Politz–Simmons approach* is also described in this chapter. This is a way to estimate the probability of a sample person being at home, which can be used to reduce the bias due to a not-at-home nonresponse.

The use of linked data sets for the analysis of nonresponse was covered in Chapter 5. Ways of re-approaching nonrespondents are discussed in this chapter. There are two reasons to conduct a re-approach of nonrespondents. The first is to

Handbook of Nonresponse in Household Surveys, By Jelke Bethlehem, Fannie Cobben, and Barry Schouten

obtain insight into the nature of the nonresponse. The second is to obtain a higher response rate and to collect information that can be used for nonresponse adjustment. An application of a fieldwork experiment that implements both the call-back approach and the basic-question approach is presented as well. Additionally linked data from external sources is used to evaluate the results of the re-approaches.

The linked data set should contain information for both nonrespondents and respondents so that a comparison can be made of respondents and nonrespondents. For example, Statistics Netherlands combines with integrates surveys with the administrative sources to which it has access. The resulting Social Statistical Database (abbreviated to SSB in Dutch) consists of administrative information on persons, households, jobs, benefits, and pensions; see Arts and Hoogteijling (2002) for more details. The population register serves as the backbone of the SSB. Work-related data from various sources are also integrated in the SSB. These data come from the Centers for Work and Income (CWI in Dutch), Tax administration, and the employment agencies (UWV in Dutch).

Uncertainty about the effect of nonresponse, especially concerning the bias of population estimators, is the impetus for constant research on methods for nonresponse reduction and adjustment. The lasting uncertainty about the influence of nonresponse was the motivation for various studies with intensive callbacks of nonrespondents e.g., Elliott et al., 2000; Keeter et al., 2000; Lynn et al., 2002; Stoop, 2001, 2004; Voogt, 2004. The studies aim at a full response in a callback survey of nonrespondents or at least a nonresponse that is missing completely at random. Stoop (2004) describes a study in The Netherlands in which about 70% of the former nonresponding households were converted to respondents.

A *callback*, or *follow-up*, of nonrespondents was first proposed by Hansen and Hurwitz (1946). They suggested investigating nonresponse in mail surveys by taking a sample of nonrespondents and trying to obtain the required information by means of a face-to-face interview. This became known as the callback approach (CBA). The aim of the CBA is to obtain as high a response rate as possible. In the CBA one may choose to use a different data collection mode, specially trained interviewers, an extended fieldwork period and/or incentives to achieve cooperation from nonrespondents.

Therefore the CBA is a rather expensive method to obtain information. Because sampling and re-approaching nonrespondents may considerably lengthen the fieldwork time, an alternative is the *basic-question approach* (BQA). This is a good method to apply when time or budget constraints arise and the CBA is not an option. The BQA can be performed as a regular part of the data collection, whereas the CBA is usually performed as a once-only experiment.

The BQA is designed on the assumption that many survey questionnaires are composed around a few basic questions. From answers to these questions the most important conclusion of the survey can be formulated. The procedure was proposed by Kersten and Bethlehem (1984), who observed that persons who refused to participate in a survey often could be persuaded to answer just a few basic questions. The main goal of the BQA is to gain insight in possible differences between respondents and nonrespondents with respect to the most important variables of the survey, as represented by the basic questions. If such differences are detected, the approach also provides information for adjusting estimates for other variables.

EXAMPLE 10.1 An Application of the BQA and the CBA

Voogt (2004) presents an example of a survey in which both the callback approach and the basic-question approach were applied. His research focused on nonresponse bias in election research. He selected a simple random sample of 995 voters from the election register of the town of Zaanstad in The Netherlands. There were two basic questions in this survey:

- Did you vote in the parliamentary election on Wednesday May 6, 1998?
- Are you interested in politics, fairly interested, or not interested?

In the first wave of the survey, people were contacted by telephone if a telephone number was available. If not, they were mailed a questionnaire. The basic-question approach was applied in a separate follow-up. All refusers were offered the possibility to answer just the two basic questions (by phone or mail).

The callback applied to those who failed to cooperate in answering the basic questions. The refusers were visited at home by interviewers. The results of the fieldwork are summarized in the table below.

Result	Cases	Percentage
Response in first wave	508	51.1%
Response in basic-question approach	196	19.7 %
Response in callback approach	224	22.5 %
Final nonresponse	67	6.7%
Total	995	100.0%

The clear conclusion from the data of this table is that a situation does not need to be dismissed as hopeless if the response is low in the first wave. With additional effort response rates can be increased substantially.

Because the researcher had access to the voting register of the town, he could establish with certainty whether all 995 people in the survey had voted. In this group 72.9% had voted. The voting behavior for the various groups is presented in the table below.

Group	% voters
Response in first wave	85.4 %
Response in basic-question approach	66.3 %
Response in call-back approach	55.8 %
Final nonresponse	53.7 %

The groups are ordered by their growing reluctance to participate. There seems to be a relationship between this reluctance and voting behavior: the more reluctant the group, the lower was the percentage of voters.

Both the CBA and the BQA re-approach a sample of nonrespondents. If the information collected in the re-approach is representative for all nonrespondents, i.e., if there is no nonresponse to the re-approach or if the nonresponse is missing completely at random (MCAR), an indication of the differences between respondents and nonrespondents can be obtained. In effect a re-approach can be regarded as a new survey. It involves a new data collection period with a new design used to collect data on nonrespondents. As such, the sequential mixed-mode design described in Chapter 4 can also be seen as a type of re-approach.

This chapter is organized as follows: The theory behind the callback approach, the basic-question approach, and the Politz–Simmons approach is presented in Section 10.2. An application of the callback approach and the basic-question approach to the Dutch Labor Force Survey (LFS) is discussed in Section 10.3. The implementation of these approaches is discussed and the results are analyzed using linked data. A summary of the main findings and some concluding remarks are given in Section 10.4.

10.2 Theory

10.2.1 THE CALLBACK APPROACH

Hansen and Hurwitz (1946) were among the first researchers to recognize that nonresponse can lead to biased estimates of population parameters. They proposed following up on nonresponse in mail surveys by taking a sample of the nonrespondents and seeking the required information by means of face-to-face interviews. If the information collected in the second phase is representative for all nonrespondents, then it is possible to obtain an indication of the difference between respondents and nonrespondents. Furthermore it is possible to correct for a nonresponse bias.

The call-back approach of Hansen and Hurvitz can also be applied in cases where the interviews in the first phase are carried out face-to-face instead of by mail. For the call-back phase specially trained interviewers can be used to re-approach the nonrespondents. Of course, this substantially increases the survey costs.

Under the fixed response model the target population consists of two strata: a stratum of respondents and a stratum of nonrespondents. Suppose that a simple random sample is selected without replacement of size n from this population. This sample is denoted by the set of indicators $a_1, a_2, \ldots, a_N$, where $a_k = 1$ means that element k is selected in the sample, and otherwise $a_k = 0$. Because it is not known beforehand to which of the two strata the elements belong, there are

$$n_R = \sum_{k=1}^{N} a_k R_k, \tag{10.1}$$

elements in the response stratum, and

$$n_{NR} = \sum_{k=1}^{N} a_k (1 - R_k), \tag{10.2}$$

elements in the nonresponse stratum, where $n = n_R + n_{NR}$.

Only the values of the n_R selected elements in the response stratum are obtained. The mean of these values is denoted by

$$\bar{y}_R = \frac{1}{n_R}\sum_{k=1}^{N} a_k R_k Y_k. \tag{10.3}$$

The bias of this estimator is equal to

$$B(\bar{y}_R) = \bar{Y}_R - \vec{Y} = \frac{N_{NR}}{N}(\bar{Y}_R - \bar{Y}_{NR}) = QK, \tag{10.4}$$

where K is the contrast and $Q = N_{NR}/N$ is the relative size of the nonresponse stratum. To apply the callback approach a simple random sample is drawn from the first phase nonrespondents. This comes down to drawing a simple random sample from the nonresponse stratum.

Formally, this approach contradicts the assumptions underlying the fixed response model. This model assumes the existence of a subpopulation consisting of elements who would never respond in a survey. However, the fixed response model should be seen as conditional on the survey design. The design of a callback survey assumes a different population (the first phase nonrespondents) with different response and nonresponse strata. The situation is depicted in Figure 10.1.

If everybody responds in the callback survey, or if the nonresponse in the callback survey is ignorable (i.e., there is no correlation with target variables), it is possible to compute unbiased estimates of nonresponse population parameters. However, this condition may not always be fulfilled in practical situations. It is possible for a population to consist of three strata: respondents, soft respondents (they cooperate in the callback approach), and hard-core nonrespondents (they never cooperate).

Now suppose that a sample of size m is selected in the callback survey. The number of respondents in this survey is denoted by m_R. Let $\bar{y}_{NR}$ denote the mean of the m_R values of the responding elements. Assume that these m_R observations

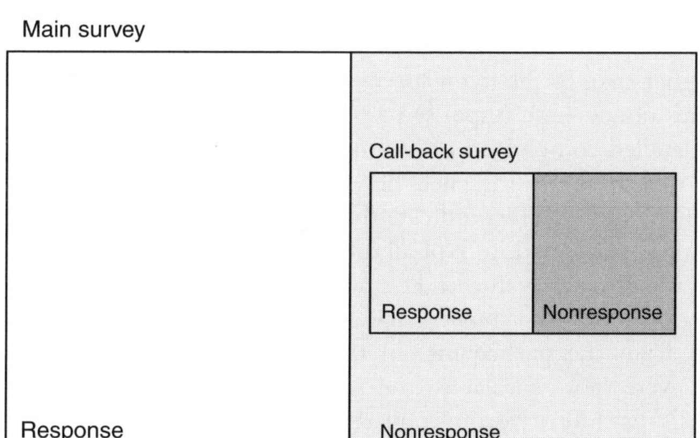

FIGURE 10.1 The callback approach.

constitute a simple random sample from the nonresponse stratum of the main survey. This mean is an unbiased estimator of the mean of the target variable in the nonresponse stratum of the main survey. Consequently

$$\frac{N_R}{N}\bar{y}_R + \frac{N_{NR}}{N}\bar{y}_{NR} \tag{10.5}$$

is an unbiased estimator of the population mean of *Y*. Unfortunately, the sizes of the response stratum and the nonresponse substrata are unknown. Therefore the quantities N_R/N and N_{NR}/N are replaced by their unbiased estimates n_R/n and n_{NR}/n. This results in the estimator

$$\frac{n_R}{n}\bar{y}_R + \frac{n_{NR}}{n}\bar{y}_{NR}. \tag{10.6}$$

Under the condition mentioned, this is an unbiased estimator.

EXAMPLE 10.2 An Application of the CBA

Stoop (2005) conducted a callback survey for the Amenities and Services Utilization Survey (AVO, in Dutch) of the Social and Cultural Planning Agency in The Netherlands. Starting with an initial sample size of 9372, the response rate turned out to be 65.1%.

A follow-up survey was carried out among the persistent refusers of the main survey. The primary goal of this follow-up survey was to discover whether refusals differed, how they differed, and what the impact of these differences was on the survey results.

Highly motivated interviewers were asked and trained for this survey. They received additional bonuses for successful interviews. They could spend, on average, 22.50 euros on each potential respondent to buy flowers or gifts, pay for their time, and invite respondents to answer questions in a public place. In fact they could spend money as deemed fit. A sample of 329 persistent refusers was re-contacted. Cooperation was obtained in 235 cases, which comes down to a response percentage of 71.4%. This response rate is even higher than in the main survey. It shows that with substantial effort, time and money, high response rates can be reached.

A detailed comparison of the main survey with the follow-up survey showed that there are differences between respondents and nonrespondents in the main survey. They indicate that refusers have a slightly lower social and cultural participation. Not all differences turned out to be significant, but this was also partly due to the small sample size of the follow-up survey.

Stoop (2005) also showed that converted refusers in the main survey differed from the participants in the follow-up survey. The converted refusers were more similar to the respondents in the main survey. The message is that more effort to convert temporary refusers in the main survey does not help to reduce the bias. Higher response rates do not necessary lead to better estimates. This study suggests that it is better to settle for a lower response rate and to spend resources getting more insight on the "real" nonrespondents, for example, by carrying out a callback survey.

The callback approach of Hansen and Hurwitz (1946) will generally be expensive. The costs per interview in the second phase will be much higher than in the first. Since the distribution of nonrespondents over the country will be less dense than that of respondents, travel costs will be relatively high. Another factor preventing application of the callback approach is time. Sampling and interviewing nonrespondents will dramatically increase the duration of the fieldwork period. Remember, the aim of the callback approach is to obtain insight into the nature of the nonresponse. To do so, there are possibly no time or money constraints.

Kersten and Bethlehem (1984) proposed the *basic-question approach* as an alternative to the callback approach. It can be applied in situations where the callback approach cannot be carried out because of time and money constraints. A similar method, *PEDAKSI*, was proposed by Lynn (2003). PEDAKSI stands for preemptive doorstep administration of key survey items. The main difference between PEDAKSI and the basic-question approach is the length of the questionnaire. Kersten and Bethlehem (1984) suggested asking only two or three basic questions, whereas Lynn (2003) used questionnaires with a larger number of questions. The theory behind the basic-question approach is described in the following section.

10.2.2 THE BASIC-QUESTION APPROACH

The basic-question approach assumes that many survey questionnaires are composed around a few basic questions. The answers to these questions are needed in order to be able to formulate the most important conclusion of the survey. If interviewers face problems in getting cooperation during the fieldwork, they can change their strategy and attempt to obtain answers to only a few basic questions once they have "a foot in the door." Another approach could be to carry out the basic-question approach afterward, by telephone or mail (i.e., for the not-at-home nonrespondents).

One way to apply the basic-question approach is to let the interviewers attempt the basic questions right after they have been confronted with a refusal for the main questionnaire. This can lead to higher nonresponse rates for the main questionnaire, as found by Van den Brakel and Renssen (1998). Therefore it may be better (but also more expensive) to re-approach the refusals after a short while with different interviewers.

The main goal of the basic-question approach is to gain insight in possible differences between respondents and nonrespondents with respect to the most important variables of the survey. If such differences are detected, the approach can provide useful information for correcting estimates for other variables.

The basic-question approach was born from the observation that people who refuse to participate can often be persuaded to answer a few basic questions. Basic questions can be distinguished in many surveys. These few questions should only be asked when it is clear that further attempts to get the questionnaire completed will be useless, namely when in the regular survey the sample element would have been processed as a nonresponse.

The basic questions are, of course, only a subset of the research variables that the survey organization would like to measure. Often the values of important research variables depend on the answers to several questions. If only a few of these questions can be asked, the value of the research variable cannot be derived.

In this section the focus is on the case of just one basic question. While it should be possible to ask more than one basic questions, it is advisable to keep to a few number of basic questions. The more questions asked, the higher is the risk of getting no information at all.

EXAMPLE 10.3 Examples of Basic Questions

Considerable insight into the nature of nonresponse can be obtained in situations where especially the survey's topic causes nonresponse. This occurs when people think the survey does not apply to them. Here are some examples of basic questions that have been used in surveys of Statistics Netherlands:

- *Housing Demand Survey:* Do you intend to move within two years?
- *Labour Force Survey:* How many people in this household have a paid job?
- *Holiday Survey:* Have you been on holiday during the last 12 months?
- *Family Planning Survey:* Based on your present circumstances, and your future expectations, how many children do you plan on having from this moment on?

The basic-question approach has helped to get as much answers as possible to important questions of a survey and seems to have worked well for several different surveys. People who refuse to cooperate can be persuaded to answer a few questions if the interviewer says "OK, I accept your refusal, but please help me fix my administration," or "OK, I will not persist but at least answer one question." Even for refusals in a telephone survey the basic-question approach may work.

To be able to compare respondents and nonrespondents, the basic question must be answered by both respondents and nonrespondents. To avoid all kinds of interviewing effects, questionnaire effects, and mode effects, the basic question must be presented to respondents and nonrespondents the same way. Therefore the basic question should be one of the first questions in a questionnaire. It is important that the answer to the basic question not be changed even when answers to subsequent questions indicate that the answer to the basic question was wrong.

Under the fixed response model, the population can be divided in a stratum U_R of N_R (potential) respondents and a stratum U_{NR} of N_{NR} nonrespondents, with $N = N_R. + N_{NR}$. In a more detailed fixed response model, the nonresponse stratum can be divided into a number L of nonresponse substrata $U_{NR,1}, U_{NR,2}, \ldots, U_{NR,L}$, according to the cause of nonresponse (e.g., refusal, noncontact and not-able). The sizes of these substrata are denoted by $N_{NR,1}, N_{NR,2}, \ldots, N_{NR,L}$, with

$$\sum_{h-1}^{L} N_{NR,h} = N_{NR}. \tag{10.7}$$

Such a more detailed model is useful if nonrespondents are expected to behave differently with respect to the basic question in different substrata.

Suppose there is one target variable Y with values $Y_1, Y_2, \ldots, Y_N$, and one basic question variable Z with values $Z_1, Z_2, \ldots, Z_N$. A simple random sample of size n is selected without replacement. There are n_R respondents and n_{NR} nonrespondents. The number of nonrespondents in substratum h is denoted by $n_{NR,h}$, for $h = 1, 2, \ldots, L$.

Not every nonrespondent answers the basic question. The number of nonrespondents in substratum h who answer the basics question is denoted by $m_{NR,h}$, for $h = 1, 2, \ldots, L$. Their total over all substrata is equal to m_{NR}.

The response means $\bar{y}_R$ and $\bar{z}_R$ are unbiased estimators of the response stratum means of the target variable and the basic question variable, respectively. Estimating the mean of the basic question variable in the nonresponse stratum is not simple. The fundamental question is whether nonrespondents who answer the basic question may be regarded as a simple random sample from all nonrespondents, at least within all nonresponse substrata. Here, it is assumed that this is the case.

Let $\bar{z}_{NR}^{(h)}$ be the mean of the available $m_{NR,h}$ values of the basic question variable in nonresponse substratum h. Then this mean is an unbiased estimator of the mean of the basic question variable in nonresponse substratum h. Consequently

$$\frac{N_R}{N}\bar{z}_R + \sum_{h=1}^{L} \frac{N_{NR,h}}{N}\bar{z}_{NR}^{(h)} \tag{10.8}$$

is an unbiased estimator. Unfortunately, the sizes of the response stratum and the nonresponse substrata are unknown. Therefore the quantities N_R/N and $N_{NR,h}/N$ are replaced by their unbiased estimates n_R/n and $n_{NR,h}/n$. This results in the estimator

$$\bar{z}_{BQ} = \frac{n_R}{n}\bar{z}_R + \sum_{h=1}^{L} \frac{n_{NR,h}}{n}\bar{z}_{NR}^{(h)}. \tag{10.9}$$

This is an unbiased estimator, provided that the nonrespondents answering the basic question are a simple random sample from all nonrespondents.

EXAMPLE 10.4 The Basic-Question Approach in the Housing Demand Survey

The basic-question approach was tested in the Dutch Housing Demand Survey in 1981 (see Kersten and Bethlehem, 1984). Excluding overcoverage, the sample size of this face-to-face survey was 82,849. The number of respondents was 58,972, and this amounts to a response percentage of 71%. When people refused to cooperate, the basic-question approach was tried at the door. In total, 8383 refusers could be persuaded to answer the basic question. This implies that 58,972 + 8,383 = 67,355 people answered the basic questions, which comes down to a response percentage of 81% (for this variable).

The basic question in the survey was: Do you intend to move within two years? The table below shows the results of these questions for two groups: the initial respondents and the refusers who answered the basic question. It is clear that there is a difference between the two groups. Initial respondents were much more inclined to move than refusers answering the basic question.

Do you intend to move within two years?	Initial respondents	Refusers answering the basic question
Yes	29.7 %	12.8 %
No	70.3 %	87.2 %
Total	100.0 %	100.0 %

The initial respondents and refusers should be asked the basic question as much as possible under the same circumstances so that they can interpret the question the same way. In the Dutch Housing Demand Survey 1981, when a second wave of fieldwork was carried out, callbacks were made for a sample of nonrespondents, including those who answered the basic question in the first wave. This provided a way to check the answers to the basic question. For 1,638 first-wave refusers, both their answers to the basic question in the first wave and to the complete interview in the second wave became available. The results are presented in the table below.

	Second wave	
First wave	Intends to move	Does not intend to move
Intends to move	8.6 %	5.1 %
Does not intend to move	12.1 %	74.1 %

The same answer was given in 8.6% + 74.1% = 82.7% of the cases. So there is a reasonable amount (but not complete) of consistency. Note that there was a time lag of three months between the two waves. It is not unlikely that people have changed their mind in this short period of fime.

Expression (10.9) gives the improved estimator for the basic question variable. But what about the other survey variables? The basic-question variable can be used to improve estimators for these variables. To that end, the basic-question variable should be treated as an auxiliary variable. If the basic-question variable is a qualitative variable, the poststratification estimator can be used. And if the basic-question variable is a qualitative variable, the ratio estimator or regression estimator can be used.

Consider the case of a qualitative basic-question variable. The expression for the poststratification estimator for a target variable Y other than a basic question is

$$\bar{y}_{PS} = \frac{1}{N}\sum_{h=1}^{L} N_h \bar{y}^{(h)}, \tag{10.10}$$

To be able to apply poststratification, the numbers N_h of population elements in the strata corresponding to the categories of the basic question variable must be known. These numbers can be estimated using the answers to the basic question (both for the initial respondents and the refusers). Next, the average of the target variable is computed for every category of the basic question variable. For this, only the data for the initial respondents can be used. By substituting these means into expression (10.10) an estimate is obtained that is, hopefully, less biased.

Poststratification works better when the target variable and the auxiliary variable have a stronger relationship. Since the basic question variable is also a target variable of the survey and target variables are often correlated, the basic-question approach is likely to produce better estimates.

Now consider the case of a quantitative basic-question variable. One way to improve the estimate for a target variable is to use a ratio estimator where the basic-question variable is used as an auxiliary variable. In this case the ratio estimator would take the form

$$\bar{y}_Q = \bar{y}_R \frac{\bar{z}_{BQ}}{\bar{z}_R}. \tag{10.11}$$

The more the values of Y and Z are proportional, the more effective is the ratio estimator.

A better approach is to use a regression estimator where the basic-question variable is treated as an auxiliary variable. This estimator would take the form

$$\bar{y}_R = \bar{y}_R - b(\bar{z}_R - \bar{z}_{BQ}), \tag{10.12}$$

where b is a vector of regression coefficients. This estimator is effective if there is a more or less linear relationship between the values of the target variable and the basic-question variable.

10.2.3 THE POLITZ AND SIMMONS APPROACH

In Chapters 1 and 4 the survey researcher had to distinguish among response types, especially contact and refusal. As a result the response mechanism was split into two sequential phases. The first phase was that of the contact attempt. The second phase was that of gaining cooperation once contact is established. Analysis of both phenomena may require different models and different auxiliary variables.

In this section a method is described for modeling the contact probabilities by estimating the probability of people being at home. This method was developed by Politz and Simmons (1949). The basic idea is to estimate these probabilities on the basis of one visit to the respondents, thereby avoiding making expensive callback visits. This technique can be used to adjust survey results obtained after a maximum of, say, three visits for a bias due to being not-at-home.

The technique of Politz and Simmons focuses on correcting a nonresponse bias due to not-at-home persons. But the problems caused by refusals are not solved. This technique will typically be applied in surveys where one can expect a relationship between being not at home and the target variables. Examples are employment, recreational activities, and mobility. In this example other types of nonresponse, like refusal, are ignored.

Taking the random response model as a starting point, each person k in the population has an unknown probability ρ_k of being at home when a contact attempt is made. In principle, the at-home probabilities can have any value in the interval [0,1], but to simplify the theory, it is assumed that these probabilities can take only one of a finite set of values $1/L$, $2/L$, ..., L/L, for a preset value of L. Thus the population is divided in L strata, where stratum h consists of all persons with at-home probability h/L, for $h = 1, 2, \ldots, L$. The total number of persons in group h is denoted by N_h, where $N_1 + N_2 + \ldots + N_L = N$. Note that this approach excludes the possible existence of a stratum of persons who are never at home so that the group has an at-home probability of 0. This situation will be discussed later.

If the stratum sizes N_1, N_2, ..., N_L are known, and it is also possible to determine without error the at-home probability ρ_k of every observed person, then poststratification can be applied to compute an unbiased estimator of the population mean of the target variable. The poststratification estimator is unbiased because within each stratum the response probabilities are constant, and therefore the stratum estimator is unbiased, resulting in an unbiased estimator of the population mean.

Unfortunately, the stratum sizes N_1, N_2, ..., N_L are not known in general. If the at-home probability can be determined for each interviewed person, it is possible to estimate the stratum sizes. Let n_h be the number of interviewed persons having at-home probability h/L. Then n_h is a random variable with expected value

$$E(n_h) = n\frac{N_h}{N}\frac{h}{L}. \tag{10.13}$$

Consequently

$$\hat{N}_h = \frac{NLn_h}{nh} \tag{10.14}$$

is an unbiased estimator for N_h.

This estimate can only be computed if the at-home probabilities are known. In practice, they are not. The approach of Politz and Simmons (1949) was to estimate these probabilities. They developed their technique for surveys in which interviewing took place at randomly chosen evenings. To estimate the probability of being at home, they asked respondents the following question: *Would you mind telling me whether or not you happened to be at home last night at just this time?*

The question was repeated for the previous four evenings. This way they obtained the number of evenings out of the last five evenings the respondent was at home. They observed that this way of obtaining the information proved to be more reliable than just asking the question: *How many nights out of the last five were you at home at just this time?*

Let t_k denote the number of times out of the last $L-1$ days the respondent was at home at just that time (ignoring the day the question was asked). Then the at-home probability can be estimated by

$$\hat{\rho}_k = \frac{t_k + 1}{L}. \tag{10.15}$$

Note that L refers to a period of L, days, including the $L-1$ previous days and the interviewing day. This is also the reason why the numerator is increased by 1, meaning the respondent was at home at the day of the interview. Also note that t_k is a random variable. The uncertainty of its value is not caused by the sampling mechanism but by independent measurement error.

Let n_h be the number of interviewed persons responding that they were at home at h of L days. For all these persons the at-home probability is estimated by h/L. Then the number of persons in the population having at-home probability h/L can be estimated by

$$\hat{N}_h = \frac{NLn_h}{nh} \qquad \text{for } h = 1, 2, \ldots, L. \tag{10.16}$$

In order to apply poststratification, the total population size must also be known. Intuitively one might be inclined to use N for this. However, note that the technique described above only applies to persons having a nonzero at-home probability. Persons with an at-home probability equal to 0 are excluded. Let N' denote the size of the population of elements with a nonzero probability being at home. Then N' can be estimated by

$$N' = \sum_{h=1}^{L} \hat{N}_h = \frac{NL}{n} \sum_{h=1}^{L} \frac{n_h}{h}. \tag{10.17}$$

By subtracting (10.17) from N, an estimate of the number of people having at-home probability 0 is obtained. Computation of this estimate might provide some insight on the adequacy of the Politz–Simmons approach. For example, a negative value, or a very large value could point at a violation of the assumptions underlying the model.

Let $\bar{y}^{(h)}$ denote the mean of the n_h values of the target variable in the stratum with at-home probability h/L. Then the Politz–Simmons estimator is defined by

$$\bar{y}_{PS} = \frac{1}{N'} \sum_{h=1}^{L} \hat{N}_h \bar{y}^{(h)} = \frac{\sum_{h=1}^{L} n_h \bar{y}^{(h)}/h}{\sum_{h=1}^{L} n_h/h}. \tag{10.19}$$

Note that this estimator can only be used to estimate the mean of the population of all people having a nonzero at home probability. The group of people having a zero at home probability is excluded. These people can never be contacted.

There are two problems that could confound the Politz–Simmons approach in practice. The first is that estimation of at-home probability becomes more complex in household surveys where a number of people could fill in the questionnaire form. Then the at-home probability of a selected person should not be estimated, but instead the probability that any eligible person is at home. This, however, requires a more detailed account of the at-home pattern.

The second problem concerns the basic assumption underlying this model, namely that interviewers make contact attempts at randomly chosen moments. This allows for unbiased estimation of at-home probabilities. However, many surveys use letters to announce the visit of an interviewer. Some people may stay at home for this visit and so increase their at-home probability. This assumption is sometimes violated because the sampled individuals are provided with telephone numbers of interviewers to make an appointment for their visit.

EXAMPLE 10.5 The Politz and Simmons Approach in the Labor Force Survey

The Politz and Simmons approach was tested in the 1981 Labor Force Survey of Statistics Netherlands. This was a household survey where all household members of age 15 and over had to be interviewed. Therefore the Politz and Simmons question was: *On how many of the past six days was someone at home around this time?*

The approach was tested in a small town on a small sample of 105 households. The results are displayed in the table below. The estimate N' for the number of households with a nonzero at-home probability turns out to be approximately equal to 4,200. This number is far higher than the total number of households in the town (3,500). So the at-home probabilities are overestimated. It should be noted that because of the small sample size, the estimate N' has a large margin.

One of the main target variables of the Labor Force Survey was the number of people in the household with a paid job. Further analysis showed that there was no relationship between at-home probabilities and the number of employed. In this case the not-at-home persons did not cause a bias.

Number of times someone was at home during the last 6 days	Frequency	At-home probability
0	1	1/7
1	2	2/7
2	0	3/7
3	6	4/7
4	9	5/7
5	10	6/7
6	77	7/7
Total	105	

10.3 Application

From July to October 2005, Statistics Netherlands carried out a follow-up survey among refusers, noncontacts, and unprocessed cases in the Dutch Labor Force Survey (LFS). In the follow-up survey two methods were applied; the *callback approach* and the *basic-question approach*. In the pilot study a sample of nonrespondents in the LFS was approached once more with a vastly shortened CATI, Web, and paper questionnaires. The sample consisted of LFS households that refused, were not processed, or were not contacted in the LFS of the months of July to October. Also an intensive follow-up of nonrespondents was conducted by a small set of selected interviewers (callback approach) and a follow-up of respondents that refused to participate in the CATI waves of the LFS. For a description of the whole pilot study, see Bethlehem et al. (2005). For a report on the follow-up of CATI panel refusers, see Banning and Schouten (2009).

The overall design of the pilot, the design of the callback approach, and the design of the basic-question approach were as discussed in Section 10.3.1. The results from the re-approaches were analyzed with the use of a linked data set. This analysis is described in Section 10.3.2; the conclusions are presented in Section 10.3.3.

10.3.1 DESIGN OF THE STUDY

10.3.1.1 Dutch Labor Force Survey The Dutch Labor Force Survey (LFS) is a continuous sample survey of residents in The Netherlands aged 15 years and older, with the exception of persons living in institutions, asylums, and old people's homes (i.e., the institutionalized population). It has a stratified two-stage sample design. In the first stage, municipalities are selected. The large municipalities are selected in this first stage with the probability equal to 1, meaning that they are always selected. The other municipalities in the sample are selected with probabilities proportional to the number of inhabitants. The second stage consists of addresses.

Every month, a sample of approximately 10,000 addresses is selected from the population register. Addresses with only residents aged 65 and older are assigned lower selection probabilities. In July and August the sample is halved because this is when the interviewers their holidays. A total of 71,000 addresses were allocated to the interviewers in 2005. The LFS is conducted as a rotating panel with five waves. The first wave is a face-to-face interview or CAPI (computer-assisted personal interview). If more than one household is found at an address, up to four households per address are interviewed, and up to eight persons per household. Proxy interviewing is allowed by the household core, namely the head of the household or his/her partner. Respondents are asked if they want to participate in four consecutive waves by telephone with a time lag of three months each.

The LFS provides monthly, quarterly, and yearly statistics on participation in the labor market, employment, unemployment, and working hours among the population aged 15 years and older. One of the key LFS items is a classification of the population of working age into three mutually exclusive and exhaustive groups—employed persons, unemployed persons, and inactive persons. The employment status is determined in a reference week. Employed persons are persons who work for more than 12 hours a week, for pay, profit, or family gain, or are not at work but have a job or business from which they were temporarily absent, for example, because of illness, holidays, education, and training. Unemployed persons are persons who are without work, who are currently available for work and are either actively seeking work in the past four weeks or have already found a job to start within the next three months. Inactive persons are those who are neither classified as employed nor as unemployed.

10.3.1.2 General Design of the LFS Pilot The pilot was motivated by Stoop (2004) and Voogt (2004), who describe studies that resulted in high response rates for former nonrespondents. Stoop (2004) re-approached a sample of 350 refusers classified by the interviewers as hard to persuade. It was not possible to link the data to administrative data, so little information is available on the nonrespondents. The study by Voogt (2004) surveyed the small town of Zaanstad in The Netherlands about voting behavior; see also Example 10.1. Voogt had access to the voting register of Zaanstad, which enabled him to confront the answers to the survey questions with the real voting behavior. He investigated both nonresponse bias and response bias in election research.

The recommendations of these studies were used as input for the LFS pilot. Besides refusals, former noncontacts and unprocessed cases were eligible for the re-approach. A random sample of approximately 1,900 LFS nonrespondents was selected for follow-up and was randomly divided over the two approaches. Due to insufficient interviewer capacity, the sample for the callback approach was not allocated to one of the interviewer regions. As a result 775 households were re-approached using the callback strategy, and 942 households using the basic-question approach.

The data set was linked to a large number of auxiliary variables. Some of these variables are related to employment status, such as having a paid job or receiving an unemployment allowance.

Eleven of the 13 interviewer districts participated in the pilot. Two districts could not participate because of a lack of available interviewers. In the participating districts two or three of the best interviewers were selected, depending on the geographical size of the district. At least two interviewers per district were needed to assign each address to an interviewer other than the interviewer who originally approached the address. In five districts a number of municipalities were excluded so that traveling distances were acceptable. During the pilot one interviewer region dropped out of the CBA. For this reason the sample size for the CBA is smaller than for the BQA. This region was removed from the analysis of the additional response in the CBA. In the BQA this region is not removed. As a consequence the analysis of the response behavior in the regular LFS using linked data gives slightly different results for the CBA and the BQA.

Although the BQA was not conducted face-to-face, for reasons of comparison households from the same districts and municipalities as for the CBA were selected

TABLE 10.1 Sizes of eligible households for the re-approach

Group	Size
Eligible LFS households	18,074
Regular response	11,275
Regular nonresponse	6,799
Eligible nonresponse	6,170
Callback approach	775
Basic-question approach	942

for the BQA. As a consequence the re-approach estimates cannot be compared directly to those in the regular LFS. However, the selected municipalities cover the full range from rural to strongly urbanized areas and make up about 75% of the population. For a comparison of estimates for the re-approach and the regular LFS, the same municipalities were selected from the LFS. Table 10.1 shows the sizes of the different groups in the pilot.

10.3.1.3 Design of the Re-approach Pilots The design features of the CBA and the BQA are presented in this section. The main features are summarized in Table 10.2 to display the differences and the similarities between the two approaches.

The features of the CBA and the BQA are discussed separately in more detail, starting with the CBA. The questionnaire and interview mode of the CBA are the

TABLE 10.2 Design features of the CBA and the BQA

Feature	Callback approach	Basic-question approach
Sample unit	All household members aged 15 +	One randomly selected person per household
Questionnaire	Original LFS questionnaire	Strongly condensed questionnaire containing only the key questions of the LFS to determine the employment situation
Mode	CAPI	Mixed mode: combination of CATI, Web and paper surveys
Fieldwork period	2 months	1 month
Control group	No	Yes
Advance letter	No	Yes
Timing	One week after being processed as nonresponse	One week after being processed as nonresponse
Incentive	Yes	No
Special interviewers	Yes	No
Eligible response types	Refusal, noncontact, unprocessed cases	Refusal, noncontact, unprocessed cases

same as in the regular LFS. This has been done to avoid mode and questionnaire effects. No advance letter was sent to nonrespondents to prevent alarming them or giving them additional reasons to refuse cooperation. The addresses in the CBA were allocated to a different interviewer than the interviewer that received the original LFS nonresponse.

A total of 28 interviewers were selected from the best CAPI interviewers. The interviewers received an *additional training* in doorstep interaction. Furthermore a special newsletter was made to inform them, and halfway through the study a meeting was organized to exchange experiences and to boost motivation. The interviewers could offer *incentives* in the form of gift vouchers as they deemed fit, and in addition the interviewers themselves could receive a bonus based on their response rate.

Paper summaries were sent to the interviewers, containing background information on the composition, gender and age of the household, the type of nonresponse and the timing, and number of previous contact attempts. The interviewers could use this information to fine-tune their contact attempts and prepare the doorstep interaction. Interviewers were, however, not obliged to adapt their strategy to the background information. It was solely provided to prepare them in dealing with the nonrespondents.

For the CBA, an *extended fieldwork period* of two months was used to enhance participation, which caused a delay in the time for the callback response to the LFS. It is known that employment has seasonal and cyclical components. Hence a delay in time can result in a difference in employment rate that is to be attributed to the month of observation and not to nonresponse bias. Unless a time series model is used for employment, it is not possible to disentangle the effects of nonresponse and time. This has not been done; instead, the time lag is accepted, and it is assumed that the time effect is small.

The design of the BQA was very different from the original LFS and different from the design of the CBA. The questionnaire for the BQA contained a small number of questions to determine the key variable of the Labor Force Survey, namely the labor force status. In the LFS these questions are asked in the beginning of the questionnaire. This way the questions were not taken out of their context, so there could be assumed that no questionnaire effects were incurred.

The design for the BQA was *mixed mode*. Addresses for which a listed landline telephone number could be linked to the survey sample were designated to CATI. The addresses without a listed landline telephone were designated to a combination of a paper questionnaire and a Web questionnaire. An advance letter was sent to the selected addresses. The letter contained a personal login to a secured website where the Web survey could be found, as well as a paper questionnaire with a return envelope. In the letter the selected person in the household was presented with a choice between the Web questionnaire and the paper questionnaire.

As mentioned before, the idea behind the BQA is that respondents who refuse to participate can often be persuaded to answer a few questions. However, asking too many questions risks getting no information at all. To reduce the response burden and thus stimulate participation, it was decided to only interview one randomly selected person per household, employing the next birthday method.

Several methods to randomly select one person within a household are identified by the literature (Kish, 1949; Salmon and Nichols, 1983; Oldendick et al., 1988; Lavrakas et al., 2000; Binson et al., 2000; Gaziano, 2005). The main objective of these procedures is to obtain a sample that is representative of the target population of the survey. Binson et al. (2000) compare the three methods that are most often used: the Kish grid, the last birthday method, and the next birthday method. Based on their findings, it was decided to use the *next birthday method* in the BQA.

In the BQA the interviewers may be allowed to ask the basic questions immediately after they receive a refusal for the main questionnaire. This may, however, lead to higher nonresponse rates for the main questionnaire; see Van den Brakel and Renssen (1998) and Van Goor and Van Goor (2003). Interviewers may switch to the BQA too soon, thereby shortening the doorstep interaction and putting less effort in converting refusers. For this reason the nonrespondents were re-approached one week after the household was processed as a nonresponse.

A fresh control sample of a 1,000 households was approached with the basic questionnaires. Ideally the basic questionnaires should also have been presented to a sample of LFS respondents. That would enable a direct comparison of the answers to the BQA from the LFS nonrespondents sample to that of the respondents. However, the experiment was designed not to intersect with the regular LFS. Therefore a fresh control sample was approached instead of LFS respondents to analyze possible mode effects and to assess the effect of the random respondent selection procedure.

Table 10.3 provides a summary of outcomes for the different groups in the BQA. Weights have been assigned to all households selected for one of the re-approaches in the LFS nonresponse to let these households represent all the eligible households; see Table 10.4. The months July through August and September through October had different selection probabilities for the CBA because of the different workloads of interviewers.

The response rate in CATI is for the basic-question approach was much higher than the response rate in the paper/Web mode. This result was obtained in both the BQA for the LFS nonrespondents and the control group. Interviewer-assisted data collection modes, in general, obtain higher response rates due to the persuasiveness of the interviewers. Griffin et al. (2001) found that offering multiple modes of response in a mailing led to a lower response rate. They suspect that offering a mode of response other than mail, in combination with a paper questionnaire,

TABLE 10.3 Different groups in the BQA

	LFS		Control	
	Sample	Response	Sample	Response
Eligible households	6,170			
Selected households	942	40%	1,000	62%
CATI	573	55%	667	80%
Paper/Web	280	23%	333	25%

TABLE 10.4 Selection of eligible households for the CBA and the BQA, weights and sizes

Group	Total size	Size selection	Weight
Eligible CBA July–Aug	3,039	391	7.77
Eligible CBA Sep–Oct	3,133	384	8.16
Eligible BQA	6,170	942	6.55
Not eligible	629	0	1.00
LFS response	11,275	0	1.00

contributes to a break in the response process and thus results in a lower overall response rate. This effect may also have been present in the LFS pilot.

The low response rates to the paper and Web questionnaires in the design of the BQA caused a strong correlation between telephone ownership and response. This finding confounds the analysis of response patterns and mode effects. Therefore the analysis is restricted to households with a listed landline telephone. In the regular LFS, 10,137 households have a listed landline telephone. Of these households 6,893 participated in the LFS, which is a response rate of 68%.

Two important aspects of the basic-question approach are the within-household respondent selection with the next-birthday method, and the possible mode effects due to the change of questionnaire and interview mode. To evaluate whether these aspects caused any differences between the response to the BQA and the response to the regular LFS that could be mistakenly attributed to the BQA, a control group has also been approached with basic questionnaires.

Cobben and Schouten (2007) analyzed whether the next-birthday respondent selection resulted in a random sample of persons within households. Their analyses indicate that there is a difference with respect to position in the household and gender. It appears that the composition of the respondents is distorted when it comes to children. However, they do not find a bias in the basic-question variable, namely employment status.

Because the implementation of the BQA involves a change in both the questionnaire and the interview modes, there is a risk of errors other than nonresponse (e.g., see Groves, 1989; Biemer and Lyberg, 2003). A change of wording and context as well as a change of mode can affect survey answers, and hence the statistics based on the survey. When mode effects are imminent, it is not possible to draw any conclusions about nonresponse error because the total survey error cannot be decomposed. For the control group, Cobben (2007) has decomposed the survey error into errors caused by nonresponse and mode effects. She found no evidence that employment status in the BQA is affected by mode effects.

10.3.2 ANALYSIS OF RESPONSE IN THE LFS AND THE RE-APPROACHES

The additional response that is obtained by the callback and basic-question methods is analyzed by two questions: What number of households participate in the re-approach strategy, and do these new households improve the composition of the response? In other words, do the new households have different characteristics than the ones in the regular LFS response. These questions are investigated separately for both the callback and the basic-question approaches by linking auxiliary variables from registers and administrative data.

10.3.2.1 Linked Data for the Analysis The samples of the LFS are drawn from the population register. In addition to the demographic auxiliary variables these samples are linked to the SSB, the Social Statistical Database of Statistics Netherlands. A number of variables in the SSB are closely related to the employment status of persons. However, the observations are made at the household level. This means that first individual auxiliary information is aggregated to the household level. The aggregation is done through the members of the core of the household, namely the head and the partner if present. Table 10.5 contains an overview of the auxiliary variables with corresponding classifications.

10.3.2.2 The Callback Approach The callback response is compared to the full callback sample of initial nonrespondents. First, bivariate tests for independence between response and the available auxiliary variables are conducted. Second, a logistic regression model is fitted to estimate the probability of response in the callback approach. Note that both the tests and the response probabilities are conditional on an initial nonresponse in the LFS.

Table 10.6 gives the p-values corresponding to Pearson's χ^2 test of independence. The p-value is the lowest significance level at which independence is rejected. The only variable that shows a significant association with response at the 5% level is the average age of the household core. However, if accounting for multiple tests the independence of callback response behavior and average age would not be rejected. The p-values for the geographic variables region and province are also small but not significant at the 5% level.

Table 10.7 gives the fit and the corresponding parameter estimates of the logistic regression model for the callback response. The resulting model has only two variables: age and region. The proportion of variance explained by the model is small, 2.5% using the definition of Nagelkerke (1991). The lack of dependence with the available auxiliary variables is a remarkable result, even for the relatively small sample size of 775 addresses. This implies that the response in the CBA was not very selective.

From Table 10.7 it is evident that the households with an average age between 35 and 54 years had a higher response rate in the CBA. Furthermore in the eastern parts of The Netherlands fewer households participated than in the other parts of the country.

Regarding the question whether the new households improve the overall composition of the response, first the p-values of bivariate tests of independence between response and several auxiliary variables were compared before and after the callback response was added to the regular LFS response. Then logistic regression models were constructed for the LFS response alone, for the combined response of LFS and callback, and for the type of response given a household responded in either the LFS or callback.

TABLE 10.5 Auxiliary variables in the analysis

Variable	Name	Categories
Average age of household core	Age	15–34 years, 35–54 years, 55 years and older
At least one member of household core is registered at the Center for Work and Income	CWI	Yes, no
Household receives a disability allowance	DaAll	Yes, no
Ethnic background of household core	Etn	Native, Moroccan, Turkish, Surinam/ Dutch Antilles, other nonwestern, other western, mixed
Aggregated gender of household core	Gender	Male, female, mixed
Type of household	HhType	Single, unmarried couple, married couple, unmarried couple with children, married couple with children, single parent, other, more than 1 household
Average house value at zipcode level (in 1,000 euros)	HsVal	Not available, 0–50, 50–75, 75–100, 100–125, 125–150, 150–200, 200–250, 250–300, 300–400, more than 400
At least one member of household core had a paid job at January 1, 2005	Job	Yes, no
Status of paid jobs in household core	JobStat	All paid job, part paid job, none paid job
Percentage of population that is nonnative on zipcode level	PercNn	Not available, 0–5%, 5–10%, 10–20%, 20% and more
Fixed landline telephone available	Phone	Yes, no
Provinces and four largest cities	Prov	Groningen, Friesland, Drenthe, Over-ijssel, Flevoland, Gelderland, Utrecht, Noor-Holland, Zuid-Holland, Zeeland, Noord-Brabant, Limburg, Amsterdam, Rotterdam, Den Haag, Utrecht City
Region of The Netherlands	Reg	North, east, west, south
At least one member of the household core is self-employed	SelfEmp	Yes, no
Household receives a social allowance	SocAll	Yes, no
Household receives an unemployment allowance	UeAll	Yes, no
Degree of urbanization	Urb	Very strong, strong, moderate, little, not

Table 10.8 shows the p-values for bivariate tests for independence between response and auxiliary variables for the LFS response and for the combined response of LFS and callback re-approach. Note that for many variables independence of response behavior is rejected at 5% or 1% levels before and after the combination of the response groups. The exceptions are average age at the household level, CWI

TABLE 10.6 Bivariate test of independence between response and nonresponse in callback and auxiliary variables

Variable	p-Value	Variable	p-Value
Urbanity	0.199	Household type	0.110
Phone	0.502	Job	0.913
Region	0.080	Disability all	0.842
Province	0.072	Unemployment all	0.656
Gender	0.793	CWI registration	0.488
Ethnicity	0.135	Self-employed	0.471
House value	0.644	Social allowance	0.893
Perc nonnative	0.625	Average age	0.048

Note: The test statistic is Pearson's χ^2.

registration, unemployment allowance, self-employed, and disability allowance. Surprisingly, the p-value for age is high. Age is often one of the strongest explanatory characteristics for response in household surveys. In the LFS, however, no strong relation to response at the household level is found for the average age.

It can be concluded from Table 10.8 that the p-values have increased for all variables after the callback response is added to the LFS response. With the callback response included, independence is not rejected at the 5% level for variables age and disability allowance. These results are the same as for the LFS response alone. However, at the 1% level, independence of gender, average house value, and paid job are not rejected, unlike the LFS response. So there are smaller deviances from independence between response behavior and household characteristics.

TABLE 10.7 Parameters in the logistic regression model for callback response with standard errors and ρ-values

Category	Parameter
Constant	−0.207
Age	
0–34	0
35–54	0.490*
> 54	0.096
Regression	
North	0
East	−0.868*
West	0.051
South	−0.332

Note: One and two asterisks denote significance at 5% and 1% level, respectively. The first category is used as a reference category

TABLE 10.8 p-Values corresponding to bivariate tests of independence using the adjusted Pearson's χ^2

	Pearson's χ^2 p-value		
Variable	LFS response	Combined	LFS versus callback
Urbanity	<0.009	<0.009	0.00
Phone	<0.009	<0.009	0.00
Region	<0.009	<0.009	0.00
Province	<0.009	<0.009	<0.009
Gender	<0.009	0.02	0.00
Ethnicity	<0.009	<0.009	0.00
House value	<0.009	0.03	0.02
Perc. nonnative	<0.009	0.00	0.00
Household type	<0.009	0.00	0.01
Job	<0.009	0.04	0.10
Disability allowance	0.00	0.16	0.54
Social allowance	<0.009	0.00	0.10
CWI registration	<0.009	0.38	0.26
Unemployment all	0.73	0.86	0.62
Self-employed	0.52	0.98	0.17
Average age	0.32	0.66	0.05
Job status	—	—	0.56
Employment status	—	—	0.32

Note: The tests were done for LFS response versus sample, combined LFS and callback response versus sample, and LFS response versus callback response.

The distributions of these characteristics in the combined response are closer to the distributions in the original sample. In general, it can be said that a more representative response was obtained by the callback re-approach.

The last column of Table 10.8 gives the p-values corresponding to weighted bivariate tests of independence of response group, LFS or callback, and the various auxiliary variables. The respondents from LFS and callback re-approach were first isolated and then it was tested whether households with different characteristics have different distributions in the two groups. As the analysis is now restricted to respondents, the bivariate tests can also be done for the target variables job status of the household and employment status of the household.

As can be seen in Table 10.8, independence for the two target variables is not rejected. Between the two response groups there is no significant difference in job and employment status at the household level. Also none of the job-related auxiliary variables leads to a rejection of independence of the response group at the 5% level. For all other auxiliary variables, independence is rejected at the 5% level. The strongest differences are in the geographical variables and the availability of a telephone number. This suggests that different respondents are obtained with the CBA.

Three logistic regression models are fitted next:

- For the probability of response in the regular LFS.
- For the probability of response in the combined response of LFS and callback re-approach.
- For the probability of being a LFS respondent given that a household responded in either the LFS or callback.

The three resulting models are shown in Table 10.9. The second column shows the parameters of the model for response in the LFS, the third column shows the parameters of the model for the response in the LFS plus the callback approach, and the fourth column shows the parameters of the model for the LFS response given response in either the LFS or callback approach.

In all cases a forward selection algorithm with a 5% significance level was applied. Hence only significant variables are shown. For the last two regressions a weighted logistic regression was done to account for the sampling of nonrespondents in the callback re-approach. For the third regression the target variables job and employment status are added.

It can be concluded from Table 10.9 that the callback re-approach does bring in new households; LFS and callback respondents are different with respect to region, having a registered phone and ethnicity. As a result the logistic regression model for the combined response has less or weaker interactions than the model for the LFS response. The only exception is the variable *Telephone*, which becomes slightly more unbalanced. However, no significant relation to the two target variables is found. In other words, if the demographic characteristics are kept fixed, then job and employment status do not have significantly different distributions. The callback nonresponse is missing at random. The results of the logistic regression suggest that the response became more representative, since the number of significant variables was reduced after the callback re-approach. This conforms with earlier results where it was found that the callback re-approach was not strongly selective.

10.3.2.3 The Basic-Question Approach The same analysis strategy as in Section 10.3.2.2 is followed. Bivariate tests are performed and then logistic regression models are constructed for the basic-question response.

The p-values for the tests of independence between response in the basic-question approach and auxiliary variables are given in Table 10.10. The p-values are computed for all households and for the households with listed telephone numbers. In the full population, the variables that show a significant association with response at the 1% level are telephone, gender, job, and social allowance. At a 5% level, independency is rejected additionally for the variables ethnicity and household type. When the analysis is restricted to households with a listed landline telephone, there are three auxiliary variables that give a significant explanation for response: average age of the household, having a job, and self-employment of a household core member.

TABLE 10.9 Parameters in (weighted) logistic regression models

	Analysis		
Category	LFS	LFS + callback	LFS versus callback
Constant	0.414*	0.723**	2.15**
Urbanity			
Very strong	0	0	
Strong	0.196*	0.309**	
Moderate	0.288*	0.160	
Little	0.312*	0.405**	
Not	0.258*	0.450**	
Telephone			
No	0	0	0
Yes	0.378**	0.447**	0.38**
Region			
North	0	0	0
East	−0.437*	0.090	−1.02**
West	−0.558**	−0.224	−1.03**
South	−0.091	0.276	−0.61
Household type			
Single	0	0	
Not married	0.049	0.091	
Married	0.087	−0.130	
Not married + children	0.134	0.200	
Married + children	0.372**	0.249*	
Single parent	0.314*	0.264	
Other	0.181	0.022	
>1 Household	0.483*	0.745**	
Ethnicity			
Native	0		0
Moroccan	−0.898**		−1.00**
Turkish	−0.372		−0.18
Surinam/Antilles	−0.040		0.22
Other nonwestern	−0.415**		−0.52*
Other western	−0.345*		0.18
Mix	0.023		0.31

Note: One and two asterisks denote significance at 5% and 1% levels, respectively. The first category is used as a reference category. Only variables significant at the 5% level are included in the models.

TABLE 10.10 Bivariate test of independence between response and nonresponse in the basic-question approach and auxiliary variables, for all households and for all households with a listed telephone number

	p-value	
Variable	All households	Listed households
Urbanity	0.42	0.58
Phone	<0.009	—
Region	0.97	0.76
Province	0.52	0.33
Gender	0.00	0.31
Ethnicity	0.04	0.93
House value	0.19	0.94
Household type	0.03	0.14
Job	0.00	0.02
Disability allowance	0.35	0.75
Social allowance	0.01	0.42
CWI registration	0.08	0.97
Unemployment allowance	0.40	0.60
Self-employed	0.12	0.01
Average age	0.06	<0.009

Note: Test statistic is Pearson's χ^2.

Next a logistic regression model is constructed for the probability of response when the basic-question approach is applied to LFS nonrespondents. Again, the analysis is performed for all households and listed telephone households. The model and regression parameters are given in Table 10.11.

The model shows significant differences in basic-question response for households with different scores on telephone, age, job, and gender characteristics. As can be seen in Table 10.11, the households with an average age between 35 and 54 years have a significantly lower response rate in the basic-question approach. Furthermore households that consist only of women or a mixture of men and women have a higher response rate, as do households where at least one person had a job at the Ist of January 2005.

Having a listed landline telephone has a large influence on the probability of response in the basic-question approach. This can to a large extent be explained by the data collection design. Based on telephone ownership, households were assigned to CATI or to a paper/Web questionnaire. The response in the CATI group was much higher, and this is reflected in the high regression parameter for telephone. The relation between response behavior and having a job is alarming, since having a job as of January 1, 2005, could be expected to relate strongly to unemployment in the summer of 2005. Conclusively, the BQA is more selective

TABLE 10.11 Parameters in the logistic regression for basic-question response with standard errors and p-values

Category	All households	Listed households
Intercept	−1.532**	1.302**
Phone		
No	0	
Yes	0.765**	
Average age		
<35		0
35–54		−1.379**
55+		−1.571**
Gender		
All male	0	
All female	0.458*	
Mix	0.470*	
Job		
No	0	
Yes	0.456**	
Self-employed		
No		0
Yes		−0.730*

Note: One and two asterisks denote significance at 5% and 1% levels, respectively. Each first category is used as a reference category

than the CBA; there is a stronger connection between the selected auxiliary variables and response.

The logistic model changes when the analysis is restricted to the households with a listed landline telephone. There is no longer a connection to having a job as of January 1 and to the gender composition of the household core. Differences in response are found, however, for age and self-employment within the households with a listed telephone. Household cores with an average age below 35 years respond better than the other households. The same is true for households of which none of the members is self-employed, which relates to the key items of the LFS. The relation between gender and BQA response behavior must therefore arise from the nonlisted households that received a paper and Web questionnaire. The bivariate relation between having a job and BQA response is still present for the listed households (see Table 10.10) but disappears when conditioning on age and self-employment.

The results in Table 10.11 indicate that the BQA tends to be more selective than the CBA. Strong correlations between response and auxiliary variables are found, some of which are job-related. The results do not necessarily imply, however, that it is useless to conduct a BQA; maybe the respondents that were lacking in the regular LFS have responded in the BQA. If the BQA is selective in the opposite direction of the LFS, then the combined response may come closer to being representative of the population.

Does the basic-question approach produce new types of households? To investigate this, the same strategy as in Section 10.3.2.2 was followed; first, the p-values of bivariate tests are evaluated, and next, logistic regression models are built to predict various response probabilities.

Table 10.12 contains the p-values of bivariate weighted tests for independence. The first column contains the p-values for the weighted test of independence between LFS response and the selected auxiliary variables. The p-values are for most variables similar in size as the p-values for the LFS response in Table 10.8. However, there are differences in some of the work-related variables and in average age. The independence tests now reject that a CWI registration and the average age are independent of the average age. Also the p-value for an unemployment allowance dropped considerably from 0.79 to 0.17. The additional regions in the BQA indicate a stronger dependence between response behavior and these variables. Nonetheless, for the majority of the auxiliary variables the p-values remain very low.

TABLE 10.12 p-Values corresponding to bivariate tests of independence using the adjusted Pearson's χ^2 test

	Pearson's chi-square p-value			
			LFS versus basic-question	
	LFS response	LFS + basic-question response	All	Listed
Urbanity	<0.009	<0,0001	0.01	0.02
Phone	<0.009	<0.009	0.80	—
Region	<0.009	0.00	0.10	0.25
Province	<0.009	<0.009	<0.009	<0.009
Gender	<0.009	<0.009	0.49	0.63
Ethnicity	<0.009	<0.009	0.65	0.52
House value	0.00	<0.009	0.79	0.28
Household type	<0.009	<0.009	0.05	0.06
Job	0.00	<0.009	0.66	0.72
Disability allowance	0.49	0.20	0.47	0.96
Social allowance	<0.009	<0.009	0.46	0.12
CWI registration	0.00	0.00	0.42	0.53
Unemployment all.	0.17	0.16	0.26	0.23
Self-employed	0.37	0.60	0.12	0.11
Average age	0.03	0.17	0.01	0.01
Employment status			0.42	0.61

Note: The tests are done for LFS response versus sample, combined LFS and basic-question response versus sample, and LFS response versus basic-question response (all households and listed households only).

The second column of Table 10.12 gives the p-values after the addition of the basic-question respondents. If the two sets are compared, many of the p-values appear to remain small for the combined responses. For the auxiliary variable average age, the p-value has increased and is now above the usual significance levels; independence between response and having average age of the household core is no longer rejected. Furthermore the p-value for self-employment has increased. Unlike for the CBA, it cannot be argued that the BQA response becomes more representative at the re-approach stage. Bivariate independence can still be rejected for many variables, including some of the (un)employment-related variables.

Again, the p-values are computed for weighted tests of independence of response group, LFS, or basic-question, given that the household responded. A distinction is made between all households and the listed households. The p-values are given in the third and fourth column of Table 10.12. The tests were also performed for the LFS key variable employment status, which could be derived from the basic questions in the condensed questionnaires.

The p-values show that for most of the auxiliary variables no significant differences are found in the univariate composition of the response of LFS and basic-question re-approach. This result remains the same even after the listed households have been isolated. So the indication is that the BQA respondents resemble the LFS respondents more than did the CBA respondents. The auxiliary variables urbanity, province, household type, and average age lead to a rejection of independence at the 5% level. The observed differences with respect to urbanity and province are not surprising. The LFS is a face-to-face survey, and contact rates in the LFS differ considerably among regions of the country because of interviewer workload, inaccessibility of addresses, and physical impediments. A re-approach using a different mode can be expected to be especially beneficial for receiving response from households that were not contacted in the LFS. Remarkably, the independence of the group of response (LFS or BQA) and having a listed phone number is not rejected. The BQA strengthens the selectiveness of response for having a listed phone. More households with a listed phone are obtained, although it would be better for the composition of the response to obtain more households without a listed phone.

Next, and analogous to Table 10.9, are the three logistic regression models that were constructed as shown in Table 10.13. The first model for the LFS response probability is the same as the first model in Table 10.9. The second model explains the response probability to the combined response of LFS and basic-question approach. The third model describes the probability that a household responded in the LFS given that the household is a respondent in either the LFS or basic-question approach. Additionally a model has been derived for the same probability but restricted to listed households.

In comparing the model for the combined response of LFS and BQA with the model for the LFS response, it can be seen that ethnicity, region, and household type are no longer significant while household cores with a job more often are respondents. The variables urbanity, phone, gender, and receiving a social allowance all remain in the model but give stronger and more significant contributions to explaining response. Still more households are found where one of the core

TABLE 10.13 Parameters in the (weighted) logistic regression models

Category	LFS response	LFS + basic-question response	LFS versus basic question: All	LFS versus basic question: Listed
Intercept	0.173	−0.008	1.211**	0.883**
Urbanity				
Very strong	0	0	0	0
Strong	0.205*	0.310**	0.139	0.066
Moderate	0.151	0.135	0.265	0.396
Little	0.420**	0.487**	0.459*	0.557*
Not	0.346**	0.457**	0.425*	0.493*
Telephone				
No	0	0		
Yes	0.371**	0.710**		
Ethnicity				
Native	0			
Moroccan	−0.563*			
Turkish	−0.331			
Surinam/Antilles	0.064			
Other nonwestern	−0.349**			
Other western	−0.205*			
Mix	0.134			
Gender				
All male	0	0		
All female	0.319**	0.516**		
Mix	0.123	0.494**		
Social allowance				
No	0	0		0
Yes	−0.407**	−0.433*		−0.630*
Average age				
<35				0
35–54				0.505**
55 +				0.556**
Job				
No		0		
Yes		0.264**		
Region				
North	0		0	
East	0.081		0.159	
West	−0,059		0.029	

(Continued)

TABLE 10.13 (*Continued*)

Category	LFS response	LFS + basic-question response	LFS versus basic-question	
			All	Listed
South	0.260		0.367	
Household type				
Single	0			
Not married	0.012			
Married	0.210			
Not married + children	0.109			
Married + children	0.366*			
Single parent	0.217			
Other	−0.459			
>1 household	0.447*			

Note: One and two asterisks denote significance at 5% and 1% levels, respectively. The first category is used as a reference category. Only the variables that are significant at the 5% level are included in the models.

members receives social allowance. Especially the selectiveness of response with respect to having a listed-phone has become more pronounced, as expected.

The strongest differences between LFS and BQA respondents are with respect to urbanity and region, as can be seen in Table 10.13. Here new respondents are obtained in the re-approach. As a consequence the dependence between response and region has disappeared. The picture is different for the listed households. If the BQA is restricted to this group, then apart from the more urban areas, more households with a social allowance are attracted as well as younger households.

10.3.2.4 R-Indicators In Chapter 7 the R-indicator was presented as a measure of representativity for the response. It is interesting to see whether the combined response for both re-approaches has become more representative with respect to a number of auxiliary variables. In Table 10.14 the R-indicators for the different groups in the re-approach study are given.

The response probabilities used to calculate the R-indicator were estimated using a fixed set of demographic and socioeconomic auxiliary variables, including job status. The results confirm the analyses in the previous sections. The additional response obtained in the basic-question approach did not lead to a more

TABLE 10.14 R-indicators for the different groups in the re-approach study of the LFS

Response	Response rate	R-indicator	Maximal bias
LFS	62%	79%	8.4%
LFS + basic-question	76%	77%	7.6%
LFS + call-back	77%	85%	4.9%

representative response subset, whereas the additional response in the callback approach did.

10.3.3 CONCLUSIONS

A comparison of the callback response and nonresponse reveals only small differences between these groups. There were differences found for average age of the household and part of the country where the household lives. Hence the composition of the response to the callback approach resembles the composition of the sample for the callback approach.

The small differences between callback response and nonresponse suggest that with respect to the available demographic, geographic, and socioeconomic characteristics, the combined response has become more similar to the sample, and this is confirmed by the analyses. A comparison of the combined response to the original LFS sample yielded smaller p-values and hence less significant differences. So the LFS and callback respondents are indeed different. When the two groups were compared directly, especially apparent were the differences concerning geographic variables, whether or not the household has a listed landline telephone, and ethnic background. The follow-up response produced an overrepresentation of the most urbanized, western parts of The Netherlands. Also households without a listed telephone were overrepresented in the follow-up. Finally, Moroccan and nonwestern households other than Moroccan and Turkish households were overrepresented in the study.

As to the key LFS questions, no significant difference was found between the job and employment status of the households in the LFS and the callback responses. This important finding held both in a bivariate comparison and in a multivariate setting. So this infers that there is no evidence that households are not missing at random in the LFS when it comes to these items.

The analysis of the callback response indicates that callback respondents are more similar to the remaining nonrespondents than to the LFS respondents. The selected refusals are the largest nonresponse group; they make up 77% of the reapproach sample. Not surprisingly, the follow-up response rate was highest for the former noncontacts, 58%. This finding conforms to that of Lynn et al. (2002), who demonstrated an intensive follow-up to be especially successful for the difficult to contact cases. However, unlike Lynn et al. (2002), this study has not found the survey estimates of employment and unemployment to be affected by the additional follow-up response.

The picture for the basic-question respondents is different. This difference is mostly due to the design of the basic-question approach. When comparing the basic-question response to the basic-question nonresponse, differences are found in telephone ownership, age, having a paid job, and gender. The results indicate that the basic-question respondents are more similar to the LFS respondents than to the remaining nonrespondents. There is still no difference in employment status between the two groups. However, new respondents are only obtained with respect to region and age. The total combined response of the LFS and the basic-question approach is slightly more representative with respect to these characteristics but not with respect to other characteristics. The improved composition of region and age again is mostly due to the conversion of former noncontacts and unprocessed cases. Region and age correlate strongly to the ability to contact households in a face-to-face survey.

The response to a Web survey could be increased by an improved design of the Web questionnaire, for instance, using the total design method (TDM) suggested by Dillman (2000). The low response rate in the paper and Web group of the basic-question approach is attenuated by the types of LFS nonrespondents assigned to this strategy. Noncontacts more often do not have a listed telephone than refusals. It is reasonable to expect lower conversion rates for refusals, especially when relatively little effort is put into converting them. Another possible explanation for the low response rates in the paper and Web group is raised by Griffin et al. (2001), who found that offering multiple modes of response in a mailing led to a lower response rate. So using only paper questionnaires could prevent a break in the response process.

In both re-approach types the conclusions about the employment status of the nonconverted nonrespondents should be drawn with care. While there is an indication in both cases that the nonrespondents resemble the converted non-respondents in terms of socioeconomic characteristics, the low conversion rate of refusals leaves ample room for differences in employment status. However, this study gives no indication that the nonresponse of the LFS may follow a not-missing-at-random data pattern for this key item.

10.4 Summary

In this chapter the callback approach and the basic-question approach are presented. The callback is more costly and has a considerably longer fieldwork period. The basic-question approach can be applied when, due to time or budget constraints, the callback approach is not an option.

The results of the callback approach and the basic-question approach to the Dutch LFS have been analyzed with linked data from external administrative sources. There were two main research questions: What households participate in the re-approach strategy, and do these new households improve the composition of the response? In addition it was analyzed whether the new households are different with respect to their employment status. In other words, it was analyzed whether the LFS nonresponse is not missing at random.

Both re-approaches were successful in obtaining additional response. The response rates in the callback and basic-question approaches were 43% and 40%, respectively. The rather passive way unlisted households are re-approached in the BQA is reflected in the response percentage: for CATI, 50% of the households participated in the survey, while in the Web and paper combination only 23% of the households responded.

The analysis of the callback response indicates that callback respondents are more similar to the remaining nonrespondents than to the LFS respondents. The survey estimates of employment and unemployment do not seem to be affected by the addition of the follow-up response.

When comparing the basic-question response to the basic-question nonresponse, differences are found with respect to telephone ownership, age, having a paid job, and gender. This is mostly due to the design of the basic-question approach. The results indicate that the basic-question respondents are more similar to the LFS respondents than to the remaining nonrespondents. There is, however, no difference in employment status between the two groups.

10.5 KEY TERMS

Basic-question: A key question of a survey. With the answers to the basic questions the most important conclusion of the survey can be formulated.

Basic-question approach: A technique to re-approach nonrespondents. A sample of the nonrespondents is drawn and the selected nonrespondents are persuaded to cooperate by offering them a much shorter questionnaire containing just a few basic questions.

Callback approach: A technique to re-approach nonrespondents. A sample of the nonrespondents is drawn and the selected nonrespondents are re-approached. In the re-approach one may choose to use a different data collection mode, specially trained interviewers, an extended fieldwork period, and incentives to achieve cooperation from the prospective respondents.

Linked data: Data from an external administrative source that can be linked to the survey sample and is available for both respondents and nonrespondents.

Mixed mode design: A design that uses multiple survey modes for the data collection of a survey.

Next-birthday method: A within-household respondent selection method that asks for the person in the household with the closest next birthday to answer the survey questions.

Politz–Simmons approach: A technique to reduce the nonresponse bias due to not-at-home reason by estimating the contact probability. This technique will typically be applied in surveys where one can expect a connection between being not at home and the target variables.

Within-household respondent selection method: A method to randomly select one person within a household. The three methods that are most often used are: the Kish-grid, the last-birthday method, and the next-birthday method.

10.6 REFERENCES

Arts, C.H., and Hoogteijling, E.M.J. (2002), Het Sociaal Statistisch Bestand 1998 en 1999. *Sociaal Economische Maandstatistiek*, 12, pp. 13–21.

Banning, R., and Schouten, B. (2009), *A Mixed-mode Follow-up of Panel Refusers in the Dutch LFS.* Report DMV-2009-08-01-RBNG-BSTN. Statistics Netherlands, The Hague.

Bethlehem, J., Cobben, F., and Schouten, B. (2005), *Projectplan Experiment Het meten van Non-responsvertekening in de EBB.* Report 09-05-TMO-BSTN, Statistics Netherlands, Voorburg.

Biemer, P.P., and Lyberg, L.E. (2003), *Introduction to Survey Quality.* Wiley, New York.

Binson, D., Canchola, J.A., and Catania, J.A. (2000), Random Selection in a National Telephone Survey: A Comparison of the Kish, Next-Birthday, and Last-Birthday Methods. *Journal of Official Statistics,* 16, pp. 53–59.

Cobben, F. (2007), *Mode effects in a Basic Question Approach for the Dutch LFS.* Discussion paper 07003, Statistics Netherlands, Voorburg.

Cobben, F., and Schouten, B. (2007), *Are You the Next to Have Your Birthday? Congratulations! You may Answer Some Questions.* Report DMV-2007-01-09-FCBN. Statistics Netherlands, Voorburg.

Dillman, D.A. (2000), *Mail and Internet Surveys: The Total Design Method.* Wiley, New York.

Elliott, M.R., Little, R.J.A., and Lewitzky, S. (2000), Subsampling Callbacks to Improve Survey Efficiency. *Journal of the American Statistical Association* 95, pp. 730–738.

Gaziano, C. (2005), Comparative Analysis of Within-household Respondent Selection Techniques. *Public Opinion Quarterly* 69, pp. 124–157.

Griffin, D.H., Fischer D.P., and Morgan, M.T. (2001), *Testing an Internet Response Option for the American Community Survey.* Paper presented at the Annual Conference of the AAPOR, Montreal, Quebec, May 17–20.

Groves, R.M. (1989), *Survey Errors and Survey Costs.* Wiley, New York.

Hansen, M.H., and Hurwitz, W.H. (1946), The Problem of Nonresponse in Sample Surveys, *Journal of the American Statistical Association* 41, pp. 517–529.

Keeter, S., Miller, C., Kohut, A., Groves, R.M., and Presser, S. (2000), Consequences of Reducing Nonresponse in a National Telephone Survey. *Public Opinion Quarterly* 64, pp. 125–148.

Kersten, H.M.P., and Bethlehem, J.G. (1984), *Exploring and Reducing the Non-response Bias by Asking the Basic Question.* Report BPA 627-84-M1. Statistics Netherlands, Voorburg.

Kish, L. (1949), A Procedure for Objective Respondent Selection within The Household. *Journal of the American Statistical Association* 44, pp. 380–387.

Lavrakas, P.J., Stasny, E.A., and Harpuder, B. (2000), A Further Investigation of the Last-Birthday Respondent Selection Method and Within-unit Coverage Error. *Proceedings of the Survey Research Methods Section, American Statistical Association,* pp. 890–895.

Lynn, P. (2003), PEDAKSI: Methodology for Collecting Data about Survey Non-Respondents. *Quality and Quantity* 37, pp. 239–261.

Lynn, P., Clarke, P., Martin, J., and Sturgis, P. (2002), The Effects of Extended Interviewer Efforts on Nonresponse Bias. In: Groves, R.M., Dillman, D.A., Eltinge, J.L.,and Little, R.J.A. (eds.), *Survey Nonresponse.* Wiley, New York, pp. 135–148.

Nagelkerke, N.J.D. (1991), Miscellanea. A Note on a General Definition of the Coefficient of Determination. *Biometrika* 78, pp. 691–692.

Oldendick, R.W., Bishop, G.F., Sorenson, S.B., and Tuchfaber, A.J. (1988), A Comparison of the Kish and Last Birthday Methods of Respondent Selection in Telephone Surveys. *Journal of Official Statistics* 4, pp. 307–318.

Salmon, C.T., and Nichols, J.S. (1983), The Next-Birthday Method of Respondent Selection. *Public Opinion Quarterly* 47, pp. 270–276.

Stoop, I.A.L. (2001), *Early, Late, Cooperative, Reluctant, Persuadable and Follow-up Respondents. A study of Nonresponse Using Fieldwork Records, Frame Data, Wave Data and a Follow-up Survey.* Paper presented at the 12th International Workshop on Household Survey Nonresponse, September 12–14, Oslo.

Stoop, I.A.L. (2004), Surveying Nonrespondents. *Field Methods* 16, pp. 23–54.

Stoop, I.A.L. (2005), *The Hunt for the Last Respondent.* PhD thesis. Faculty of Social Sciences, Utrecht University, Utrecht.

Van den Brakel, J., and Renssen, R. (1998), A Field Experiment to Test Effects of an Incentive and a Condensed Questionnaire on Response Rates in The Netherlands Fertility and Family Survey. *Research in Official Statistics* 3, pp. 55–63.

Van Goor, A., and Van Goor, H. (2003), De Bruikbaarheid van de Methode van de Centrale Vraag voor het Vaststellen van Non-responsvertekeningen: Vier Empirische Tests. *Sociologische Gids* 50, pp. 378–391.

Voogt, R.J.J. (2004), *Nonresponse Bias, Response Bias and Stimulus Effects in Election Research.* PhD thesis. Faculty of Social Sciences, University of Amsterdam, Amsterdam.

10.7 EXERCISES

Exercise 10.1 Which statement about the callback approach is not correct?

(a) In the callback approach, nonrespondents should be re-approached by the interviewer that originally obtained the nonresponse.

(b) The response in the callback approach can be used for the adjustment of a nonresponse bias if the nonresponse in the callback approach is missing completely at random.

(c) The response in the callback approach can be used for adjustment of nonresponse bias if everyone responds in the callback survey.

(d) The callback approach increases the length of the fieldwork.

Exercise 10.2 Which statement about the basic-question approach is not correct?

(a) Ideally the basic question must be answered by both respondents and nonrespondents so that the respondents and nonrespondents can be compared.

(b) The basic question must be presented to respondents in a situation that resembles the nonresponse situation as much as possible to avoid interviewer effects, questionnaire effects, and mode effects.

(c) The basic question should be one of the first questions in the questionnaire.

(d) The basic-question approach should be applied immediately after a nonresponse is obtained.

(e) It is important that the answer to the basic question not be changed when the answers to subsequent questions indicate that the answer to the basic question is wrong.

(f) The basic-question approach should be applied when the callback approach is not an option due to time or budget constraints.

Exercise 10.3 What is the main reason the callback approach may not work in a practical situation?

(a) The nonresponse in the second phase is too low.

(b) The nonrespondents in the second phase resemble the first-phase respondents too much.

(c) Re-approached nonrespondents may show aggressive behavior.

(d) The callback approach will only work in face-to-face surveys.

Exercise 10.4 When applying the basic-question approach, where should the basic questions ideally be placed in the questionnaire?

(a) The questionnaire should start with the basic questions.

(b) The questionnaire should end with the basic questions.

(c) The basic questions should be somewhere in the middle of the questionnaire.

(d) It does not matter where the basic questions are.

Exercise 10.5 Under which circumstances is estimating contact probabilities useful?

(a) When making callback visits is expensive.

(b) When questions may also be answered by other members of the household.

(c) When there exists a clear relationship between the target variables of the survey and refusal patterns.

(d) When there exists a clear relationship between the target variables of the survey and being at home or not.

Exercise 10.6 In what situation would the Politz and Simmons approach be applicable to participation probabilities?

(a) When respondents are able to predict their own response behaviour in similar surveys.

(b) When respondents know how people in general feel about the survey.

(c) When respondents would again be respondents when re-approached.

(d) When respondents are able to guess how often they have responded over the last few years.

CHAPTER ELEVEN

The Use of Response Propensities

11.1 Introduction

The introduction of the propensity score method has drawn much attention in the survey methodological literature to the idea of response probabilities (Rosenbaum and Rubin, 1983). The propensity score method derives from evaluation studies that estimate average treatment effects. In treatment effect studies there usually are two groups: one group that receives the treatment, and one group that serves as a control group and does not receive the treatment. The statistic of interest is the effect of the treatment. However, to measure this effect without bias, it is necessary to remove all possible differences in outcomes that arise from the different composition of the treatment and the control groups. For this purpose the propensity score is introduced as a way to balance the composition of the two groups.

The idea of using the propensity score method in survey methodology was introduced by Harris Interactive (for e.g., see Taylor et al., 2001). Harris Interactive used the propensity score method to solve the problems of undercoverage and self-selection in self-selecting Internet panels. Schonlau et al. (2004) and Lee (2006) elaborate on the technique of the propensity score method with in this context. It begins with the availability of a reference survey that does not have the same problems of undercoverage and self-selection. Ideally this should be a probability-based survey with full response, or where the response is missing completely at random (MCAR). This reference survey is used as a benchmark for the Internet

Handbook of Nonresponse in Household Surveys, By Jelke Bethlehem, Fannie Cobben, and Barry Schouten

respondents by balancing the composition of the attributes of the Web respondents so that they are similar to the composition of the attributes of the reference survey respondents.

Therefore these attributes, as measured in both surveys, reflect the differences between Internet respondents and sample elements that would not respond to an Internet panel request. Duffy (2005) uses behavioral, attitudinal, and sociodemographic information for this purpose. These attributes are sometimes called *webographic* or *psychographic variables.*

The propensity score is determined as the conditional probability that a sample element responds to the Internet panel, given the attributes. The actual value of the propensity score has no meaning, but it is assumed that sample elements with approximately the same propensity score are similar with respect to the attributes that are used in estimating the propensity scores. Based on this assumption, the self-selection bias and the bias due to undercoverage are removed from the Internet panel. Hence the approach of Harris Interactive depends on having a good reference survey.

There are some disadvantages to this method. Bethlehem (2007) shows that the variance of the estimates based on this type of survey is dominated by the smaller of the two. Thus, if the sample size of the reference survey is small, the variance of the estimates for the target variables in the Internet panel is large, regardless of the size of the internet panel. In addition, it is not realistic to assume that the reference survey does not experience problems with undercoverage or nonresponse. Furthermore the panel is usually an Internet survey, and the reference survey is telephone or face-to-face. As such, there is also a risk of differential measurement errors, or mode effects.

The methods presented in this chapter use response probabilities, or response propensities, for the nonresponse bias adjustment without the use of a reference survey. First, the response propensity is introduced in Section 11.2.1. In Section 11.2.2 the adjustment weighting approach to nonresponse and the usual methods to estimate population characteristics based on a survey with nonresponse are outlined. These methods can be modified to incorporate response propensities. In Section 11.2.3 a number of modifications to nonresponse adjustment methods are described that use response propensities. A number of methods is applied to the General Population Survey (GPS). This is the topic of Section 11.3.

11.2 Theory

11.2.1 THE RESPONSE PROPENSITY

Let $k = 1, 2, \ldots, N$ be the population of interest. The first-order inclusion probabilities are denoted by π_k for $k = 1, \ldots, N$. The selection indicator is denoted by a_k, which is 1 if element k is selected in the sample; otherwise, it is 0. Let $d_k = 1/\pi_k$ be the design weight.

A sample of size n is selected from the population. For every sample element i there is a vector of auxiliary variables, denoted by $X_i = (X_{i1}, X_{i2}, \ldots, X_{ip})'$. This information comes, for instance, from the population register. In addition there is a survey item Y that is only observed for respondents.

It is assumed that every element k in the population has a nonzero, unknown response probability, denoted by ρ_k. This corresponds to the random response model. If element k is selected in the sample, a random mechanism is activated that results with probability ρ_k in response and with probability $(1 - \rho_k)$ in nonresponse.

The response probabilities ρ_k are theoretical quantities. The definition of response probabilities is not straightforward. It involves at some stage a decision on how to deal with the dependence of the response probabilities on the circumstances under which the survey is being held. For example, the number and timing of contact attempts and the interviewer characteristics. If these circumstances change, it is very likely that the individual response probabilities also change.

In addition, response probabilities vary over time. However, the more conditions are imposed on the response probabilities, the more fixed they become. The fixed response model arises as a special case of the random response model when the response probabilities are viewed conditional on very detailed circumstances. No variation is left, and the response probabilities become deterministic. In the case where no conditions are imposed, the response behavior is stochastic, which is the case in the random response model.

The response probability ρ is a latent variable, it is not observed, but instead the corresponding response indicator R is observed. The binary response indicator R_k equals 1 if element k responds to the survey request; otherwise, it is 0. A vector R of response indicators can be introduced, namely $R = (R_1, R_2, \ldots, R_N)'$, where $P(R_k = 1) = \rho_k$ and $P(R_k = 0) = 1 - \rho_k$.

These probabilities can be estimated for the sample elements ($a_k = 1$). By using an appropriate model based on auxiliary information, sample-based estimates of the response probabilities can be computed:

$$\hat{\rho}_k = \hat{\rho}(X_k) = P(R_k = 1 | a_k = 1, X_k) \tag{11.1}$$

for $k = 1, 2, \ldots, n$; $\hat{\rho}_k = \hat{\rho}(X_k)$ is referred to as the *response propensity*. The response propensity is the estimated response probability conditional on the sample and the individual characteristics X_k. The response propensities can be estimated by, for example, a logit or a probit model.

11.2.2 TRADITIONAL NONRESPONSE ADJUSTMENT METHODS

11.2.2.1 Full Response Case Horvitz and Thompson (1952) show that when the inclusion probabilities π_k are known and nonzero, it is always possible to unbiasedly estimate the population mean by

$$\bar{y}_{HT} = \frac{1}{N} \sum_{k=1}^{N} \frac{a_k Y_k}{\pi_k}. \tag{11.2}$$

The Horvitz–Thompson estimator can be used as a general tool to construct unbiased estimators for many sampling designs. The estimator (11.2) uses the auxiliary information from the sampling design through the inclusion probabilities π. Only selected elements are included ($a_k = 1$), and the value for the survey item Y_k is weighted by the inverse of the first-order inclusion probability π_k. This ensures that elements that are overrepresented in the survey, meaning that they have a large inclusion probability, receive a smaller weight, and vice versa. Let $d_k = 1/\pi_k$ be the design weight. Since $a_k = 0$ for elements not in the sample, it is possible to rewrite (11.2) as

$$\bar{y}_{HT} = \frac{1}{N}\sum_{i=1}^{n} d_i y_i, \tag{11.3}$$

where y_i is the sample value of Y_i. Adjustment weighting replaces this estimator by a new estimator

$$\bar{y}_w = \frac{1}{N}\sum_{i=1}^{n} w_i y_i, \tag{11.4}$$

where $w_i = g_i \times d_i$ and g_i is the correction weight, often referred to as g-weight, produced by a weighting adjustment method. It can easily be seen that the Horvitz–Thompson estimator (11.3) is obtained by $g_i = 1$ for all i.

11.2.2.2 The Horvitz–Thompson Estimator for Nonresponse By introducing nonresponse into the sampling theory based on the random response model, another type of randomness enters into the framework. Besides the randomness of the sample selection, every element has an unknown individual response probability ρ_k. So elements are equipped with known inclusion probabilities and unknown response probabilities.

The randomization theory, or design-based theory, now becomes a quasi-randomization theory (Oh and Scheuren, 1983). This leads to the two-phase approach for nonresponse adjustment (Särndal et al., 1992). First, a sample is selected from the target population. The selection is followed by nonresponse. This theory leads to a modified version of the Horvitz–Thompson estimator

$$\bar{y}_{HT}^{r} = \frac{1}{N}\sum_{i \in r} \frac{d_i y_i}{\rho_i}, \tag{11.5}$$

where r is the response subset. The Horvitz–Thompson estimator is extended to two phases: sample selection, as reflected in the design weight, and response, as reflected in the response probability.

However, the response probabilities are unknown in practice. Särndal (1981) proposes to insert the estimated response probabilities $\hat{\rho}_i$ into (11.5) so that

$$\hat{\bar{y}}_{HT}^{r} = \frac{1}{N}\sum_{i \in r} \frac{d_i y_i}{\hat{\rho}_i}. \tag{11.6}$$

Hence $g_i = \hat{\rho}_i^{-1}$. Bethlehem (1988) suggests replacing the unknown response probabilities ρ_i by the Horvitz–Thompson estimator for the mean response probability, namely

$$\bar{\rho}_{HT} = \frac{1}{N}\sum_{k=1}^{N} d_k R_k \tag{11.7}$$

Hence $g_i = \bar{\rho}^{-1}$. Then the modified Horvitz–Thompson estimator in case of nonresponse can be expressed as

$$\bar{\bar{y}}_{HT}^{r} = \frac{1}{N}\sum_{i\in r} \frac{d_i y_i}{\bar{\rho}_{HT}}. \tag{11.8}$$

Both modifications of the Horvitz–Thompson estimators, (11.6) and (11.8), use auxiliary information from the design stage, and in addition estimator (11.6) uses auxiliary information for the estimation of ρ.

11.2.2.3 The GREG Estimator for Nonresponse

The GREG estimator uses auxiliary information from the design stage through the inclusion probabilities. In addition the GREG estimator uses auxiliary information to make the sample representative with respect to these variables. Hence the quality of the auxiliary information is important.

According to Särndal and Lundström (2005), an ideal auxiliary variable has three features: it explains the response behavior well, it explains the survey variables well, and it identifies the most important domains for publication of the survey statistics. Särndal and Lundström (2008) note that in practice, even the best auxiliary variables will not be capable of completely removing the nonresponse bias. So they propose an indicator to compare and rank auxiliary variables in their effectiveness of adjusting for nonresponse bias. The indicator that Särndal and Lundström (2008) propose is mainly based on the first feature, explaining response behavior. Schouten (2007) proposes an algorithm to select the most influential auxiliary variables for nonresponse adjustment. His approach takes into account both the relationship with the response behavior and the survey items (features 1 and 2).

Bethlehem (1988) introduced the generalized regression (GREG) estimator to adjust for the nonresponse bias. The GREG estimator calibrates weighted sample estimates for quantitative variables to known population parameters. It is derived from the standard regression estimator, and it incorporates the adjustment for auxiliary variables as a modification of the weights. The GREG estimator can cover several auxiliary variables, unequal weights, transformations of variables, and interactions between variables.

For the case of nonresponse Bethlehem (1988) introduces the following modified version of the general regression estimator:

$$\bar{y}_{GR}^{r} = \bar{y}_{HT}^{r} + \left(\bar{X} - \bar{x}_{HT}^{r}\right)'\hat{\beta}_{r}^{(s)} \tag{11.9}$$

where $\bar{y}^r_{HT}$ is the modified Horvitz–Thompson estimator of the survey variable mean, estimator (11.5), $\bar{x}^r_{HT}$ is the analogue of $\bar{y}^r_{HT}$, and $\hat{\beta}^{(s)}_r$ is defined as

$$\hat{\beta}^{(s)}_r = \left(\sum_{i=1}^{n} d_i R_i X_i X_i' \right)^{-1} \left(\sum_{i=1}^{n} d_i R_i X_i Y_i \right). \tag{11.10}$$

Under the condition that there exists a vector λ' such that $\lambda' X_i = 1$, the general regression estimator in the case of nonresponse can be expressed as

$$\bar{y}^r_{GR} = \bar{X}' \hat{\beta}^{(s)}_r . \tag{11.11}$$

This estimator can be expressed as a weighted sum of observations on the survey variable based on the subset of respondents, or

$$\bar{y}_w = \frac{1}{N} \sum_{i \in r} w_i Y_i, \tag{11.12}$$

where again $w_i = g_i \times d_i$. The *g*-weights can be expressed as

$$g_i = 1 + (\bar{X} - \bar{x}^r_{HT})' \left(\sum_{i \in r} d_i X_i X'_i \right)^{-1} X_i. \tag{11.13}$$

For the case where the auxiliary information X is available for all elements in the population, $\bar{X}$ equals

$$\bar{X} = \frac{1}{N} \sum_{k=1}^{N} Xk, \tag{11.14}$$

and where the auxiliary information is only available for the sample elements, $\bar{X}$ is equal to

$$\bar{X} = \frac{1}{N} \sum_{i=1}^{n} d_i X_i. \tag{11.15}$$

X may also be a combination of the two: that is, population and sample level information. Consequently the GREG estimator in (11.9) can also be expressed as

$$\bar{y}^r_{GR} = \frac{1}{N} \sum_{i \in r} d_i g_i Y_i. \tag{11.16}$$

This shows that the GREG amounts to a weighting procedure.

11.2.3 NONRESPONSE ADJUSTMENT METHODS BASED ON THE RESPONSE PROPENSITY

The propensity score method was introduced by Rosenbaum and Rubin (1983). They describe the propensity score *e(X)* as the conditional probability of assignment to a particular treatment given a vector of observed auxiliary variables X. Adjustment for the one-dimensional propensity score $e(X)$ proved to be sufficient to remove the bias because of all the observed auxiliary variables X. So instead of adjusting for all observed auxiliary variables, adjustment for the univariate propensity score turned out to be good enough.

There are two important assumptions under which the propensity score method can be employed. The first one is the missing-at-random (MAR) assumption. This assumption is also known as the *conditional independence* assumption, *selection on observables*, the *unconfoundedness* assumption, or the *ignorable treatment* assumption.

For the survey response the propensity score is defined as the conditional probability that a sample element k with characteristics X_k responds to the survey, namely the response probability $\rho(X_k)$. In that case, the "treatment" is survey response R and the "outcome" is the values of the survey variables.

By the MAR assumption, the survey variables are independent of the response behavior conditional on X. This is denoted by

$$Y \perp R|X, \tag{11.17}$$

where $\perp$ indicates orthogonality, or independence. In other words, within subpopulations defined by values of the observed characteristics, there is random assignment. This implies that all variables that affect either the survey variable or the response behavior must be observed, meaning there is no selection on unobservables.

The second assumption is the *matching assumption*. It states that

$$0 < \rho(X) < 1. \tag{11.18}$$

This assumption ensures that for each value of X there are both sample elements that respond and sample elements that do not respond. Sample elements with a response probability equal to 0 or 1 cannot be compared because for these values there are no counterparts. For elements that always respond there is no need to match, but elements with $\rho(X) = 0$ must be discarded, and so they fall outside the population.

Rosenbaum and Rubin (1983) show that the conditional independence assumption given X implies conditional independence given $\rho(X)$. Put differently,

$$Y \perp R|X \Rightarrow Y \perp R|\rho(X). \tag{11.19}$$

The propensity score method thus balances the response behavior on observed auxiliary variables. Rosenbaum and Rubin (1983) provide a proof that

$$X \perp R|\rho(X), \tag{11.20}$$

which implies that sample elements with the same response propensity have the same distribution of the auxiliary variables used in the estimation of the response probability, regardless of whether or not they responded.

The attractiveness of the propensity score method lies in the fact that matching can be based on the one-dimensional response propensity instead of all the auxiliary variables used in the estimation of the response probabilities. The true response probabilities are not known in practice. Therefore they have to be estimated. In section 11.2.1 the response propensities $\hat{\rho}(X_i)$ are introduced as the estimated response probabilities.

The response probabilities have to be estimated in such a way that the balancing property (11.20) of the propensity score method still holds for the response propensities. The response propensity can be used to adjust for nonresponse in different ways. Two approaches to use response propensities are distinguished. The first approach uses the response propensities directly in the estimation of the survey variable. This approach can be applied in two distinct ways: as *response propensity weighting* and as a *response propensity GREG estimator.*

The second approach makes indirect use of the response propensities. Strata are formed based on the response propensities, but in the actual weighting the response propensities are not used. This approach is referred to as *response propensity stratification*.

In the direct approach, the response propensities are used as weights in the calculation of the estimates. Therefore this approach depends on the correct specification of the response propensity model. In the response propensity stratification approach, the final estimates depend less on the accuracy of the model that is used to calculate the response propensities because the response propensities are used to form the strata. Hence the propensity score $\rho(X)$ is smoothed.

In both the response propensity weighting and the response propensity stratification approach, the available auxiliary information is used to model the relationship between the response and the auxiliary variables. The response propensity GREG estimator involves the relationship between the survey item and the auxiliary information as well. The different approaches are presented in the following sections.

11.2.3.1 Response Propensity Weighting Kalton and Flores-Cervantes (2003) describe the method of logistic regression weighting. The response probabilities are estimated with a logistic regression model and inserted in the modified Horvitz–Thompson estimator as described in (11.6). Särndal (1981) and Little (1986) have also applied this approach.

Kalton and Flores-Cervantes (2003) note that when the auxiliary information used in the logistic regression model is a set of categorical variables, and there are no interactions in the model, logistic regression weighting is similar to raking modeling or multiplicative weighting (see Chapter 8). It is, however, more flexible in that it can include continuous variables, unlike raking modeling. Kalton and Flores-Cervantes (2003) also note that logistic regression weighting, unlike raking modeling, does not ensure that the weighted sample marginal distributions conform to the population marginal distributions.

Following the suggestion of Särndal (1981) as presented in (11.6), the response propensity can be inserted in the modified Horvitz–Thompson estimator as

$$\hat{\bar{y}}_{HT}^{r} = \frac{1}{N}\sum_{i \in r} \frac{d_i y_i}{\hat{\rho}(X_i)}. \tag{11.21}$$

This estimator is referred to as the *response propensity weighting* estimator. The only difference from (11.6) is that the response propensity is used instead of the response probability.

As presented here, response propensity weighting does not ensure that the distribution of the auxiliary variables that are used in the model for the response probabilities are calibrated to the population. Therefore the same variables should be included in the GREG model described in (11.16).

11.2.3.2 Response Propensity Stratification Another way of using the response propensities is to stratify the sample based on the response propensities. Within strata that have approximately the same value of $\hat{\rho}(X_i)$ the probability that $R_i = 1$ then does not depend on the values of X. This conclusion follows directly from the balancing property (11.20) of the propensity score method. This property implies that strata should be defined based on the response propensity; within these strata the response behavior is the same for respondents and nonrespondents.

Suppose that the sample is stratified into L strata based on the response propensities. Cochran (1968) suggests that it is enough to use five strata, meaning $L = 5$. The strata are denoted by $s_1, s_2, \ldots, s_L$. For each stratum a dummy variable is introduced, denoted by $Z_1, Z_2, \ldots, Z_L$. For an element in a certain stratum h, the corresponding dummy variable Z_h is assigned the value 1, and all other dummy variables are set to 0.

The sample size of stratum h is denoted by n_h. These sample sizes are random variables and not fixed numbers. Poststratification assigns the same weight to all elements in the same stratum. The design weight is defined as $d_i = N/n$. The correction weight g_i for element i in stratum h is defined as

$$g_i = \frac{n_h}{n_{r,h}}, \tag{11.22}$$

where $n_{r,h}$ is the number of responding sample elements in stratum h.

Consequently the poststratification estimator can be expressed as

$$\bar{y}_{PS}^{\hat{\rho}(X)} = \frac{1}{N}\left(\frac{n_1}{n_{r,1}}\sum_{i \in s_1} Y_i + \frac{n_2}{n_{r,2}}\sum_{i \in s_2} Y_i + \cdots + \frac{n_F}{n_{r,F}}\sum_{i \in s_L} Y_i\right) = \frac{1}{N}\sum_{h=1}^{L} n_{r,h}\bar{y}_r^{(h)}, \tag{11.23}$$

where $\bar{y}_r^{(h)}$ is the (unweighted) response mean for the survey variable in stratum h.

This poststratification estimator calibrates to the sample level instead of the population level due to the use of the response propensities. An additional step is required to adjust for the sampling design. Estimator (11.23) is referred to as the *response propensity stratification* estimator.

11.2.3.3 Response Propensity GREG Estimator In the response propensity weighting approach, the design weights are updated for the second phase in the sampling process by dividing them by the response propensities. The available information is used to model the relationship between the response and the auxiliary variables. This approach can be extended by regarding the relationship between the survey variable and the auxiliary information as well.

Särndal and Lundström (2005) show how the GREG estimator described in (11.16) can be extended to account for the second phase of nonresponse:

$$\bar{y}^r_{GR} = \frac{1}{N}\sum_{i\in r}\frac{d_i}{\rho_i}g_{i\rho}y_i. \tag{11.24}$$

The g-weights in (11.24) are

$$g_{i\rho} = 1 + (\bar{X}-\bar{x}^r_{HT})'\left(\sum_{i\in r}\frac{d_i}{\rho_i}X_iX'_i\right)^{-1}X_i. \tag{11.25}$$

Compared to the g-weights from the full response case, the distinction is that the modified version $\bar{x}^r_{HT}$ of the Horvitz–Thompson estimator for the mean of the auxiliary variables is used and that the design weights d_i are updated for the second phase, the response phase, by dividing them by the response probabilities ρ_i.

Based on a suitable model using auxiliary information, the unknown response probabilities ρ_i are replaced by their estimates $\hat{\rho}(X_i)$ in (11.24) so that

$$\bar{y}^r_{GR} = \frac{1}{N}\sum_{i\in r}\frac{d_i}{\hat{\rho}(X_i)}g_{i\rho}y_i, \tag{11.26}$$

and likewise

$$g_{i\hat{\rho}} = 1 + (\bar{X}-\bar{x}^r_{HT})'\left(\sum_{i\in r}\frac{d_i}{\hat{\rho}(X_i)}X_iX'_i\right)^{-1}X_i. \tag{11.27}$$

This estimator is referred to as the *response propensity GREG estimator.*

The auxiliary information X_i in (11.25) not necessarily is the same auxiliary information X_i as in (11.26). In (11.26) the relationship with the response behavior is modeled, whereas in (11.25) the relationship with Y_i is regarded.

11.3 Application

The methods presented in the previous section are applied here to the General Population Survey (GPS). For the GPS, persons were selected by a stratified two-stage sample. In the first stage, municipalities were selected within regions with probabilities proportional to the number of inhabitants. In the second stage, an equal probability sample was drawn in each selected municipality. Sampling frames were the population registers of the municipalities. The sampling design was such

TABLE 11.1 Fieldwork results of the General Population Survey

Result	Frequency	Percentage
Sample size	32,019	100.0%
Response	18,792	58.7%
Nonresponse	13,227	41.3%

that each person had the same probability of being selected (a so-called self-weighting sample). A detailed description of the GPS is given in Chapter 1. In Table 11.1 the survey results for the GPS are displayed.

The theory described in Section 11.2.3 applies for the true response propensities. These true response propensities, however, are not known in practice. Therefore they have to be estimated. This has to be done in such a way that the balancing property of the propensity score still holds. Hence, the estimated response probabilities also have to balance the distribution of the auxiliary variables between the respondents and the nonrespondents.

In Section 11.3.1 it is described how the response propensities are estimated. Then, in Section 11.3.2 the balancing property of the response propensity stratification is checked. In Section 11.3.3 the different response propensity methods are applied to the GPS.

11.3.1 ESTIMATING RESPONSE PROPENSITIES

Imbens and Rubin (2010) devote a whole chapter to the estimation of the propensity score. They describe how the model can be built with first- and second-term interactions. Next they describe how the adequacy of the specification of the model for the propensity score can be assessed by inspection of the estimated propensity score.

The model is built without quadratic terms or interactions. This is mainly done because all auxiliary variables are categorical variables. Furthermore Schouten (2004) shows that including interaction terms does not improve the model fit.

The response propensities are modeled by a logistic regression model. Other models can be used too, for instance, a probit model. Dehija and Wahba (1999) conclude that the choice of model does not particularly influence the results. Therefore estimation of response probabilities is limited to the logit model in this section.

The response indicator R_i is the dependent variable, and all the auxiliary variables are candidates for the independent variables X_i in the response propensity model. The logistic regression model can be expressed as

$$\log\left(\frac{R_i}{1-R_i}\right) = \text{logit}(\rho(X_i)) = X'_i\beta. \tag{11.28}$$

Suppose that there are P independent variables selected in the model. Then β is a $P \times 1$-vector of coefficients that correspond to the different auxiliary variables in

X_i. The response propensities $\hat{\rho}(X_i)$ are obtained by

$$\hat{\rho}_i = \hat{\rho}(X_i) = \frac{\exp(X'_i\beta)}{1+\exp(X'_i\beta)}. \quad (11.29)$$

The response propensity is the estimated response probability, conditional on the sample and the individual characteristics X_i. Hence $\hat{\rho}_i = \hat{\rho}_j$ if $X_i = X_j$, meaning persons i and j have the same probability of response if they are in the same strata defined by X.

To estimate the response probabilities, a multivariate logistic model is built. First, the bivariate relationship between the auxiliary variables and the response indicator is evaluated using Cramér's V. The results are displayed in Table 11.2. Next, a multivariate response model is constructed by starting from the most significant variable and stepwise including less significant variables until no more significant relationships with the response indicator remain.

The significance is tested with a Wald-test. The χ^2-value is given for each individual variable and for the full model. Furthermore the Nagelkerke pseudo-R^2 (Nagelkerke, 1991) is reported as a measure of the model fit. The results for the final model are given in Table 11.3 and Table 11.4. In Table 11.3 one asterisk denotes significance at a 5% level; two asterisks denote significance at a 1% level. More

TABLE 11.2 Cramér's V statistic for bivariate tests of independence between the auxiliary variables and response behavior

Auxiliary variable	V
Region of the country	0.163
Degree of urbanization	0.153
Has listed phone number	0.150
Percentage nonnatives in neighborhood	0.138
Percentage nonwestern nonnatives in neighborhood	0.133
Average house value in neighborhood	0.115
Type of nonnative	0.112
Type of household	0.106
Size of the household	0.099
Marital status	0.097
Is married	0.097
Is nonnative	0.087
Has social allowance	0.077
Age in 13 classes	0.061
Has an allowance	0.061
Children in household	0.056
Has a job	0.037
Age in 3 classes	0.030
Has disability allowance	0.021
Gender	0.011
Has unemployment allowance	0.000

TABLE 11.3 Logistic regression model for the response propensity

Variable	Category	β
Intercept		−0.203
Region	Woodlands	0.011
(Reference Greenfields)	Lowlands	−0.275**
	Highlands	−0.008
	Metropolis	−0.721**
Urbanization	Strong	−0.079
(Reference very strong)	Fairly	−0.000
	Little	−0.008
	Not	0.083
Listed phone	Yes	0.460**
(Reference no)		
Average house value	50–75 thousand	−0.039
(Reference < 50 thousand)	75–100 thousand	−0.009
	100–125 thousand	
	125–150 thousand	0.097
	150–200 thousand	0.114
	200–250 thousand	0.148
	250–300 thousand	0.175
	300–350 thousand	0.186
	350–400 thousand	0.094
	400–450 thousand	0.122
	450–500 thousand	0.085
	> 500 thousand	0.221
Ethnic background	First-generation nonwestern	−0.634**
(Reference native)	First-generation western	−0.187**
	Second-generation nonwestern	−0.244
	Second-generation western	0.006
Type of household	Couple without children	−0.323
(Reference single)	Couple with children	−0.439
	Single parent	−0.341
	Other	−0.744*
Size of household	2	0.342
(Reference 1)	3	0.411
	4	0.597
	5 or more	0.739*
Marital status	Married	0.323**
(Reference not married)	Widowed	0.087
	Divorced	0.176**

(*Continued*)

TABLE 11.3 (*Continued*)

Variable	Category	β
Has social allowance	Yes	−0.188**
(Reference no)		
Has a job	Yes	0.141**
(Reference no)		
Age	35–54 years	−0.109**
(Reference 18–34 years)	55 years and older	−0.107*
Gender	Female	0.093**
(Reference male)		

details about the response propensity model can be found in Chapter 5. Based on the logistic regression model presented in Table 11.4, the response propensities $\hat{\rho}(X_i)$ can be calculated as described in (11.29).

11.3.2 BALANCING PROPERTY

The purpose of the balancing property described in (11.19) is to obtain response propensities that balance the distribution of the auxiliary variables between the respondents and the nonrespondents. This means that within subpopulations that have the same value for the response propensity, the distribution of the auxiliary variables for the respondents should be the same as the distribution of the auxiliary

TABLE 11.4 Multivariate model for the response propensities

Variable	Wald χ^2
Having a listed phone	242.0
Region	164.8
Ethnic background	93.7
Size of household	52.6
Age (3 categories)	11.0
Has a job	23.7
Marital status	74.1
Gender	14.7
Type of household	23.6
Has a social allowance	8.6
Average housevalue	25.3
Degree of urbanization	16.3
pseudo-R^2	0.042
χ^2	1805.62
df	40

variables for the nonrespondents. Notice that this is a different purpose than obtaining an estimated value of the response probability that is closest to the true value of the response probability.

The balancing property should be checked to determine whether the response probability is adequately modeled. Ideally this is done by comparing elements with exactly the same response propensity. This can only be done if the response propensity takes on a small number of values, and thus if the auxiliary variables take on a small number of values. Since this is rarely the case in practice, the procedure is performed in strata based on the response propensity.

Suppose that the sample is stratified into L strata based on the response propensities. The strata are denoted by $s_1, s_2, \ldots, s_L$. For each stratum a dummy variable is introduced, denoted by $Z_1, Z_2, \ldots, Z_L$. For an element in a certain stratum h, the corresponding dummy variable Z_h is assigned the value 1, and all other dummy variables are set to 0.

Within each of these strata, the adequacy of the response propensity will be evaluated by checking whether

$$R_i \perp X_i | Z_h \tag{11.30}$$

for all $h = 1, 2, \ldots, L$.

The response behavior within strata has to be homogeneous, meaning the response propensity should not vary too much within strata. If this is not the case, and the response propensity varies within a stratum, then the balancing condition cannot be checked because a difference in the distribution of the auxiliary variables between the respondents and the nonrespondents can also be caused by the varying response propensity itself.

There are a number of ways to construct strata. The number of observations can be balanced over the different strata. Another option is to let the width of the strata depend on the range of the response propensities. Then the strata will be more homogeneous with respect to the response propensities. By balancing the observations, there is a risk of a wide range of response propensities within one stratum, and hence heterogeneous response behavior within one stratum. Therefore it is better to construct strata based on the values of the response propensity.

Also the number of strata has to be determined. Imbens and Rubin (2010) give an iterative procedure to determine the number of strata in which the auxiliary variables are balanced between the respondents and the nonrespondents. With this procedure, the stratum boundaries are chosen in such a way that within strata the response propensities do not vary too much; in other words, the response behavior within strata is homogeneous. Cochran (1968) suggests that it is enough to use five strata.

Once the strata are constructed, the balancing condition can be checked by testing for a bivariate relationship between the response indicator and the auxiliary variables within each of the strata. This can be done, for instance, by a χ^2-based test like Cramér's V statistic.

For the analysis of the GPS, it was decided to follow Cochran's rule of thumb of five strata. In addition strata were constructed that were homogeneous with respect to the response behavior. Table 11.5 shows the range of the response

TABLE 11.5 Response propensity strata

Stratum	Range	n_h	$n_{r,h}$
1	0.10–0.24	303	63
2	0.24–0.38	1,913	609
3	0.38–0.52	5,385	2,504
4	0.52–0.66	14,690	8,777
5	0.66–0.80	9,728	6,839
Total		32,019	18,792

propensities divided into five strata of equal width. The minimum value that is attained for the response propensity is 0.10. The maximum value is 0.80. Within these values, intervals of equal width are formed. So, more precisely, stratum 1 ranges from 0 to 0.24 and stratum 5 from 0.66 to 1. The number of sample elements in stratum h is denoted n_h; the number of responding elements in stratum h is denoted $n_{r,h}$.

As can be seen in Table 11.5, the number of responding elements increases as the response propensity increases, whereas the number of sample elements in the strata does not necessarily increase. The response propensity strata are displayed graphically in Figure 11.1. In addition to the five strata, a kernel density line was estimated. This represents the continuous distribution of the response propensity in the sample. The higher the number of strata, the more the distribution of the strata will resemble the continuous distribution. Figure 11.2 shows a histogram for 25 strata. Notice that the distribution of the strata resembles the estimated kernel density much more than the stratification with five response propensity strata. However, Cochran (1968) suggests that five strata are enough for stratification purposes.

The next step in the analysis is to the check the balancing condition. Therefore tests of independence between response behavior and the auxiliary variables in the

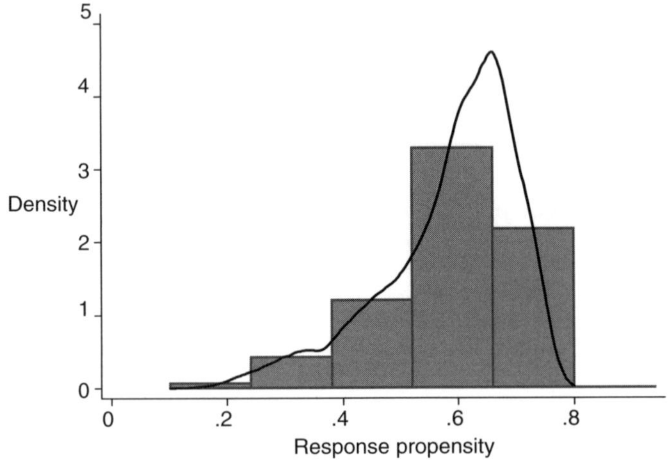

FIGURE 11.1 Response propensities divided into 5 strata.

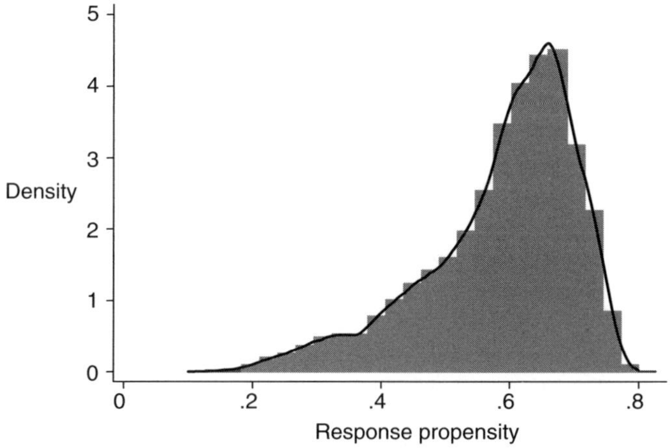

FIGURE 11.2 Response propensities divided into 25 strata.

different strata are performed. The value of Cramér's *V* statistic is reported for the test of independence between response and the auxiliary variable in the entire data set (the overall value), as well as for the test of independence between response and the auxiliary variable in the five different response propensity strata. The results for these tests are given in Table 11.6.

Cramér's *V* statistic obtains a value of 1 for complete dependence, and 0 for complete independence. Values of Cramér's *V* that are closer to 0 (compared to the overall value) suggest a decreased dependence between response and the auxiliary variable in the respective stratum.

It can be concluded from Table 11.6 that the balancing condition is satisfied in strata 2 to 5. For several auxiliary variables in the first stratum, the values for Cramér's *V* are higher than for the entire data set (in bold). This may be due to the small number of observations in the first stratum. In the first stratum the number of respondents is smaller than in the higher strata because the first stratum consists of those sample elements with the lowest response propensities.

The higher values for Cramér's *V* suggest that with respect to the variables concerned, the response propensity stratification did not result in a more balanced distribution between respondents and nonrespondents. These variables are: degree of urbanization, average house value, ethnic background, household size, marital status, social allowance, age in three categories, and gender.

There are a number of solutions to this problem. The first is to delete observations for which the response propensity is a very low or very high value. Both in the lower and the higher regions of the response propensities, the distribution of respondents and nonrespondents in strata is skewed. In lower strata, there are a lot of nonrespondents and few respondents; in the higher strata, there are a lot of respondents and a few nonrespondents. Also in methods that use the response propensity for matching, it is common to delete these observations, since there is a low probability of matching. Another solution would be to use the algorithm by Imbens and Rubin (2010), since this algorithm is aimed at constructing strata in such a way that the final stratification satisfies the balancing condition.

TABLE 11.6 Cramér's V statistic for the independence of response and auxiliary variables within the different response propensity strata

Variable	Stratum	Cramér's *V*	Variable	Stratum	Cramér's *V*
Region		0.163	Household size		0.100
	1	0.133		**1**	**0.106**
	2	0.063		2	0.031
	3	0.041		3	0.035
	4	0.020		4	0.020
	5	0.014		5	0.042
Degree of urbanization		0.153	Marital status		0.097
	1	**0.179**		**1**	**0.101**
	2	0.071		2	0.052
	3	0.039		3	0.009
	4	0.015		4	0.024
	5	0.024		5	0.017
Listed phone		0.150	Social allowance		0.077
	1	0.004		**1**	**0.080**
	2	0.035		2	0.003
	3	0.006		3	0.003
	4	0.025		4	0.011
	5	0.007		5	0.005
House value		0.116	Has a job		0.038
	1	**0.180**		1	0.023
	2	0.058		2	0.016
	3	0.039		3	0.012
	4	0.024		4	0.006
	5	0.031		5	0.035
Ethnic background		0.112	Age (3 categories)		0.030
	1	**0.117**		**1**	**0.150**
	2	0.028		2	0.037
	3	0.044		3	0.010
	4	0.016		4	0.008
	5	0.021		5	0.040
Household type		0.106	Gender		0.011
	1	0.106		**1**	**0.147**
	2	0.016		2	0.033
	3	0.021		3	0.006
	4	0.031		4	0.009
	5	0.034		5	0.004

11.3.3 APPLICATION TO GPS DATA

In this section the estimated response propensities described in Section 11.2 are used for the adjustment for nonresponse bias in the GPS data set. The methods of response propensity weighting and response propensity stratification are applied. The results for these two methods are presented in Table 11.7. In the third column the response means for the different target variables are shown. The standard errors are calculated as $\sqrt{p(1-p)/n}$, with p the percentage of the category of the target variable and n the number of respondents ($n = 18{,}972$). The standard errors for the response mean are displayed in column 4.

In columns 5 and 7 the estimates obtained by response propensity weighting respectively, stratification are displayed. The standard errors for the estimates are obtained with a nonparametric bootstrap method; see Efron and Tibshirani (1993).

The nonparametric bootstrap calculates the standard error of the estimated target variables by drawing a number $b = 1, 2, \ldots, B$ of so-called bootstrap samples. These are samples drawn independently, and with replacement from the original data set, of the same size n as the original data set. On every bootstrap sample b the estimates for the target variables are calculated. Thus B replications of the estimates are obtained, $\hat{Y}_b^{BT}$, $b = 1, 2, \ldots, B$. The standard error for the empirical distribution of these B replications is an estimate for the standard error of the estimates for the target variables:

$$s_R^{BT} = \sqrt{\frac{1}{B-1}\sum\nolimits_{b=1}^{B}\left(\hat{Y}_b^{BT} - \hat{\bar{Y}}^{BT}\right)^2}, \tag{11.31}$$

where

$$\hat{\bar{Y}}^{BT} = \frac{1}{B}\sum\nolimits_{b=1}^{B}\hat{Y}_b^{BT}. \tag{11.32}$$

is the average estimate for target variable Y. In the approximations $B = 1{,}000$.

In column 9 of Table 11.7 the results from applying the GREG estimator are displayed. The standard errors for the estimates with the GREG estimator are calculated by a first-order Taylor approximation of the linear regression estimator.

The response mean is provided so that the effect of the nonresponse bias adjustment can be considered. For instance, the percentage estimate for persons who definitely do not want to move within one year has decreased after the adjustment. So it appears that persons who do not want to move are overrepresented in the response. The same holds for persons with a PC in the household, persons in very good or good health, persons with a newspaper subscription, persons who are active in a club, persons who are at least a little interested in politics, persons with a Roman Catholic or Protestant upbringing, and persons that are employed.

The most striking effect of the nonresponse bias adjustment is found for persons who own a house. Among the respondents, the percentage of persons who own a house is 62.5%. After adjustment for the nonresponse bias, the estimated percentage of persons who own a house decreases to 58.8% with response propensity stratification, and even a little lower with response propensity weighting (58.4%) and the

TABLE 11.7 Results from application of response propensity weighting and stratification to the GPS (estimates in %)

Variable	Category	Response mean	se	Propensity weighting	se	Propensity stratification	se	GREG	se
PC in household	Yes	57.8	0.36	55.2	0.38	55.5	0.36	55.3	0.32
	No	42.6	0.36	44.8	0.38	44.6	0.36	44.7	0.32
Want to move within a year	Definitely not	72.3	0.33	70.7	0.33	70.9	0.32	70.7	0.32
	Possibly	15.2	0.26	15.7	0.27	15.6	0.26	15.7	0.27
	Cannot find something	1.7	0.09	1.8	0.10	1.8	0.09	1.8	0.10
	Definitely	8.4	0.20	9.3	0.20	9.2	0.19	9.3	0.22
	Is going to move	2.5	0.11	2.5	0.12	2.5	0.11	2.5	0.12
General health condition	Very good	22.4	0.30	21.8	0.32	21.9	0.30	21.8	0.30
	Good	55.4	0.36	54.7	0.39	54.8	0.37	54.8	0.37
	reasonable	13.1	0.25	13.6	0.25	13.6	0.23	13.6	0.25
	Varied	6.6	0.18	7.0	0.18	7.0	0.17	7.0	0.19
	bad	2.5	0.11	2.8	0.12	2.8	0.11	2.8	0.13
Has newspaper subscription	Yes	65.6	0.35	62.8	0.36	63.1	0.34	62.8	0.33
	No	34.4	0.35	37.2	0.36	36.9	0.34	37.2	0.33
Is active in a club	Yes	44.4	0.36	42.8	0.38	43.0	0.36	42.8	0.36
	No	55.6	0.36	57.2	0.38	57.0	0.36	57.2	0.36
Is interested in politics	Very interested	11.6	0.23	11.9	0.24	11.9	0.23	11.9	0.24
	Fairly interested	42.4	0.36	41.8	0.40	41.9	0.38	41.9	0.36
	Little interested	29.1	0.33	28.6	0.35	28.7	0.34	28.5	0.33
	Not interested	16.9	0.27	17.7	0.28	17.6	0.27	17.7	0.28

Job level	Very low	3.6	0.14	3.8	0.14	3.8	0.14	3.8	0.14
	Low	13.3	0.25	13.0	0.26	13.1	0.25	13.0	0.24
	Middle	22.8	0.31	21.9	0.32	22.1	0.31	22.0	0.28
	High	11.2	0.23	11.0	0.25	11.0	0.23	11.0	0.22
	Academic	4.2	0.15	4.1	0.15	4.1	0.14	4.2	0.15
	No job	44.8	0.36	46.2	0.37	46.0	0.36	46.1	0.23
Level of education	Primary	20.0	0.29	21.3	0.29	21.1	0.28	21.2	0.29
	Junior secondary	9.2	0.21	9.2	0.22	9.2	0.21	9.2	0.21
	Prevocational	17.4	0.28	17.0	0.29	17.0	0.28	17.0	0.28
	Senior secondary	6.9	0.19	7.1	0.19	7.1	0.18	7.1	0.19
	Post secondary	28.0	0.33	26.9	0.33	27.1	0.32	26.9	0.32
	Higher professional	13.3	0.25	13.3	0.26	13.3	0.25	13.3	0.25
	University	5.2	0.16	5.3	0.17	5.3	0.16	5.4	0.17
Owns a house	Yes	62.5	0.35	58.4	0.37	58.8	0.35	58.3	0.30
	No	37.5	0.35	41.6	0.37	41.2	0.35	41.7	0.30
Religious denomination	None	37.2	0.35	38.5	0.37	38.3	0.36	38.7	0.34
	Roman Catholic	33.7	0.34	32.0	0.36	32.2	0.35	31.9	0.31
	Protestant	22.3	0.30	21.1	0.32	21.2	0.31	21.1	0.28
	Islam	1.6	0.09	2.7	0.10	2.6	0.09	2.7	0.10
	Other	5.3	0.16	5.7	0.17	5.7	0.16	5.7	0.18
Employment situation	Works 12 hours or more	55.2	0.36	53.9	0.37	54.0	0.36	53.9	0.23
	Works less than 12 hours	6.1	0.17	5.8	0.18	5.9	0.17	5.8	0.17
	Does not work	38.7	0.36	40.3	0.35	40.1	0.35	40.3	0.21

GREG estimator (58.3%). Persons who own a house seem to be more inclined to respond to the surveys than persons who do not own their own house.

The results for the job level and the level of education are less clear. Persons with a very low job level or no job seem to be underrepresented. The same holds for persons with a primary or senior secondary school education, as well as for persons with a university degree.

In general, the estimates for the three different nonresponse bias adjustment methods are very similar. The three methods always agree on the direction in which the estimate must be adjusted, and the GREG estimator usually adjusts the most, followed by response propensity weighting. All estimates differ significantly from the response mean but fall in the confidence intervals of the estimates made with other methods.

The standard errors for response propensity weighting are slightly higher than for the other methods. This is because with response propensity weighting, the response propensities are directly used in the estimation of the target variables. The GREG estimator has the lowest standard errors. This is also one of the advantages of the GREG estimation, namely variance reduction.

It can be concluded that the adjustment technique had a small impact on the size of the adjustment. Apparently the information used for the adjustment is considered most important. It is remarkable how close the estimates based on response propensity stratification are to the estimates based on the GREG estimator. The response propensity stratification method based on five strata is able to remove most of the nonresponse bias relative to the other estimators.

11.4 Summary

The response propensity is the conditional probability that a person responds to the survey request, given the available background characteristics. In order to compute response propensities, auxiliary information for all sample elements is needed.

The attractiveness of methods based on response propensities is that instead of adjusting for all observed auxiliary variables, only the univariate response propensity score can be used for the adjustment. The purpose is to obtain response propensities that balance the distribution of the auxiliary variables between the respondents and the nonrespondents. In that case the balancing property holds. It means that within subpopulations that have the same value for the response propensity, the distribution of the auxiliary variables for the respondents is the same as the distribution of the auxiliary variables for the nonrespondents.

The response propensities can be used in the estimation of the target variables directly, by using the response propensities as weights, or indirectly, by forming strata based on the response propensities. In the direct approach, the response propensities are used as weights in the calculation of the estimates. Therefore this approach depends on the correct specification of the response propensity model. In the response propensity stratification approach, the final estimates depend less heavily on the accuracy of the model that is used to calculate the response propensities because the response propensities are used to form the strata. However, the available observations in each stratum are determined by the response propensities.

The response propensity strata have to be constructed so that within the strata the response behavior is homogeneous. Cochran (1968) suggested that it is enough to use five strata. Once the strata are constructed, the balancing condition can be checked for by testing for a bivariate relationship between the response indicator and the auxiliary variables within each of the strata.

Upon application of the response propensity methods to the GPS, the corresponding estimates were close to the estimates obtained by the GREG estimator. This result is remarkable in that only five strata were used for the response propensity stratification method. The standard errors for the two response propensity based methods were slightly higher than for the GREG estimator.

Use of the response propensities alone, as in response propensity weighting and response propensity stratification, does not ensure that the distribution of the characteristics of sample elements will be calibrated to known distributions in the population. To ensure that all the auxiliary variables will be reproduced, the response propensities have to be used in combination with the GREG estimator.

11.5 KEY TERMS

Balancing condition Within subpopulations that have the same value for the response propensity, the distribution of the auxiliary variables for the respondents is the same as the distribution of the auxiliary variables for the nonrespondents.

Propensity score method The method that originates from evaluation studies to estimate average treatment effects. In treatment effect studies there usually are two groups: one that receives the treatment, and one group that serves as a control group and does not receive the treatment. The statistic of interest is the effect of the treatment. However, to measure this effect without bias, it is necessary to remove all possible differences in outcome that arise from the different compositions of the treatment and the control groups. For this purpose the propensity score is used to balance the composition of the two groups.

Propensity score A type of probability measure introduced by Rosenbaum and Rubin (1983) as the conditional probability of assignment to a particular treatment given a vector of observed auxiliary variables X.

Response probability A theoretical quantity that gives the probability that a person will respond to a survey.

Response propensity GREG estimator A generalized regression estimator to adjust for a nonresponse bias where the weights are updated for nonresponse using the response propensities. This method makes direct use of the response propensities.

Response propensity stratification A poststratification method to adjust for nonresponse bias, where the strata are based on the response propensities. This method indirectly uses the response propensities.

Response propensity weighting A weighting method to adjust for nonresponse bias, where the weights are based on the response propensities and directly used in the estimation of the target variable.

Response propensity The estimated response probability $\hat{\rho}(X_i)$ for person i conditional on the sample and the individual characteristics X_i.

11.6 REFERENCES

Bethlehem, J.G. (1988), Reduction of Nonresponse Bias through Regression Estimation. *Journal of Official Statistics* 4, pp. 251–260.

Cochran, W. (1968), The Effectiveness of Adjustment by Subclassification in Removing Bias in Observational Studies. *Biometrics* 24, pp. 205–213.

Dehija, R., and Wahba, S. (1999), Causal Effects in Non-Experimental Studies: Re-Evaluating the Evaluation of Training Programs. *Journal of the American Statistical Association* 94, pp. 1053–1062.

Duffy, B., Smith, K., Terhanian, G., and Bremer, J., (2005), Comparing Data from Online and Face-to-Face Surveys. *International Journal of Market Research* 47, pp. 615–639.

Efron, B., and Tibshirani, R.J. (1993), *An Introduction to the Bootstrap.* Chapman and Hall/CRC, Boca Raton, FL.

Horvitz, D.G., and Thompson, D.J. (1952), A Generalization of Sampling Without Replacement from a Finite Universe. *Journal of the American Statistical Association* 47, pp. 663 –685.

Imbens, G., and Rubin, D.B. (2010), *Causal Inference in Statistics.* Forthcoming.

Kalton, G., and Flores-Cervantes, I. (2003), Weighting Methods. *Journal of Official Statistics* 19, pp. 81–97.

Lee, S. (2006), Propensity Score Adjustment as a Weighting Scheme for Volunteer Panel Web Surveys. *Journal of Official Statistics* 22, pp. 329–349.

Little, R. (1986), Survey Nonresponse Adjustments for Estimates of Means. *International Statistical Review* 54, pp. 139–157.

Nagelkerke, N.J.D. (1991), Miscellanea, A Note on a General Definition of the Coefficient of Determination. *Biometrika* 78, pp. 691–692.

Oh, H.L., and Scheuren, F.J., (1983), Weighting Adjustment for Unit Nonresponse. In: Madow, W.G., Olkin, I.,and Rubin, D.B. (eds.), *Incomplete Data in Sample Surveys,* Vol. 2, pp. 143–184.

Rosenbaum, P.R., and Rubin, D.B. (1983), The Central Role of the Propensity Score in Observational Studies for Causal Effects. *Biometrica* 70, pp. 41–55.

Särndal, C.E. (1981), Frameworks for Inference in Survey Sampling with Application to Small Area Estimation and Adjustment for Non-Response. *Bulletin of the International Statistical Institute* 49, pp. 494- 513.

Särndal, C.E., and Lundström, S. (2005), *Estimation in Surveys with Nonresponse.* Wiley, Chichester, UK.

Särndal, C.E., and Lundström, S. (2008), Assessing Auxiliary Vectors for Control of Nonresponse Bias in the Calibration Estimator. *Journal of Official Statistics* 24, pp. 167–191.

Särndal, C.E., Swensson, B.,and Wretman, J. (1992), *Model Assisted Survey Sampling.* Springer, New York.

Schonlau, M., Zapert, K., Simon, L.P., Sanstad, K.H., Marcus, S.M., Adams, J., Spranca, M., Kan, H., Turner, R., and Berry, S.H., (2004), A Comparison between Responses from a Propensity-Weighted Web Survey and an Identical RDD Survey. *Social Science Computer Review* 22, pp. 128–138.

Schouten, B. (2004), Adjustment for Bias in the Integrated Survey on Living Conditions (POLS) 1998. Discussion paper 04001. Statistics Netherlands, Voorburg.

Schouten, B. (2007), A Selection Strategy for Weighting Variables under a Not-Missing-At-Random Assumption. *Journal of Official Statistics* 23, pp. 51–68.

Taylor, H., Bremer, J., Overmeyer, C., Siegel, J.W., and Terhanian, G. (2001), The Record of Internet-Based Opinion Polls in Predicting the Results of 72 Races in the November 2000 U.S. Elections. *International Journal of Market Research* 43, pp. 127–136.

11.1 EXERCISES

Exercise 1.1 The aim of this exercise is to perform a propensity weighting adjustment as described by expression (11.21) using the data from the GPS. The data can be found on the website. First, estimate response probabilities using a binary logistic regression model and all the available auxiliary information. In SPSS you can use the forward selection strategy based on the Wald criterion to do this. Save the estimated response propensity as *rprob* and label the variable *Response propensity*. Compute weights by taking the inverse of the estimated response probabilities *rprob*.

Use **Transform | Compute** to introduce a new variable *propwght.* Its value is obtained for a respondent by *propwgt* = 1/*rprob*.

To be scaled properly, the weights have to be divided by average response probability of the respondents. Consequently the weights will average to 1 over the response. First, select the respondents with **Data | Select cases**. Then compute the average weight of the respondents with **Analyze | Descriptive statistics | Descriptives**. Last, use **Transform | Compute** again to adjust the values of *propwgt*.

Weight the response data using **Data | Weight cases**. Be sure that only respondents are selected. Compute weighted estimates of the two target variables "Owns a house" and "PC in household" by **Analyze | Descriptive Statistics | Frequencies**. What are the estimates? Compare them to the response means.

Exercise 11.2 The aim of this exercise is to perform a response propensity stratification adjustment as described by (11.23) using the data from the GPS. This comes down to poststratification based on strata constructed on the basis of response probabilities.

First, a categorical variable *propclas* is needed. This is obtained by aggregating individuals into five classes, based on the estimated response probabilities *rprob*. To determine these classes, look at the distribution of the estimated response probabilities, after plotting a histogram of the estimated response probabilities for the complete, unweighted sample. **Graphs | Histogram** can be used for this.

Classes can be formed to divide the individuals equally over the response probabilities, meaning every class contains the same number of individuals. An alternative is to form classes according to the response probabilities, whereby every class has the same width of the response probabilities. In this exercise, classes with the same width in response propensities are constructed. The first class comprises sample persons with a response propensity between 0 and 0.2, the second class between 0.2 and 0.4, the third between 0.4 and 0.6, the fourth between 0.6 and 0.8, and the last between 0.8 and 1. The last class, however, is empty because the maximum response propensity is < 0.8. So there will be only four classes.

Use **Transform | Recode | Into different variables** to recode the variable *rprob* into the four desired classes. Recode to an output variable *propclas* and assign the label "propensity class." The resulting variable *propclas* will be the stratification variable for the nonresponse adjustment.

First compute the four weights. As the weight to the sample, use

$$d_i = \frac{n_h/n}{r_h/r};$$

for the weight of an element in class h, use the estimator

$$\bar{y}_{PS} = \frac{1}{n}\sum_{h=1}^{L} n_h \bar{y}^{(h)},$$

to find the target variables "Owns a house" and "Has a PC." Here, n_h is the sample size of class h, n is the sample size, r_h is the response size in class h, r is the response size, and $\bar{y}^{(h)}$ is the response mean of the target variable in class h. To compute the poststratification estimator, compute these quantities first. To this end, complete the table below.

h	n_h	r_h	w_h
1			
2			
3			
4			
Total	32,019	18,792	

Derive the class sizes n_h and r_h by **Analyze | Descriptive Statistics | Crosstabs,** by crossing the response indicator *response* with the variable *propclas*. Compute the weights by hand and import them in SPSS by constructing a new weight variable *propwgt2* using **Transform | Compute** (you have to do this four times, for every category of *propclas*).

Finally, estimate the target variables by activating the weights with **Data | Weight cases**, respondent selection in **Data | Select cases**, and weighted estimation in **Analyze | Descriptive Statistics | Frequencies**. What are the estimates? Are they different from the first approach? Also compare them to the response means.

CHAPTER TWELVE

Analysis and Adjustment Accounting for the Cause of Nonresponse

12.1 Introduction

Nonresponse adjustment methods assign weights to respondents so that the weighted respondents represent the population as close as possible with respect to known auxiliary information. The nonresponse is assumed to be missing-at-random (MAR), which leads to less biased estimates for population characteristics.

The distribution of the auxiliary information for the subsample of respondents is calibrated to known auxiliary information. The success of the nonresponse adjustment depends on the relevance of the auxiliary information (see Chapter 9 for the selection of auxiliary variables). An ideal auxiliary variable has three features: it explains well the response behavior, it explains well the target variables of the survey, and it identifies the most important domains for publication of the survey statistics.

The calibration framework embodies a large number of conventional non-response adjustment methods, including linear weighting (see Chapter 8). This method uses the generalized regression (GREG) estimator. The GREG is based on

Handbook of Nonresponse in Household Surveys, By Jelke Bethlehem, Fannie Cobben, and Barry Schouten

the relationships among the response behavior, the survey items and the auxiliary variables. Another method that has become popular in adjusting for the nonresponse bias is the propensity score method. This method is based on estimating response probabilities, or response propensities (see Chapter 11), and does not belong to the calibration framework.

These nonresponse adjustment methods do not distinguish between different types of respondents. Different types of response, however, result from distinctive phenomena. Also the effects of the response types on the bias may vary. This can be seen by looking at the nonresponse bias of the estimated response mean of target variable Y under the fixed response model:

$$B(\bar{y}_R) = \frac{N_{NR}}{N}(\bar{Y}_R - \bar{Y}_{NR}), \tag{12.1}$$

where N is the total number of population elements and N_{NR} is the number of nonrespondents (see also Chapter 2).

When a close look is taken at the two main nonresponse types, it follows that this bias can be attributed to two causes: noncontact and refusal. These types of nonresponses are nested because sample elements can only refuse once contacted and noncontacted elements either would have refused or participated would they have been contacted; see Figure 12.1.

Equation (12.1) then becomes

$$\begin{aligned} B(\bar{y}_R) = {} & \frac{N_{NC}}{N}\left(\frac{N_{R|NC}}{N}\left(\bar{Y}_R - \bar{Y}_{R|NC}\right) + \frac{N_{RF|NC}}{N}\left(\bar{Y}_R - \bar{Y}_{RF|NC}\right)\right) \\ & + \frac{N_C}{N}\frac{N_{RF|C}}{N_C}\left(\bar{Y}_R - \bar{Y}_{RF|C}\right), \end{aligned} \tag{12.2}$$

where $\bar{Y}_R$ is the respondent mean for target variable Y and N_{NC}/N is the fraction of noncontacts in the population. These noncontacts can be divided into elements

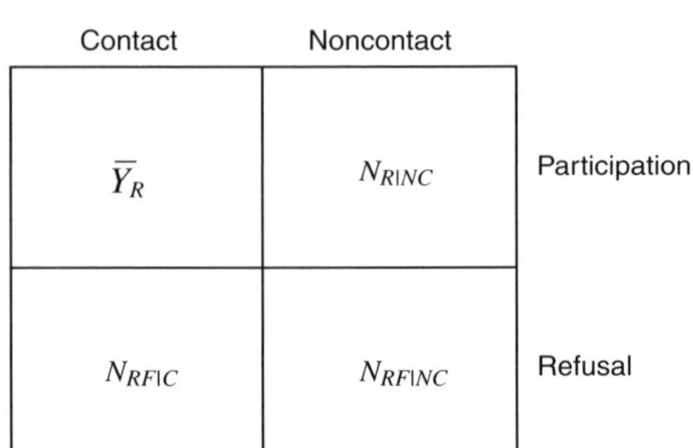

FIGURE 12.1 Nesting of the two main causes of nonresponse.

that would have responded when contacted whose number is denoted by $N_{R|NC}$, and sample elements that would have refused participation once contacted, denoted by $N_{RF|NC}$. $\bar{Y}_{R|NC}$ is the mean for the first group, and $\bar{Y}_{RF|NC}$ for the second.

Participation and refusal are conditional on being contacted. The last term in equation (12.2) reflects this nesting of refusals within contact, with N_C the number of contacted elements, $N_{RF|C}$ the number of refusals among the contacted elements, and $\bar{Y}_{RF|C}$ the mean for target variable Y for the elements that refused participation. If the different groups display a different relationship with Y, for instance, if $\bar{Y}_{R|NC} \neq \bar{Y}_R$, and $\bar{Y}_{RF|NC} \neq \bar{Y}_R$, the effects may cancel out. When the composition of the nonresponse varies over years, and the nonresponse types are not accounted for in the adjustment for nonresponse bias, the estimates for the target variables in the survey may differ due to a change in the composition of the response instead of real changes in society.

The literature describes several methods that take into account the different nonresponse types in the analysis of nonresponse. Durrant and Steele (2007) use a multilevel, multinomial logistic regression model for contact and participation. The multinomial model allows for different variables and coefficients in the equations for the distinct response types. A model is fitted for all households, for both contact and participation. However, this model does not allow for correlation between the response types. In addition it does not take into account that households can only participate when they are contacted first.

Multilevel models are used to analyze interviewer effects. This type of model accounts for the effect of clustering sample elements within interviewers. By combining the multilevel model with the multinomial model, as Durrant and Steele (2007) have done, it becomes possible to allow for a correlation between the unobserved interviewer influences on both contact and participation.

Lepkowski and Couper (2002) employ a nested logit model to analyze contact, and participation conditional on contact. The nested logit model does account for the sequential nature of the response process by separately modeling contact and participation, where participation is only modeled for those households that were successfully contacted.

Both the multinomial model and the nested logit model assume independence among the response types. It is, however, more realistic to assume that the different response probabilities, for instance, the contact probability and the participation probability, are dependent. A dependency between response types can be caused by misclassification. This occurs, for example, when households that have a low participation probability pretend to be absent (or plan to be absent, in case of an appointment) to avoid being confronted with the survey request. These households will be classified as noncontacts, whereas in fact they refuse participation. Another reason for a dependency between processes arises when there are unobserved variables that affect different propensities in the response process.

The nested logit model accounts for the sequential nature of the response process by modeling each of the distinct stages using only the observed indicators for each response type. This allows the equations to be estimated separately. However, unobserved variables are not accounted for in the models. As a result the underlying propensities are not defined if their corresponding indicators are missing. Nicoletti and Peracchi (2005) introduce a model that defines the propensities also when the corresponding indicators are not observed, by explicitly modeling the censoring of the indicators. The model that they use is based on the bivariate probit model with sample selection (Van de Ven and Van Praag, 1981).

Groves and Couper (1998) present nested models that distinguish between contact and participation. They make the assumption that conditional on the available background information, participation and contact are independent processes. This assumption is referred to as the conditional independence assumption and it corresponds to the missing-at-random assumption. The approach of Groves and Couper (1998) involves constructing weights for the estimation of survey items where the nonresponse due to noncontact and refusals can be accounted for. They employ a weighted logistic regression technique to update the design weights, namely the inverse inclusion probabilities. They carry out a weighted logistic regression twice, first to adjust the design weights for nonresponse due to noncontact. In the second weighted logistic regression model they model the participation stage, where the initial design weights are replaced by the updated weights from the first model. The final weights reflect the selection in the two stages of the response process. A similar sequential weight adjustment approach is described by Iannacchione (2003).

The sequential weight adjustment approach, however, does not take into account the relationship between the response behavior and the survey variables. The survey variables can be introduced into the bivariate probit model with sample selection described by Nicoletti and Peracchi (2005), if the survey variable is bivariate. In case of a continuous survey variable, the sample selection model proposed by Heckman (1979) can be applied. In fact the bivariate probit model with sample selection is an adapted version of the sample selection model.

This chapter is outlined as follows: In Section 12.2.1 models to analyze different types of nonresponse are presented. The nested logit model and the bivariate probit model with sample selection are discussed. The difference between these models with respect to the sequential nature of the response process and the correlation between the different response types is outlined. These models can be refined to include interviewer effects by means of the multilevel model. This approach is the topic of Section 12.2.1.4.

In Section 12.2.2 it is discussed how these nonresponse analysis models can be used to adjust for selective nonresponse. The sequential procedure to adjust the design weights for the different types of nonresponse is described. Additionally the sample selection model proposed by Heckman (1979) is presented. Issues regarding the identification and the estimation of this model are considered, as well as a number of methods to estimate the sample selection model. Furthermore it is shown how these models can be extended to account for the different stages in the response process and to adjust the survey items for nonresponse bias at the same time. Some of the methods are illustrated in Section 12.3, where they are applied to the GPS. Section 12.4 ends the chapter with conclusions and suggestions for further research.

12.2 Theory

12.2.1 METHODS FOR NONRESPONSE ANALYSIS

12.2.1.1 Introduction It is generally acknowledged in the survey literature that it is important to distinguish among different types of response. Moreover an important aspect of the response process is its sequential nature. The aim of this section is to provide an overview of the different models that are being used to analyze nonresponse. It is discussed to what extent the models distinguish among different response types, and how they account for the sequential nature of the response process.

The nested logit model and the bivariate probit model with sample selection are presented, respectively, in Section 12.2.1.2 and Section 12.2.1.3. The nested logit model and the bivariate probit with sample selection focus on the analysis of nested response processes, whereby they allow for different assumptions regarding the dependence of the various processes. It is possible to extend these models to allow for interviewer effects. A model that is particularly useful for analyzing interviewer effects is the multilevel model, which is introduced in Section 12.2.1.4. It is, however, less straightforward to change the underlying assumptions of the model regarding the correlation between the processes.

There are a number of reasons to allow for a correlation between the different processes. In general, a correlation between processes arises when there are factors that affect both processes, but these factors are not included in the model. These factors are often referred to as *unobservables* and their influence can be accounted for by allowing the equations to be correlated. Another reason for the processes to be correlated arises due to misclassification. For instance, if a sample element does not want to participate in the survey and pretends to be absent when contacted, the sample element will be misclassified as a noncontact whereas in fact it is refusing to participate.

In Figure 12.2 the response process is displayed as a sequence of different events (with corresponding probability): being processed (probability ξ_i), contacted (γ_i), able to participate (θ_i), and willing to participate (ζ_i). Notice that not every sample element passes through every stage of the response process. Whether or not sample elements proceed to a next stage depends on the outcome of the preceding stage.

In Table 12.1 the indicators for the different stages in the response process are summarized: being processed U, contacted C, able to participate L and willing to participate P. These indicators are equal to 1 in case a sample element proceeds to the next stage (or participates in the survey), and zero if they drop out of the response process at that stage. If a sample element dropped out in an earlier stage, the subsequent indicators are not observed. Since the other characteristics for the sample elements are known, this missingness is referred to as *censoring*.

In the nested logit model the underlying probabilities are not defined for the censored indicators. Other models, for instance, the bivariate probit model with sample selection, assume that although the indicators are censored, the underlying

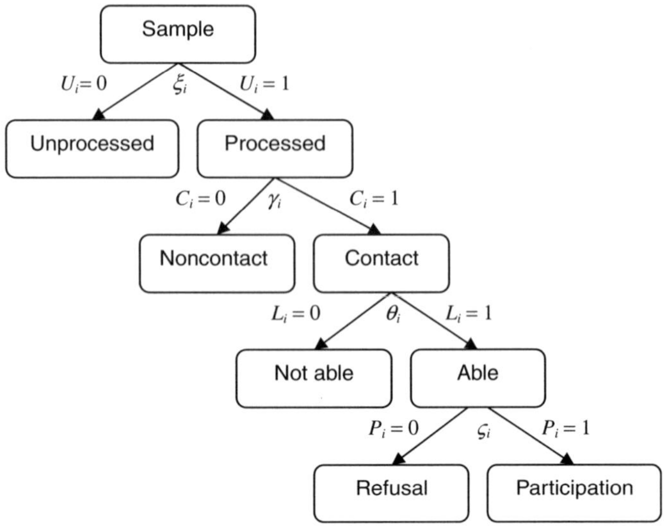

FIGURE 12.2 Nested representation of the elaborate response process.

probabilities are still defined. The different models will be described for the two main causes of nonresponse: noncontact and refusal. This simplified nested response process then reduces to Figure 12.3.

An extension of the models to more types of nonresponse is straightforward. For ease of notation the elaboration is restricted to contact and participation.

12.2.1.2 The Nested Logit Model The nested logit model consists of separate logistic regression models for each of the stages in the response process. The different stages are analyzed on the subset of observations that is available for each of the processes. The estimated probabilities are referred to as propensities. The contact propensity $\hat{\gamma}(X_i)$ is the estimated contact probability, conditional on the sample and the individual characteristics X_i. Likewise the participation propensity $\hat{\zeta}(X_i)$ is the estimated probability of participation, conditional on the sample and characteristics X_i.

The logit model for the contact propensity is described as

$$\log\left(\frac{\gamma_i}{1-\gamma_i}\right) = \text{logit}(\gamma(X_i^C)) = X_i^{C'}\beta^C \tag{12.3}$$

TABLE 12.1 Probabilities in the response process

Process	Indicator	Values	Probability
Being processed	U	1 processed, 0 unprocessed	ξ
Contact	C	1 contact, 0 no contact	γ
Being able	L	1 able, 0 not able	θ
Participation	P	1 participation, 0 refusal	ζ

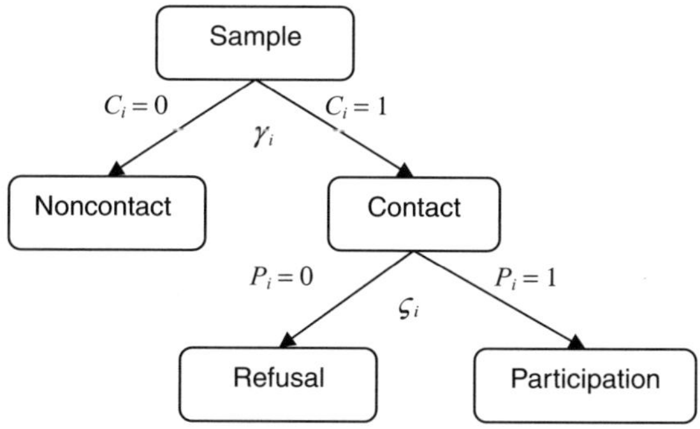

FIGURE 12.3 Simplified nested response process.

for $i = 1, 2, \ldots, n^c$ with n^c the number of contacted elements. X^C is a $n^c \times J$-vector of variables, J is the number of variables in the model and β^C is a $J \times 1$-vector of coefficients corresponding to the variables in X^C.

In the simplified nested response process displayed in Figure 12.3, $n^c = n$, but in the more elaborate response process the contact stage is preceded by the process stage and hence only processed sample elements are contacted. The contact stage is, however, the first stage in the simplified response process and therefore the indicator C is defined for all sample elements.

Only sample elements for which $C = 1$ proceed to the next stage. For those elements the logit model for the participation propensity is defined as

$$\log\left(\frac{\zeta_i}{1-\zeta_i}\right) = \operatorname{logit}(\zeta(X_i^{(P|C=1)})) = X_i^{(P|C=1)'}\beta^{(P|C=1)} \tag{12.4}$$

for $i = 1, 2, \ldots, n^P$. The superscript $(P|C = 1)$ indicates that sample elements can participate conditional on being contacted. n^P is the number of sample elements for which the contact indicator P is defined, so the number of sample elements for which $C_i = 1$.

With models (12.3) and (12.4) the propensities in the simplified response are obtained, being the contact propensity $\hat{\gamma}(X_i)$ and the participation propensity $\hat{\zeta}(X_i)$. Passing through the response process, the number of available observations decreases; that is, $n = n^c > n^P$. In addition the number of conditions under which the stages in the response process are being analyzed increases.

The different stages in the response process are nested. Therefore the combination of these models is referred to as the *nested logit model*. The assumption underlying the nested logit model is that the different response types are not correlated, given the covariates X^C and $X^{P|C=1}$. Hence, in the general setting, it is assumed that the probability of drop out in stage k is independent of drop out in stages 1 to $k-1$ given the covariates, meaning $\zeta(X_i) = P(P_i = 1|C_i = 1; X_i) = P(P_i = 1|X_i)$. As a consequence equations (12.3) and (12.4) can be estimated separately.

12.2.1.3 The Bivariate Probit Model with Sample Selection For the estimation of response propensities, both logit and probit models can be used. It does not really matter what functional form is used: probit or logit. In fact even the linear model can be used as an approximation.

The logistic distribution can be approximated by a normal distribution with mean 0 and variance $\pi/3$, the logistic distribution only has slightly more weight in the tails. In the former section the logistic model is used to estimate the propensities in the response process. The logit model has as an advantage that the cumulative distribution function is simpler to evaluate, whereas the probit involves an unevaluated integral. Furthermore the inverse transformation of the logit model is interpretable as the log-odds, whereas the inverse of the probit transformation does not have a direct interpretation. The estimated coefficients $\hat{\beta}$ will be different in the two models, but roughly proportional so that the probit and the logit model produce rather similar results.

An essential difference between the two models that is important in the context of nonresponse is the specification of the error term. The error term of the logistic model follows a logistic distribution. The error term in the probit model follows a standard normal distribution. The different distributions of the error term lead to different assumptions. The conventional logistic distribution function of the logit model does not include a correlation between the different logit models in the nested logit model. The error term distribution has to be customized to allow for a correlation with other error terms, but in the standard case there is none.

Consequently the equations in the nested logit model are forced to be independent from each other. In the probit model the error terms may follow a joint distribution that allows for a correlation between the equations. The difference with the nested logit model is that the propensities are now defined for all sample elements, regardless of whether the corresponding indicators are observed.

The sample elements for which the indicators in one stage of the response process are not observed are *censored*, conditional on the outcome of the preceding stage (or stages). This censoring can be regarded as a form of *sample selection.*

A model that can be used to introduce a correlation between the different response types, and that accounts for the censoring of unobserved households, is the bivariate probit model with sample selection (Van de Ven and Van Praag, 1981).

The bivariate probit model with sample selection is also used by Nicoletti and Peracchi (2005). They apply the model to predict future contact and participation propensities in a panel survey. Jenkins et al. (2006) apply an extended version of the model to analyze patterns of consent, namely patterns in respondents' agreement to link additional data to the survey. They use a multivariate probit model for four bivariate outcomes, where two of the outcomes are subject to incidental truncation. Truncation occurs if no information at all is observed (i.e., no auxiliary variables nor the dependent variable), whereas censoring implies that only the dependent variable is not observed.

The bivariate model for the simplified response process displayed in Figure 12.3 can be described in terms of the latent variable regression models for the contact propensity $\hat{\gamma}$ and the participation propensity $\hat{\zeta}$. The bivariate probit model is defined as

$$\begin{aligned} y_i^* &= X_i^C \beta^C + \varepsilon_i^C, \\ \zeta_i^* &= X_i^P \beta^P + \varepsilon_i^P, \end{aligned} \tag{12.5}$$

for $i = 1, 2, \ldots, n$. Note that in this bivariate probit model the propensities are defined for all n, contrary to the nested logit model in (12.3) and (12.4) where the participation propensity is only defined for those sample elements that are contacted.

The latent variables y^* and ζ^* are not observed. Only the corresponding indicators C and P are observed. These indicators can be defined as

$$C_i = \begin{cases} 1, & \text{if } y_i^* > 0, \\ 0, & \text{otherwise,} \end{cases} \tag{12.6}$$

$$P_i = \begin{cases} 1, & \text{if } \zeta_i^* > 0, \\ 0, & \text{otherwise.} \end{cases} \tag{12.7}$$

The sample selection that occurs due to censoring of sample elements is introduced in the model by acknowledging that additionally P_i is only observed if $C_i = 1$, meaning if $y_i^* > 0$.

To estimate this model, a distributional assumption on the error terms in (12.5) has to be made. A conventional distribution is the bivariate standard normal distribution:

$$\begin{pmatrix} \varepsilon_i^C \\ \varepsilon_i^P \end{pmatrix} \sim N\left(0, \begin{bmatrix} 1 & \zeta_{cp} \\ \zeta_{pc} & 1 \end{bmatrix}\right), \tag{12.8}$$

where $\zeta_{cp} = \zeta_{pc}$ is the correlation between the contact and the participation equation. The variances are set equal to 1 because, otherwise, the model is only identifiable up to scale; see Dubin and Rivers (1989).

The bivariate model with sample selection can be estimated by maximum likelihood estimation (MLE; see Greene, 2002). Van der Klaauw and Koning (2003) show that MLE performs well even when the distributional assumption is not correct.

The log-likelihood for model (12.5) can be constructed by dividing the sample elements into three mutually exclusive groups: those that are not contacted ($C_i = 0$), those that are contacted but refuse participation ($C_i = 1$ and $P_i = 0$), and those that are contacted and participate ($C_i = 1$ and $P_i = 1$). Each of these groups contributes to a specific part of the likelihood. Thus the log-likelihood for the sample of n observations is

$$\begin{aligned} L(\beta^C, \beta^P, \xi_{cp}) &= \textstyle\sum_{i=1,\ldots,n} \big((1 - C_i) \times \log P(C_i = 0 | X_i^C) \\ &\quad + C_i(1 - P_i) \times \log P(C_i = 1; P_i = 0 | X_i^C, X_i^P) \\ &\quad + C_i P_i \times \log P(C_i = 1; P_i = 1 | X_i^C, X_i^P)\big). \end{aligned} \tag{12.9}$$

Cobben (2009) gives an expression for the conditional probabilities in (12.9). Maximizing the likelihood results in estimates for the conditional probabilities. Again, (12.5) can be extended to more response stages. With k stages, (12.8) becomes a k-variate normal distribution and (12.9) will have $k + 1$ terms.

12.2.1.4 The Multilevel Model Surveys can be conducted using interviewers, or by means of self-administration. In interviewer-assisted surveys, the interviewer communicates with the respondent either face-to-face or over the telephone. In these types of surveys, the interviewer plays an important role in the response process. Biemer and Lyberg (2003) provide a discussion on the role of the interviewer. They distinguish between standardized interviewing and conversational interviewing. In standardized interviewing, the interviewer has to follow a protocol that tries to ensure that the interviewers behave in a standard way, thus reducing the risk of measurement error in responses caused by the interviewer. In conversational interviewing, interviewers are freer to interact with respondents, to optimize the survey situation for the respondent in order to minimize measurement error in all error sources and not just the interviewer's error.

Whatever style of interviewing, the interviewers will affect the response process in the survey, as well as the answers provided to the survey. An *interviewer effect* occurs when the interviewers systematically affect the probability that sample elements participate, or are contacted. Also the interviewers can affect the survey answers, for instance, when the interviewer thinks positive of voting and sample elements, therefore overreport voting behavior. This is a *social desirability bias*.

Since multiple sample elements are assigned to one interviewer, the effect of the interviewer on the response process and the survey answers is clustered within interviewers. To analyze interviewer effects and to use the lowest level of available information (i.e., the sample element level), this clustering has to be accounted for. Similarly an analysis of respondent effects, with the inclusion of interviewer data, should take into account the clustering and thus the hierarchical structure of the data too.

In this section it is shown how the models in the previous sections can be refined to include interviewer variance into the analysis models. The type of model that can account for interviewer variance appears in the literature under a variety of different names: hierarchical regression model (Hox, 1994), random effects model (Agresti, 2002), variance component model (Anderson and Aitkin, 1985), and multilevel model (Hox, 1995; Durrant and Steele, 2007). The term that is most commonly used in the survey literature is the *multilevel model*.

The multilevel model can be used to describe different forms of nesting. For instance, the nesting of persons within families, pupils within schools, or interviewers within geographical areas as described in Durrant and Steele (2007). It is described here how the model can be used to analyze the effect of interviewers on data quality. In survey sampling, the sample elements are drawn randomly from the sampling frame. In case of a face-to-face survey, clusters of elements are assigned to interviewers based on geographic locations. The characteristics of the interviewer are then confounded with the characteristics of the respondents, and there is a correlation between the characteristics of the interviewer and the respondents. For instance, persons that live in highly urbanized areas will be interviewed by an urban interviewer. This may improve the response, since similarity leads to liking (see the compliance principles discussed in Chapter 3), but it can also introduce an unwanted interviewer effect. In case of a CATI survey using a centralized system, sample elements are distributed randomly to interviewers. The confounding of characteristics does not apply to this situation, but the interviewers still affect the response process and survey answers.

A detailed description of the multilevel model can be found in Hox (1995). The logistic regression model for the response indicator R and response propensity $\hat{\rho}(X_i)$ can be extended to include interviewer effects. This involves introducing an extra index for the interviewers and a corresponding random term for the interviewer effect, which is denoted by ε. Suppose that there are $m = 1, 2, \ldots, M$ interviewers. Then the multilevel model becomes

$$\text{logit}(\rho(X_{im})) = X_{im}^{R}\beta + \varepsilon_m \tag{12.10}$$

for $i = 1, \ldots, n$ and for $m = 1, \ldots, M$. The random effect for interviewer m, ε_m, is included in the equation. Sample elements with the same interviewer are thus assigned the same interviewer effect. It is assumed that $\varepsilon_m \sim NID(0, \sigma_m^2)$, where NID denotes normally and independently distributed.

12.2.1.5 Advantages and Disadvantages The models described in the previous sections can be employed to analyze response behavior. The underlying assumptions of the models with respect to the correlation structure of the distinct stages in the response process are different. Furthermore they differ in the way that they account for the sequential character of the response process. Table 12.2 summarizes the characteristics of the different models.

The multinomial model allows for a distinction among the different response types in a sense that the dependent variables for each of the response types, and the corresponding coefficients, are allowed to be different. This type of model, however, does not acknowledge the sequential nature of the response process, nor does it allow for a dependency among response types.

The nested logit model does account for the sequential nature of the response process. Nevertheless, because the different indicators are only defined on subsets of the sample, the truncation that arises due to unobserved indicators is not accounted for. Hence the underlying assumption of the nested logit model is that the processes are independent from each other.

TABLE 12.2 Characteristics of the nonresponse analysis models

Model	Distinguish response types	Sequential nature of response process	Dependence assumption
Multinomial	Yes	No	No
Nested logit	Yes	Yes	No
Bivariate probit with sample selection	Yes	Yes	Yes

The bivariate probit model with sample selection treats the censoring of sample elements as a form of sample selection. It makes the censoring explicit by acknowledging the fact that certain response types are not observed, conditional on previous stages in the response process. The underlying propensities are being estimated based on their observed indicators, but unlike the nested logit model, the propensities are defined for every sample element regardless of whether the corresponding indicator is observed. In addition the joint distribution of the error terms in the equations allows for a correlation and therefore allow for a dependency between the different stages in the response process.

The multilevel model tackles another type of nesting, namely that of the clustering of sample elements within interviewers. Therefore this model is a refinement of the other models to allow for interviewer effects. The clustering causes confounding of characteristics from the interviewer and the sample elements. When analyzing interviewer effects, the multilevel model distinguishes between these two levels of information. Furthermore, when the data are analyzed on the respondent level and data on the interviewers are included in the models, the cluster effect has to be accounted for. Otherwise, the standard errors are much lower than they should be, leading to results that appear more significant than they are.

The most elaborate nonresponse analysis model thus is the bivariate model with sample selection refined with interviewer effects. In this model the response types are allowed to be correlated and defined for all sample elements, and the effects of the respondents and the interviewers are separated.

12.2.2 ALTERNATIVE METHODS FOR NONRESPONSE ADJUSTMENT

12.2.2.1 Introduction The objective of nonresponse adjustment methods is to reduce the bias in estimated survey variables for errors due to selective nonresponse. The influence of the different causes of nonresponse on survey target variables can be a reason to use more advanced methods. In the previous section, a number of models were discussed that describe the relationship between the characteristics of sample elements and their response behavior. In this section, these models are extended to adjust for a nonresponse bias.

To describe the relationship between response behavior and auxiliary information, it is sufficient to include demographic or socioeconomic characteristics that are not survey specific but are approximately the same in all surveys. However, the relation between auxiliary information and the target variable Y can be different for each variable, and for different surveys. For instance, the topic of the survey can influence survey participation, and the name of the survey alone may cause nonresponse. In that case, changing the name of the survey will influence the response rate. For example, the Dutch "Fertility Survey" had a low response rate. Changing the name of the survey to "Family Planning Survey" increased the response.

Furthermore there are situations where distinct causes of nonresponse have a different influence on the survey variables. In those situations especially, the composition of the response influences the estimated survey variables, so it is important to model and take into account any specific relationships.

EXAMPLE 12.1 Relationship Between Employment and Noncontact

Daalmans et al. (2006) analyzed the relationship between an increasing response rate and estimates for the number of employed persons in the Dutch Labor Force Survey (LFS) in 2004. In 2004 the composition of the response changed because of a number of changes in the fieldwork strategy. Especially the noncontact rate has decreased because the main objective of the new fieldwork strategy was to optimize the probability of finding sample elements at home. Accordingly, interviewers are instructed to visit all their addresses for the first time in the first half of the month. Contact attempts are preferably made at the end of the day, in the early evening, or on Saturdays. When nobody is found at home, the subsequent attempt should be made at a different time

Daalmans et al. (2006) estimated number of employed persons as a function of the number of contact attempts made. The estimates are made using the LFS with a GREG estimator. When the number of attempts increased, the estimated number of employed persons increased, as in the figure below.

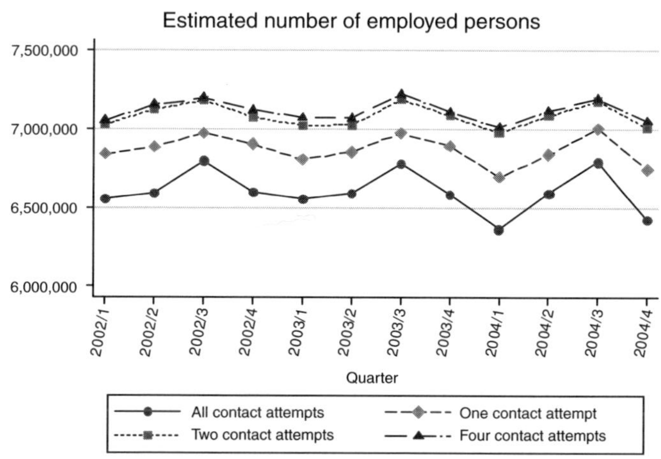

Daalmans et al. (2006) concluded that there is a very strong relation between contact and employment, and that hard-to-contact persons more often are employed.

There may be a correlation between the survey variables and the different response types. Again, if the response types display a different relationship with the survey variable Y, the effect on the nonresponse bias is different, as shown in equation (12.1).

Another issue that arises is the number and variety of survey variables. A survey usually consists of a large number of survey variables. The effect of nonresponse can be different for each of these variables, and ideally one would want to adjust each of the survey variables separately, namely use different adjustment models for different survey variables. This is not desirable in practice. The adjustment weights calculated by the linear regression estimator can be described in terms of the *X*-variables, thereby not relying on a specific target variable but using a fixed set of *X*-variables for every survey variable. Hence it is possible to use only one weighting model for all the survey variables, which is more practical than a different weighting model for each survey variable.

In this section two weighting adjustment methods are discussed that take into account different response types and the sequential character of the response process. The sequential weight adjustment method is presented, and the sample selection model (with as a special case the bivariate probit model with sample selection). The sample selection model additionally allows for a correlation between the response types, and a correlation between response types and survey items. Section 12.2.2.2 describes the sequential weight adjustment technique. In Section 12.2.2.3 the sample selection model is applied to the situation of survey nonresponse. Issues regarding the identification and the estimation of the sample selection model are also discussed.

12.2.2.2 The Sequential Weight Adjustment Approach Groves and Couper (1998) and Iannacchione (2003) describe the sequential weight adjustment method for a two-stage response process consisting of contact and participation conditional on contact. The nested logit models described in Section 12.2.1.2 are used to construct nonresponse adjustment weights. Here the sequential weight adjustment method is described and the advantages and disadvantages of the method are outlined.

The method consists of sequentially fitting a number of logistic regression models. First, a logistic model for the contact probability is constructed for all sample elements:

$$\text{logit}(\gamma(X_i^C)) = X_i^{C'}\beta^C \tag{12.11}$$

for $i = 1, \ldots, n$, with X_i^C assumed to contain a constant term. The model parameters β^C are estimated by MLE. The contact propensity can be computed as

$$\hat{\gamma}_i = \frac{\exp(X_i^C\hat{\beta}^C)}{1+\exp(X_i^C\hat{\beta}^C)} = \frac{1}{1+\exp(-X_i^C\hat{\beta}^C)}. \tag{12.12}$$

Based on this regression, sample elements that have successfully been contacted ($C_i = 1$) receive as a weight $w_i = 1/\hat{\gamma}_i$ so that they represent the sample. In the second step a weighted logistic regression is fitted for the participation propensity using only the subsample of contacted sample elements:

$$\text{logit}(\zeta(X_i^{P|C=1})) = X_i^{(P|C=1)'}\beta^{(P|C=1)} \tag{12.13}$$

for $i = 1, \ldots, n^c$, where n^c are the sample elements that have been contacted for which $C_i = 1$. Again, $\beta^{P|C=1}$ can be estimated by MLE and the participation propensity can be calculated as

$$\hat{\zeta}_i = \frac{1}{1 + \exp(-X_i^{P|C=1}\hat{\beta}^{P|C=1})}. \tag{12.14}$$

The final weights w_i^* are obtained by multiplying w_i with

$$1 + \exp(-X_i^{P|C=1}\hat{\beta}^{P|C=1}), \tag{12.15}$$

which leads to the following expression for w_i^*:

$$w_i^* = w_i(1 + \exp(-X_i^{P|C=1}\hat{\beta}^{P|C=1})) = \frac{w_i}{\hat{\zeta}_i} = \frac{1}{\hat{\gamma}_i\hat{\zeta}_i}. \tag{12.16}$$

Both Groves and Couper (1998) and Iannacchione (2003) have used a weighted logistic regression model in the first step, to account for the survey design by including the design weights d_i (the inverse of the inclusion probability, $d_i = \pi_i^{-1}$).

The weights that are obtained by the sequential weights adjustment procedure can be used in the regular nonresponse adjustment methods, for instance, in the Horvitz–Thompson estimator or the propensity score adjustment methods; see Chapter 11. The sequential weight adjustment method thus accounts for the sequential nature of the response process. Furthermore it allows for a distinction between the different response types. However, when the estimated propensities are directly used as weights, the variance of the estimator can become large. This effect increases with the number of nested processes (Little and Vartivarian, 2005). Another disadvantage of the method is that it does not allow for correlation between response types, the same as the nested logit model. Moreover the participation propensity is only defined for the subset of sample elements that have been successfully contacted.

This disadvantage can be overcome by using probit models instead of logit models. The probit model with sample selection that is presented in Section 12.2.1.3 defines an underlying, latent participation propensity that is also defined for sample elements that are not contacted. Furthermore the error distribution of the probit model allows for correlation between the distinct stages.

Because of these properties of the bivariate probit model with sample selection, it is not necessary to explicitly weight the observations for the different types of response, as in the nested logit model approach described above. Nicoletti and Peracchi (2005) have applied the bivariate probit model with sample selection described in Section 12.2.1.3.

To adjust for nonresponse bias, the relationship with the target variable(s) of the survey should also be included in the models. Nicoletti and Peracchi (2005) do not include this relationship; their sole focus is the response process. Therefore in the next section an extension of the probit model with sample selection that includes the relationship with the survey target variable(s) is discussed.

12.2.2.3 The Sample Selection Model The sample selection model was first proposed by Heckman (1979) to determine women's wages and labor force supply; see also Heckman (1974) and Gronau (1974). The outcome that the researchers were interested in concerned women's wages. It was argued that the observed wages for women were not a random sample of women's wages. The selection was the decision of a woman to become active in the labor force. If the decision of a woman to participate in the labor force depends on characteristics that also influence her wage, then there is a selectivity bias. When all variables that determine the participation decision are observed, these can simply be inserted in the wage equation.

Conditional on these variables, the wages for nonworking and working women are the same. This corresponds to the MAR assumption, or the conditional independence assumption. If, however, there are unobservable characteristics, and these characteristics influence both the decision to participate in the labor force and the wage, then there is a relationship between the selection (the participation decision) and the outcome (women's wages). In this case it is not sufficient to include the observed characteristics in the wage equation because through the unobserved characteristics the wage is also influenced by the decision to participate in the labor force.

In the context of survey nonresponse, the sample selection arises due to self-selection of respondents, either explicit (refusal) or implicit (not able, noncontact) and in some cases also due to actual sample selection (unprocessed cases). Both types are referred to as *sample selection.*

First, the sample selection model for survey nonresponse is described in the simplified form where only the final response and the outcome for the target variable are considered; see Figure 12.4. Such a sample selection model to fit the various causes of nonresponse (noncontact, refusal, nonprocessed, not able, etc.) was generalized by Cobben (2009).

The sample selection model consists of two equations, as shown in Figure 12.4. The first equation models the response propensity. The outcome of the first

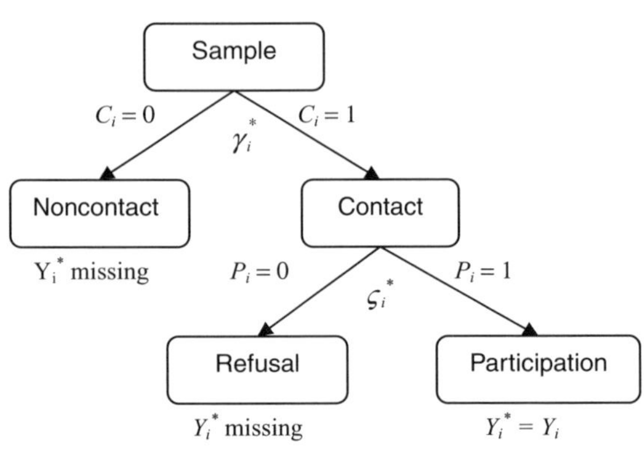

FIGURE 12.4 Sample selection model for nonresponse.

equation determines whether the survey target variable is observed. This equation is therefore referred to as the *selection equation*. The second equation accounts for the censoring of sample elements that do not respond, and models the target variable. Therefore the second equation is referred to as the *regression equation*.

The sample selection model for survey nonresponse consist of two equations

$$\begin{aligned} \rho_i^* &= X_i^R \beta^R + \varepsilon_i^R, \\ Y_i^* &= X_i^Y \beta^Y + \varepsilon_i^Y, \end{aligned} \tag{12.17}$$

for $i = 1, \ldots, n$. Both equations are latent variable regression equations, meaning ρ_i^* is not observed but instead the binary response indicator R_i is observed. If $R_i = 1$, Y_i^* is observed. Otherwise, Y_i^* is missing, meaning it is censored. Thus Y_i can be defined as

$$Y_i = \begin{cases} Y_i^*, & \text{if } R_i = 1, \\ \text{missing}, & \text{if } R_i = 0. \end{cases} \tag{12.18}$$

In (12.17) and (12.18), Y_i^* is a quantitative variable. The bivariate model with sample selection described in the previous section is a modification of the sample selection model for discrete outcomes. In the case of a binary categorical survey variable Y_i^*, such as having or not having a job, the bivariate probit model with sample selection is the appropriate model to use.

In the sample selection model the error term distribution is assumed to be bivariate normal with variance σ^2, which leads to the following joint distribution of the error terms:

$$\begin{pmatrix} \varepsilon_i^R \\ \varepsilon_i^Y \end{pmatrix} \sim N\left(0, \begin{bmatrix} 1 & \zeta\sigma \\ \zeta\sigma & \sigma^2 \end{bmatrix}\right), \tag{12.19}$$

with ζ the correlation between the two error terms. The population variance of Y_i^* thus equals σ^2. The population variance of Y_i may be different due to selection.

The aim of the sample selection model is to adjust the target variables of the survey for nonresponse bias. The Horvitz–Thompson estimator $\bar{y}_{HT}$ is an unbiased estimator for the population mean of a survey variable Y:

$$\bar{y}_{HT} = \frac{1}{N} \sum_{k=1}^{N} d_k Y_k. \tag{12.20}$$

In the case of nonresponse, this estimator may be biased. So the modified Horvitz–Thompson estimator $\bar{y}_{HT}^r$ deals with nonresponse bias by adjusting the inclusion probabilities for the second phase of nonresponse:

$$\bar{y}_{HT}^r = \frac{1}{N} \sum_{i \in r} \frac{d_i Y_i}{\rho_i}. \tag{12.21}$$

In the sample selection model the Horvitz–Thompson estimator can be modified by replacing Y_i with the expected value of Y_i given the model and $R_i = 1$. This estimator is denoted by $\bar{y}_{HT}^{sel}$:

$$\bar{y}_{HT}^{sel} = \frac{1}{N}\sum_{i=1}^{n} d_i\hat{E}[Y_i|R_i = 1, X_i^R, X_i^Y]. \tag{12.22}$$

Estimator (12.22) can be calculated by estimating the parameters of the sample selection model. There are a number of methods that can be used. Vella (1998) provides an overview of the different methods. The most commonly used methods are maximum likelihood estimation and the two-step estimator proposed by Heckman (1979). Both methods depend heavily on the underlying distributional assumptions. To avoid distributional assumptions, the literature suggests using semiparametric estimation methods. For example, Gallant and Nychka (1987) discuss a general semiparametric estimation strategy.

First consider the expected value of Y_i conditional on $R_i = 1$, and X_i^R, X_i^Y:

$$\begin{aligned} E[Y_i|R_i = 1, X_i^R, X_i^Y] &= X_i^Y\beta^Y + E[\varepsilon_i^Y|R_i = 1, X_i^R, X_i^Y] \\ &= X_i^Y\beta^Y + E[\varepsilon_i^Y|\varepsilon_i^R > -X_i^R\beta^R, X_i^R, X_i^Y]. \end{aligned} \tag{12.23}$$

From this conditional expectation it can seen that there is a selection bias if ε_i^Y and ε_i^R are dependent, or if X_i^R and X_i^Y are correlated, meaning $E[\varepsilon_i^Y|\varepsilon_i^R] \neq 0$. A test for the presence of a selection bias involves testing whether the correlation coefficient $\zeta = 0$. If $\zeta = 0$, there is no selection bias and the two equations of the sample selection model can be estimated separately.

Maximum likelihood estimation (MLE) amounts to finding specific parameter values that optimize the probability of finding the data that have been observed. The likelihood function is a function of the parameters, conditional on the observed data, so that $L(\beta^R, \beta^Y, \sigma^2, \zeta; R, Y)$. It is usually simpler to maximize the log of the likelihood, denoted by $\ell(\beta^R, \beta^Y, \sigma^2, \zeta; R, Y)$. For the sample selection model (12.17) the log-likelihood function can be derived as in the bivariate probit model with sample selection; see equation (12.9). There are two mutually exclusive groups, respondents $R_i = 1$ and nonrespondents $R_i = 0$. For sample elements that respond, Y_i is observed, so their contribution to the likelihood is $P(R_i = 1; Y_i = y_i|X_i^R, X_i^Y)$. For nonresponding sample elements, Y_i is not observed, and their contribution to the likelihood equals $P(R_i = 0|X_i^R)$. Thus the log-likelihood for the sample of n observations is

$$\begin{aligned} L(\beta^R, \beta^Y, \sigma^2, \zeta; R, Y) &= \sum_{i=1}^{n}((1 - R_i)\log P(R_i = 0|X_i^R) \\ &\quad + R_i\log P(R_i = 1; Y_i = y_i|X_i^R, X_i^Y)), \end{aligned} \tag{12.24}$$

where the conditional probabilities are defined as

$$P(R_i = 0|X_i^R) = \int_{-\infty}^{0} P(\rho_i^* = r|X_i^R)dr = \Phi(-X_i^R\beta^R) \tag{12.25}$$

and

$$\begin{aligned} P(R_i = 1; Y_i = y_i|X_i^R, X_i^Y) &= \int_{-\infty}^{0} P(\rho_i^* = r; Y_i^* = y|X_i^R, X_i^Y)dr \\ &= \Phi\left(\frac{-X_i^R\beta^R - \dfrac{\zeta}{\sigma}(y - X_i^Y\beta^Y)}{\sqrt{1-\zeta^2}}\right) \\ &\quad \times \frac{1}{\sigma}\varphi\left(\frac{y - X_i^Y\beta^Y}{\sigma}\right). \end{aligned} \tag{12.26}$$

The estimates for the coefficients that are obtained by MLE are consistent and efficient, but not necessarily unbiased. The results become inconsistent if the underlying distributional assumptions are violated. Furthermore full likelihood estimation may become cumbersome if the number of parameters is large.

Heckman (1979) developed a *two-step procedure* that avoids some of the complications of full MLE. Again, consider the conditional expectation of Y_i :

$$\begin{aligned} E[Y_i|R_i = 1, X_i^R, X_i^Y] &= X_i^Y\beta^Y + E[\varepsilon_i^Y|\varepsilon_i^R > -X_i^R\beta^R, X_i^R, X_i^Y] \\ &= X_i^Y\beta^Y + \zeta\sigma\frac{\varphi(-X_i^R\beta^R)}{1-\Phi(-X_i^R\beta^R)} \\ &= X_i^Y\beta^Y + \zeta\sigma\frac{\varphi(X_i^R\beta^R)}{\Phi(X_i^R\beta^R)}. \end{aligned} \tag{12.27}$$

The term $\varphi(X_i^R\beta^R)/\Phi(X_i^R\beta^R)$ is referred to as the *inverse Mills ratio*, denoted by λ_i. In this context, λ_i is a monotone decreasing function of the probability that a sample element responds to the survey request. The interpretation of the inverse Mills ratio λ_i is that observations with a higher response probability have a lower probability of introducing selection bias. The first step of the Heckman two-step procedure involves estimation of β^R by applying a probit model to the selection equation. The log-likelihood for the probit model is

$$L(\beta^R) = \sum_{i=1}^{n}((1-R_i)\log\Phi(-X_i^R\beta^R) + R_i, \log(1-\Phi(-X_i^R\beta^R)). \tag{12.28}$$

The ML estimate for β^R is denoted $\hat{\beta}^R$. The inverse Mills ratio can be computed with the estimated values by

$$\hat{\lambda}_i = \frac{\varphi(X_i^R \hat{\beta}^R)}{\Phi(X_i^R \hat{\beta}^R)}. \tag{12.29}$$

The rationale behind the two-step estimator arises from regarding the conditional expectation of ε_i^Y as an omitted variable from the regression equation in the sample selection model. The selection bias is caused by this omission. Therefore, in the second step of Heckman's procedure, the omitted variable $\hat{\lambda}_i$ is included in the regression equation to adjust for the selectivity. The regression equation can be estimated straightforwardly by OLS, based on the subset of observations for which $R_i = 1$. The regression equation becomes

$$Y_i = X_i^Y \beta^Y + \sigma\zeta\hat{\lambda}_i + \varepsilon_i^{Y^*}. \tag{12.30}$$

From this equation it can be seen that testing for sample selectivity bias reduces to testing whether $\zeta = 0$. The two-step procedure provides consistent estimators, but it ignores the fact that the error term $\varepsilon_i^{Y^*}$ now is heteroskedastic because of the inclusion of the estimated inverse Mills ratio. The standard errors in the regression equation have to be adjusted to account for the first step (e.g., see Greene, 1981). As a consequence, under the same distributional assumptions, MLE is more efficient than the two-step estimator.

Consider the case where $X_i^R = X_i^Y$; that is, the same variables are used in the selection equation and the outcome equations. Identification of the parameters β^Y in the two-step method now depends on the nonlinearity of the inverse Mills ratio. The inverse Mills ratio is not linear in $X_i^R \beta^R$, but the function that maps $X_i^R \beta^R$ into the inverse Mills ratio is linear for certain ranges of $X_i^R \beta^R$. If $X_i^R = X_i^Y$ and the inverse Mills ratio is linear in $X_i^R \beta^R$. Then there is perfect collinearity in the second step:

$$Y_i = X_i^Y \beta^Y + \sigma\zeta\, X_i^Y \hat{\beta}^R + \varepsilon_i^{Y^*}. \tag{12.31}$$

As a consequence the parameters are not identifiable without additional assumptions or restrictions. For this reason it is common to make an exclusion restriction. This involves including at least one of the regressors in X_i^R that is not included in X_i^Y. The exclusion restriction only identifies $P(Y_i | X_i^Y)$ if the instrumental variable predicts well whether or not Y_i is observed, but the variable must be unrelated to the value of Y_i.

Usually there are substantive arguments to include more information in the selection equation, for instance, paradata that describes the cause of nonresponse but is unrelated to survey variables. A good source is interviewer information. Identification of the sample selection models, however, hinges on the assumption of a bivariate normal distribution of the error terms. This causes a serious lack of robustness against misspecification.

The two methods discussed, MLE and Heckman's two-step estimator, are fully parametric and have been widely criticized for their dependence on the normality assumptions (Maddala, 1991; Johnston and DiNardo, 1997; Vella, 1998). A lot of research has been devoted to the relaxation of the distributional assumptions. Dubin and Rivers (1989) discuss and present alternative parametric methods to relax the normality assumption, for instance, using a logit rather than a probit form of the regression equation. Vella (1998) presents a survey on estimation methods and discusses departures from the fully parametric models in the direction of relaxation of the distributional assumptions. Maddala (1991) proposes to use generalized distributions, or semiparametric methods. Nonparametric estimation methods are presented in Das et al. (2003). Gallant and Nychka (1987) propose a semiparametric method that appears to be very practical. They use a Hermite series approximation for the joint distribution of the error terms.

Most of the literature describes models for continuous outcome variables. Cobben (2009) discusses how these models can be adjusted to include categorical survey variables.

12.2.2.4 Advantages and Disadvantages The nonresponse analysis models described in Section 12.2.1 are important in that they motivate the evaluation of different response types, as explored in a lot of the recent survey literature (e.g., Groves et al., 2002; Bethlehem and Schouten 2004; Stoop, 2005; Durrant and Steele, 2007). In fact, until now most of the literature on different response types has focused on methods to analyze nonresponse.

The methods described in Section 12.2.1 can be used to adjust for the nonresponse bias. The nested logit model allows for a sequential weight adjustment approach as suggested by Groves and Couper (1998) and Iannacchione (2003). However, this approach uses only the relationship between the response and background variables. It does not take into account the relationship with the survey items.

For discrete bivariate variables, the bivariate probit model with sample selection can be used to adjust for nonresponse bias. Standard statistical software is available to analyze situations where there is one selection equation and a bivariate survey item. The sample selection model proposed by Heckman (1979) describes a model that can be used for continuous survey items. It is shown how this model can be estimated for the situation of survey nonresponse based on one selection equation. Two estimation methods are described, maximum likelihood estimation and Heckman's two-step estimator. The parametric estimation methods are highly dependent on the distributional assumption of the models. Therefore also semiparametric procedures are available, such as described by Gallant and Nychka (1987). These are, however, not frequently applied in practice.

There are no standard ways to apply the sequential response process to several response types. Nevertheless, the bivariate probit model with sample selection and the sample selection model can be extended to multiple selection equations. Furthermore a lot of the survey items are categorical. Cobben (2009) describes how

TABLE 12.3 Fieldwork results of the General Population Survey

Result	Frequency	Percentage
Sample size	32,019	100.0 %
Response	18,792	58.7%
Nonresponse	13,227	41.3%
Unprocessed cases	*2,456*	*7.7%*
Noncontact	*1,847*	*5.8%*
Not-able	*1,034*	*3.2%*
Refusal	*7,890*	*24.6%*

In Italic: the different types of nonresponse.

the sample selection model can be adapted to fit those situations. There as yet is no standard statistical software for these methods.

12.3 Application

12.3.1 NONRESPONSE ANALYSIS WITH DIFFERENT TYPES OF RESPONSE

The application of the nonresponse analysis methods described in Section 12.2.1 to the General Population Survey (GPS) is covered in this section. All analyses are run in Stata.

Persons for the GPS were selected by means of a stratified two-stage sample. In the first stage, municipalities were selected within regions with probabilities proportional to the number of inhabitants. In the second stage, an equal probability sample was drawn in each selected municipality. Sampling frames were the population registers of the municipalities. The sampling design was such that each person had the same probability of being selected (a so-called self-weighting sample). Table 12.3 gives the survey results for the GPS.

Ultimately about 59% of the sampled persons responded. Refusal was by far the largest cause of nonresponse (59%). In 14% of the cases, no contact could be established with the sampled persons. And 19% of the cases were not processed in the field. Reasons for this type of nonresponse were lack of capacity (high workload

TABLE 12.4 Composition of responses due to noncontact and refusal

Result	Frequency	Percentage
Sample size	32,019	100.0 %
Response	18,792	58.7%
Nonresponse	13,227	41.3%
Noncontact	*4,303*	*13.4%*
Refusal	*8,924*	*27,8%*

of the interviewer) and interviewer not being available (illness or holiday). Only 8% of the nonresponse was due to people not able to answer the questions because of illness, a handicap, or language problems.

When the different types of nonresponse were restricted to noncontact and refusal, the unprocessed cases had to be merged with noncontact, and the not-able cases with refusal. The composition of the response is given in Table 12.4.

The subset of observations that were successfully contacted consisted of $n^c = 27{,}716$. The nested logit model consisted of two separate logit models, one for contact and one for participation. The contact model was fitted for all the observations. The participation model was fitted for the subset of observations about the persons contacted.

First, a bivariate analysis of the auxiliary variables with contact and participation was performed. The values for Cramér's V statistic are listed in Table 12.5. The auxiliary variables are ordered according to their value for the overall response. To contrast the analysis with different response types and the situation where only response and nonresponse are considered, the values for Cramér's V and response are shown in the fourth column of Table 12.5.

TABLE 12.5 Tests of bivariate dependence between the auxiliary variables with contact, participation, and response

	Cramér's V		
Auxiliary variable	Contact	Participation	Response
Region of the country	0.205	0.092	0.163
Degree of urbanization	0.186	0.086	0.153
Has listed phone number	0.138	0.107	0.150
Percentage nonwestern nonnatives in neighborhood	0.144	0.088	0.138
Percentage nonnatives in neighborhood	0.163	0.081	0.133
Average house value in neighborhood	0.109	0.088	0.115
Ethnic background	0.098	0.089	0.112
Type of household	0.122	0.067	0.106
Size of the household	0.114	0.066	0.099
Marital status	0.130	0.058	0.097
Is nonnative	0.088	0.058	0.087
Has social allowance	0.058	0.064	0.077
Age in 13 classes	0.095	0.071	0.061
Has an allowance	0.034	0.055	0.061
Children in household	0.053	0.038	0.056
Has a job	0.012	0.053	0.037
Age in 3 classes	0.084	0.044	0.030
Has disability allowance	0.001	0.025	0.021
Gender	0.025	0.001	0.011
Has unemployment allowance	0.003	0.002	0.000

For contact, the strongest bivariate relationship is found for the region of the country ($V = 0.205$), followed by the degree of urbanization ($V = 0.186$), and then the percentage of nonnatives in the neighborhood ($V = 0.163$). For participation, the strongest relationship is found for having a listed phone number ($V = 0.107$), followed by region of the country ($V = 0.092$), and then ethnic background ($V = 0.089$). Compared to contact, these values are low. For contact, the bivariate relationship with having a listed phone number is stronger, at $V = 0.138$.

In general, the values for Cramér's *V* are highest for contact. When it comes to participation, Cramér's *V* values are much lower. Hence, it appears that making contact is more strongly related to the characteristics of the sample element than participation.

Next separate multivariate logistic models were built for contact and participation. The same strategy as described in Chapter 5 was followed. First, based on Cramér's *V*, the strongest candidate for the model was selected. Stepwise, the logistic regression model with one additional variable was evaluated. The final model was found if none of the additional variables has a significant relation with the independent variable. The results are shown in Tables 12.6 through 12.8.

The relative significance, as adjusted for the effects of other variables in the model, was evaluated by the *p*-value of the Wald test for the joint significance of all

TABLE 12.6 Contributions of the auxiliary variables to the final models for contact, participation, and response

Contact		Participation		Response	
Variable	Wald χ^2	Variable	Wald χ^2	Variable	Wald χ^2
Region	135.40	Listed phone	149.84	Listed phone	242.04
Listed phone	127.57	Region	113.14	Region	164.79
Age in 13 categories	81.95	Marital status	75.87	Ethnic background	93.65
Urbanization	43.17	Age in 13 categories	72.45	Marital status	74.12
Type of household	42.84	Is nonnative	72.02	Size of household	52.58
Average house value	41.19	Age in 3 categories	63.59	Average house value	25.29
Has children	35.51	Ethnic background	58.48	Has a job	23.70
Marital status	31.02	Average house value	44.50	Type of household	23.61
Gender	21.37	Size of household	43.70	Urbanization	16.28
Percentage of non-natives	19.62	Has a job	14.40	Gender	14.72
Has a job	12.00	Has social allowance	11.30	Age in 3 categories	10.97
Is nonnative	8.97	Gender	3.91	Has a social allowance	8.62

TABLE 12.7 Logistic regression model for the contact probability

Variable	Category	β	se
Intercept		0.8868**	0.2028
Region	Woodlands	0.1424	0.0762
(Reference Greenfields)	Lowlands	−0.2063**	0.0721
	Highlands	0.0279	0.0728
	Metropolis	−0.9491**	0.1009
Urbanization	Strong	−0.0718	0.0837
(Reference very strong)	Fairly	0.1126	0.0882
	Little	−0.0074	0.0895
	Not	0.3465**	0.0977
Percentage of nonnatives	5–10%	0.0310	0.0481
(Reference <5%)	10–15%	−0.0571	0.0582
	15–20%	−0.1228	0.0699
	20–30%	−0.1940**	0.0707
	30–40%	−0.2256*	0.0923
	40–50%	−0.2860*	0.1186
	>50%	−0.3722**	0.1394
Listed phone	Yes	0.4452**	0.0394
(Reference no)			
Marital status	Married	0.2880**	0.0570
(Reference not married)	Widowed	0.2856**	0.0962
	Divorced	0.0401	0.0726
Type of household	Couple without children	0.3688**	0.0591
(Reference single)	Couple with children	0.0219	0.0834
	Single parent	Collinear	
	Other	−0.0700	0.1534
Average house value	50–75 thousand	0.3062*	0.1511
(Reference < 50 thousand)	75–100 thousand	0.3643*	0.1474
	100–125 thousand	0.3602*	0.1483
	125–150 thousand	0.3489*	0.1492
	150–200 thousand	0.3047*	0.1480
	200–250 thousand	0.2214	0.1517
	250–300 thousand	0.2167	0.1588
	300–350 thousand	−0.0300	0.1674
	350–400 thousand	−0.1479	0.1831
	400–500 thousand	0.1223	0.1982
	> 500 thousand	0.2067	0.2306
Age in 13 categories	20-24 years	−0.4124**	0.1161
(Reference 18–19 years)	25-29 years	−0.3040**	0.1167

(Continued)

TABLE 12.7 (*Continued*)

Variable	Category	β	se
	30-34 years	−0.2629*	0.1186
	35-39 years	−0.2021	0.1208
	40-44 years	0.0090	0.1253
	45-49 years	−0.0118	0.1267
	50-54 years	0.0618	0.1303
	55-59 years	−0.0135	0.1361
	60-64 years	0.2272	0.1442
	65-69 years	0.1183	0.1450
	70-74 years	0.3128*	0.1533
	75 +	0.2099	0.1460
Is nonnative	Yes	−0.1444**	0.0482
(Reference no)			
Has children	Yes	0.4805**	0.0806
(Reference no)			
Gender	Female	0.1631**	0.0353
(Reference male)			
Has a job	Yes	0.1430**	0.0413
(Reference no)			
Pseudo R^2		0.0780	
χ^2-value		1,982.01	
df		49	

categories of a variable. The null hypothesis of the Wald test was that all coefficients corresponding to the categories of a variable are equal to zero—against the alternative hypothesis that at least one of the coefficients is not equal to zero. If $p > 0.05$, the variable was not added to the model.

The contributions of the Wald χ^2-values are shown in Table 12.6. The estimated coefficients ß with the corresponding standard errors (se) are reported in Tables 12.7 and 12.8 for all categories except the reference category. An asterisk indicates that a category is significant at a 5% level, and two asterisks at a 1% level. The sign of the coefficient ß determines whether the value of the dependent variable is increasing or decreasing with respect to the reference category. For the general model fit there are additionally reported the Nagelkerke pseudo R^2, the total χ^2, and the degrees of freedom (df) of the model.

The final logistic model for contact is displayed in Table 12.7, and for participation in Table 12.8.

The region of the country and having a listed phone are clearly the most influential variables in the models for contact and participation. Likewise in the response model these variables give the strongest contributions to the model.

TABLE 12.8 Logistic regression model for the participation probability

Variable	Category	β	se
Intercept		1.1674**	0.1905
Listed phone	Yes	0.4064**	0.0332
(Reference no)			
Region	Woodlands	−0.0420	0.0497
(Reference Greenfields)	Lowlands	−0.2872**	0.0474
	Highlands	−0.0378	0.0484
	Metropolis	−0.3900**	0.0581
Ethnic background	First generation nonwestern	Collinear	
(Reference native)	First generation western	0.4839**	0.0923
	Second generation nonwestern	0.3608	0.2579
	Second generation western	0.7092**	0.0933
Average house value	50–75 thousand	−0.2958	0.1695
(Reference < 50 thousand)	75–100 thousand	−0.2978	0.1650
	100–125 thousand	−0.1763	0.1646
	125–150 thousand	−0.1517	0.1647
	150–200 thousand	−0.0932	0.1637
	200–250 thousand	−0.0280	0.1653
	250–300 thousand	−0.0056	0.1686
	300–350 thousand	−0.0332	0.1751
	350–400 thousand	0.0701	0.1885
	400–500 thousand	−0.0921	0.1913
	> 500 thousand	0.0406	0.2125
Age in 13 categories	20–24 years	−0.3789**	0.0995
(Reference 18–19 years)	25–29 years	−0.4543**	0.0995
	30–34 years	−0.4398**	0.1011
	35–39 years	0.3099**	0.0604
	40–44 years	0.2821**	0.0615
	45–49 years	0.1258*	0.0596
	50–54 years	Collinear	
	55–59 years	−0.0837	0.0779
	60–64 years	0.0042	0.0783
	65–69 years	0.1028	0.0793
	70–74 years	Collinear	
	75 years +	−0.2038**	0.0745
Size of household	2	−0.2022**	0.0485
(Reference 1)	3	−0.2813**	0.0548
	4	−0.1443*	0.0582
	5 or more	−0.0160	0.0671

(*Continued*)

TABLE 12.8 (*Continued*)

Variable	Category	β	se
Has social allowance	Yes	−0.2392**	0.0711
(Reference no)			
Marital status	Married	0.3727**	0.0458
(Reference not married)	Widowed	0.0558	0.0719
	Divorced	0.2570**	0.0645
Is nonnative	Yes	−0.6424**	0.0757
(Reference no)			
Has a job	Yes	0.1276**	0.0336
(Reference no)			
Age	35–54 years	−0.8453**	0.1060
(Reference 18–34 years)	55 years +	−0.7314**	0.1196
Gender	Female	0.0541*	0.0274
(Reference male)			
Pseudo R^2		0.0270	
χ^2-value		941.99	
df		42	

Age in 13 categories is influential for both contact and participation. However, in the response model age in three categories is selected. The degree of urbanization appears only in the model for contact, and for response. Hence the degree of urbanization seems to be unrelated to participation.

The type of household seems to be related to contact, but not to participation. Still does turn up in the response model. The average house value plays a role in both contact and participation, so it is also included in the response model. Having children is only related to contact and is not selected in the response model. Marital status is an influential variable when it comes to participation, but less so for contact. It is included in the response model. Gender shows only a weak relation to both contact and participation. Also in the response model the contribution of gender to the model is small.

The percentage of nonnatives is only related to contact. Having a job is related to both contact and participation and also turns up in the response model. Being nonnative is related to both contact and participation, but is not selected in the response model. The size of the household is not related to contact, but gives a considerable contribution to the participation model and is also selected in the response model. The same holds for having a social allowance.

In Table 12.7 the pseudo R^2 for the contact model is not very high, only 7.8% of the variance in contact is explained by the model.

The results are not surprising and confirm what is known from the literature on nonresponse. In Lowlands, the probability of contact is lower than in Greenfields. Areas with very low urbanization have a higher contact probability than highly urbanized areas. The contact probability decreases with an increasing percentage of nonnatives in the neighborhood. Having a listed phone increases the contact probability. Married and widowed persons have a higher probability of contact than unmarried persons, and couples with children have a higher contact probability than singles. Neighborhoods with a very low (< 50 thousand euros) average house value have a low probability of contact. Compared to the 18 to 19-year-olds, persons aged 20 to 34 have a lower probability of contact, and persons aged 70 to 74 a higher probability. Natives have a higher probability of contact than nonnatives. Having children implies a higher contact probability, so does having a job. Women have a slightly higher contact probability than men.

In Table 12.8 the pseudo R^2-value for the participation model ($R^2 = 0.0270$) is even lower than for the contact model. That implies that there is a lot of variation left that is not explained by the model. This may be an advantage; see Chapter 7.

Having a listed phone implies a higher participation probability than not having a listed phone. In Lowlands and Metropolis, the participation probability is lower than in Greenfield. First- and second-generation Western nonnatives have a higher probability of participation than natives. This result is different from the response model. There, first-generation Western nonnatives have a lower response probability than native persons. This could be caused by the low contact probability of nonnatives. Compared to 18 to 19-year-old persons, 20 to 34-year-old persons have a lower probability of participation, and 35 to 49-year-old persons have a higher probability of participation. Persons aged 75 and older also have a low probability of participation. Households consisting of 2, 3, or 4 persons have a lower participation probability than single person households. Most probably, the contact probability may be high for these types of households, but once contacted, the probability of participation is lower than for single person households. Persons with a social allowance and nonnatives have even less probability of participation. Persons with a job and women have a high probability of participation. Married and divorced persons have a higher probability of participation than unmarried persons and persons aged 0 to 34-years have a higher probability of participating than those aged 35 and over.

The conclusions based on the two logit models do not seem to differ a lot from the conclusions based on the response model presented in Chapter 5. Only with respect to ethnic background, other conclusions are reached. Hence there is not much indication of a correlation between contact and participation. In the nested logit models discussed above, no correlation between contact and participation is modeled. Therefore the bivariate probit model with sample selection, as described in Section 12.2.1.3 is applied to the GPS data. The same models as for the nested logit model are used (see Table 12.6). The results of fitting the bivariate probit model with sample selection to the GPS data are displayed in Table 12.9.

The results of these models are not different from the results of the nested logit model. The parameter estimates for the logit model and the probit model are roughly proportional (Johnston and DiNardo, 1997).

The estimated correlation $\hat{\zeta}_{cp}$ between the contact and the participation model equals 0.067. This is close to 0. If the correlation between contact and participation

TABLE 12.9 Results of fitting a bivariate probit model with sample selection to contact and participation in the GPS

Contact				Participation			
Variable	Category	β	se	Variable	Category	β	se
Intercept		0.5456**	0.1153	Intercept		0.6765**	0.1522
Region	Woodlands	0.0822*	0.0390	Listed Phone	Yes	0.2546**	0.0260
(Reference Greenfields)	Lowlands	−0.1041**	0.0374	(Reference no)			
	Highlands	0.0227	0.0380	Region	Woodlands	−0.0250	0.0299
	Metropolis	−0.5409**	0.0552	(Reference Greenfields)	Lowlands	−0.1777**	0.0304
Urbanization	Strong	−0.0464	0.0459		Highlands	−0.0235	0.0291
(Reference very strong)	Fairly	0.0488	0.0477		Metropolis	−0.2561**	0.0609
	Little	−0.0105	0.0484	Ethnic background	First-generation nonwestern	collinear	
	Not	0.1645**	0.0517	(Reference native)	First-generation western	0.2987**	0.0570
Percentage of nonnatives	5–10%	0.0137	0.0252		Second-generation nonwestern	0.227	0.1591
(Reference < 5%)	10–15%	−0.0299	0.0313		Second-generation western	0.4384**	0.0573
	15–20%	−0.0687	0.0383	Average housevalue	50–75 thousand	−0.1758	0.1052
	20–30%	−0.1100**	0.0395	(Reference < 50 thousand)	75–100 thousand	−0.1752	0.1034

	30–40%	−0.1341*	0.0527		100–125 thousand	−0.0996	0.1033
	40–50%	−0.1761*	0.0695		125–150 thousand	−0.0854	0.1032
	>50%	−0.2361**	0.0846		150–200 thousand	−0.0498	0.1024
Listed Phone	Yes	0.2415**	0.0219		200–250 thousand	−0.0107	0.1029
(Reference no)					250–300 thousand	0.0007	0.1048
Marital status	Married	0.1576**	0.0315		300–350 thousand	−0.0179	0.1073
(Reference not married)	Widowed	0.1710**	0.0516		350–400 thousand	0.0409	0.1143
	Divorced	0.0280	0.0411		400–500 thousand	−0.0514	0.1177
Type of household	Couple without children	0.2111**	0.0331		> 500 thousand	0.0313	0.1305
(Reference single)	Couple with children	0.0081	0.0461				
	Single parent	collinear					
	Other	−0.0409	0.0897				
Average housevalue	50–75 thousand	0.1836*	0.0902	Age in 13 categories	20–24 y	−0.2327**	0.0608
(Reference < 50 thousand)	75–100 thousand	0.2182*	0.0878	(Reference 18–19 years)	25–29 y	−0.2757**	0.0598
	100–125 thousand	0.2198*	0.0882		30–34 y	−0.2672**	0.0604
	125–150 thousand	0.2103*	0.0885		35–39 y	0.1861**	0.0375
	150–200 thousand	0.1824*	0.0878		40–44 y	0.1691**	0.0372
	200–250 thousand	0.1401	0.0896		45–49 y	0.0762*	0.0363
	250–300 thousand	0.1385	0.0928		50–54 y	collinear	
	300–350 thousand	0.0087	0.0975		55–59 y	−0.0557	0.0485
	350–400 thousand	−0.0529	0.1059		60–64 y	0.0014	0.0479
	400–500 thousand	0.0941	0.1126		65–69 y	0.0612	0.0487
	> 500 thousand	0.1433	0.1293		70–74 y	collinear	

(Continued)

TABLE 12.9 *(Continued)*

Contact				Participation			
Variable	Category	β	se	Variable	Category	β	se
Age in 13 categories	20–24 y	−0.2283**	0.0627		75 +	−0.1259**	0.0458
(Reference 18–19 years)	25–29 y	−0.1659**	0.0629	Size of household	2	−0.1155**	0.0344
	30-34 y	−0.1368*	0.0641	(Reference 1)	3	−0.1611**	0.0396
	35-39 y	−0.1067	0.0651		4	−0.0783*	0.0411
	40–44 y	0.0028	0.0672		5 or more	0.0001	0.0455
	45–49 y	−0.0051	0.0681	Has social allowance	Yes	−0.1482**	0.044
	50–54 y	0.0278	0.0696	(Reference no)			
	55–59 y	−0.0109	0.0731	Marital status	Married	0.2289**	0.0295
	60–64 y	0.1113	0.0769	(Reference not married)	Widowed	0.0377	0.0467
	65–69 y	0.0529	0.0775		Divorced	0.1564**	0.0394
	70–74 y	0.1522	0.0812	Is nonnative	Yes	−0.3995**	0.0479
	75 +	0.029	0.078	(Reference no)			
Is nonnative	Yes	−0.0818**	0.0269	Has a job	Yes	0.0792**	0.0208
(Reference no)				(Reference no)			
Has children	Yes	0.2711**	0.045	Age	35–54 years	−0.5077**	0.0629
(Reference no)				(Reference 18–34 years)	55 years and older	−0.4341**	0.072
Gender	Female	0.0857**	0.019	Gender	Female	0.0352**	0.0174
(Reference male)				(Reference male)			
Has a job	Yes	0.0759**	0.0224				
(Reference no)							

ζ_{cp} is 0, the log-likelihood for the probit model with sample selection is equal to the sum of the log-likelihoods for the probit model for participation and the probit model for contact. This can be tested with a likelihood-ratio test. The χ^2-value for this test equals 0.12, with a p-value of 0.7279. The outcome of the bivariate probit model with sample selection is thus not much different from the outcome of fitting the models for contact and participation separately. Hence, given the variables used in the analysis and the chosen model, there is no indication that the two processes are correlated. This result is in line with Lynn and Clarke (2002) and Nicoletti and Peracchi (2005).

12.3.2 A SEQUENTIAL WEIGHT ADJUSTMENT FOR NONRESPONSE

The sequential weight adjustment discussed in Section 12.2.2.2 consists of four steps. First, a logit model is fitted to estimate the contact probability. Based on the estimated contact probability $\hat{\gamma}$, the weights $w_i = 1/\hat{\gamma}_i$ are constructed. These are then used in a weighted logistic regression to estimate the participation probability. The estimated participation probabilities $\hat{\zeta}$ are used to construct the final weight, which is $w_i^* = w_i/\hat{\rho}_i = 1/\hat{\gamma}_i\hat{\zeta}_i$.

The same models for contact and participation as described in Section 12.3.1 are applied for the sequential weight adjustment. The results of this approach are presented in Table 12.10. In the third column of Table 12.10 the response mean is given; the sequential weight adjusted estimates are in column 4, and the estimated target variables based on the GREG-estimator are in column 5.

TABLE 12.10 Response means and estimates for the target variables in the GPS based on the sequential weight adjustment approach and the GREG estimator

Variabele	Category	Response mean	Sequential weight adjustment	GREG
PC in household	Yes	57.8	55.2	55.3
	No	42.6	44.8	44.7
Wants to move within a year	Definitely not	72.3	70.7	70.7
	Possibly	15.2	15.6	15.7
	Cannot find something	1.7	1.8	1.8
	Definitely	8.4	9.4	9.3
	Is going to move	2.5	2.5	2.5
General health condition	Very good	22.4	21.8	21.8
	Good	55.4	54.7	54.8
	Reasonable	13.1	13.6	13.6
	Varied	6.6	7.1	7.0
	Bad	2.5	2.8	2.8

(Continued)

TABLE 12.10 *(Continued)*

Variabele	Category	Response mean	Sequential weight adjustment	GREG
Has newspaper subscription	Yes	65.6	62.9	62.8
	No	34.4	37.1	37.2
Is active in a club	Yes	44.4	42.8	42.8
	No	55.6	57.2	57.2
Is interested in politics	Very interested	11.6	11.9	11.9
	Fairly interested	42.4	41.9	41.9
	Little interested	29.1	28.5	28.5
	Not interested	16.9	17.7	17.7
Job level	Very low	3.6	3.8	3.8
	Low	13.3	13.1	13.0
	Middle	22.8	22.0	22.0
	High	11.2	11.0	11.0
	Academic	4.2	4.2	4.2
	No job	44.8	46.0	46.1
Level of education	Primary	20.0	21.3	21.2
	Junior secondary	9.2	9.1	9.2
	Prevocational	17.4	17.0	17.0
	Senior secondary	6.9	7.0	7.1
	Postsecondary	28.0	26.9	26.9
	Higher professional	13.3	13.3	13.3
	University	5.2	5.4	5.4
Owns a house	Yes	62.5	58.3	59.3
	No	37.5	41.7	41.7
Religious denomination	None	37.2	38.4	38.7
	Roman Catholic	33.7	32.1	31.9
	Protestant	22.3	21.2	21.1
	Islam	1.6	2.7	2.7
	Other	5.3	5.7	5.7
Employment situation	Works 12 hours or more	55.2	54.0	53.9
	Works less than 12 Hours	6.1	5.8	5.8
	Does not work	38.7	40.2	40.3

When compared, the differences between the estimates obtained by the sequential weight adjustment approach (column 4) and the estimates obtained by the GREG estimator (column 5) are not unambiguous. Yet the estimates do not differ much. Some estimates from the sequential weight adjustment column adjust the response mean more than the GREG estimator, for instance, on having a PC in the household, the general health condition, and the level of education. Other estimates are closer to the response mean than the estimates based on the GREG estimator, for instance, the variables on having a newspaper subscription and on religious denomination.

12.3.3 SAMPLE SELECTION MODEL TO ADJUST FOR NONRESPONSE

The sample selection model as described in Section 12.2.2.3 was applied for overall response for the target variables *PC in household, owns a house, has a newspaper subscription*, and *is active in a club*. The sample selection model then reduces to the bivariate model with sample selection. This is the same model that was applied in Section 12.3.1 for contact and participation in the GPS. The relationship between the auxiliary variables and the response was modeled in the selection equation. No distinction was made for the different types of nonresponse. The auxiliary variables and the different target variables were modeled by the outcome equation. See (12.17) for the two equations. The response model used was similar to the response model of Chapter 5. The models for the target variables consisted of the five variables that showed the strongest bivariate relationships with the target variables. Hence for each target variable a different model was used. The models for the target variables are summarized in Table 12.11.

The results are presented in Table 12.12. In the third column the response means are given. The fourth column presents the results from applying the sample selection model to the GPS. The last column shows the estimates from the GREG estimator.

TABLE 12.11 Variables in the outcome equations for the different target variables

Target variable	Model
PC in household	Age in 13 categories, age in 3 categories, size of household, type of household, has a job
Has newspaper subscription	Average house value, age in 13 categories, percentage of nonwestern nonnatives in neighborhood, percentage of nonnatives in neighborhood, has a listed phone
Is active in a club	Percentage of nonwestern nonnatives in neighborhood, average house value, ethnic background, has a listed phone, extent of urbanization
Owns a house	Average house value, percentage of nonwestern nonnatives in neighbor hood, percentage of nonnatives in neighborhood, type of household, size of household

TABLE 12.12 Results from applying the sample selection model to the GPS

Target variable	Category	Response mean	Sample selection model	GREG
PC in household	Yes	57.8	55.0	55.3
	No	42.6	45.0	44.7
Has newspaper subscription	Yes	65.6	62.8	62.8
	No	34.4	37.2	37.2
Is active in a club	Yes	44.4	42.8	42.8
	No	55.6	57.2	57.2
Owns a house	Yes	62.5	58.5	59.3
	No	37.5	41.5	41.7

TABLE 12.13 Estimated correlation between response and target variables in the GPS

Target variable	$\hat{\zeta}$	χ^2-Value	p-Value
PC in household	−0.8	186.8	0.0000
Has newspaper subscription	−0.8	113.2	0.0000
Is active in a club	−0.4	10.6	0.0011
Owns a house	−0.8	213.7	0.0000

The sample selection model allows for a correlation ζ between the selection equation (response) and the outcome equation (target variable). Hence this correlation is also estimated in the model. A zero correlation implies that there is no selection bias due to nonresponse. The results for the different target variables are given in Table 12.13.

As can be seen in Table 12.13, there is a negative correlation between the four bivariate target variables in the GPS and response. This implies that the values for the target variables in the response should be adjusted downward for the nonresponse bias. This effect is evident in the estimates shown in Table 12.12. Compared to the response mean, the estimates for the categories "yes" of the target variables are lower for the sample selection estimates. The same holds for application of the GREG estimator. Being active in a club shows the smallest correlation to the response. Hence also the adjustment made to the estimated percentage of persons active in a club is smaller than the adjustments made to the other target variables.

The estimates based on the sample selection model and the estimates based on the GREG estimator are very close. No standard errors for the sample selection model were calculated, but it seems that the sample selection model adjusted the estimates a little bit more than the GREG estimator. For all target variables, except having a newspaper subscription, the estimates from the sample selection model are further away from the response mean than the GREG estimator.

12.4 Summary

The usual nonresponse analysis and nonresponse adjustment methods do not distinguish among the different causes of nonresponse. This chapter presented different ways to analyze nonresponse and to adjust for the nonresponse bias that do account for different types of nonresponse.

For the nonresponse analysis, the multinomial model, the nested logit model, the bivariate probit model with sample selection, and the multilevel model were discussed. These models have different underlying assumptions about the correlation structures at distinct stages in the response process. Furthermore they differ in the way that they account for the sequential character of the response process.

The multinomial model allows for distinctions to be made among the different response types. That is the dependent variables for each of the response types, and the corresponding coefficients, are allowed to be different. This type of model, however, does not acknowledge the sequential nature of the response process, nor does it allow for a dependency between response types.

The nested logit model does account for the sequential nature of the response process. Nevertheless, because the different indicators are only defined on subsets of the sample, the truncation that arises from unobserved indicators cannot be accounted for. Hence the underlying assumption of the nested logit model is that the processes are independent of each other.

The bivariate probit model with sample selection treats the censoring of sample elements as a form of sample selection. It makes the censoring explicit by acknowledging the fact that certain response types are not observed, conditional on previous stages in the response process. The underlying propensities are estimated based on their observed indicators, but unlike the nested logit model the propensities are defined for every sampled person regardless of whether the corresponding indicator is observed. In addition the joint distribution of the error terms in the equations allows for a correlation and therefore allows for a dependency between the different stages in the response process.

The multilevel model is based on another type of nesting, namely, the clustering of sample elements within interviewers. This model is a refinement of the other models to allow for interviewer effects. The clustering confounds the characteristics of the interviewer and the sampled persons. The multilevel model distinguishes between two levels of information. When the data are analyzed at the respondent level and data on the interviewers are included, the clustering effect has to be accounted for in the models. Otherwise, the estimated standard errors would be much lower than they are and lead to false conclusions as significance levels are met too soon.

In Section 12.3.1, the nested logit model and the bivariate probit model with sample selection were applied to the GPS data set, and a distinction was made between contact and participation. The results from the nested logit were contrasted with the nonresponse analysis in Chapter 5, where no further distinction between response types was made. The two logit models yielded outcomes that do not differ much from those using the response model of Chapter 5. Hence, given the variables used in the analysis and the chosen model, there is no indication that the two processes are correlated. This is further confirmed by the bivariate probit model with sample selection, in which the correlation between contact and participation is estimated. The estimated correlation $\hat{\zeta}_{cp}$ between the contact and the participation models turned out to be 0.067. This is close to 0. The χ^2-value for a likelihood ratio test was 0.12 with a p-value of 0.7279. Hence the outcome of the bivariate probit model with sample selection was not so different from the outcome of fitting the models for contact and participation separately.

The methods for nonresponse analysis were applied to adjust for the nonresponse bias. The nested logit model allowed for a sequential weight adjustment as suggested by Groves and Couper (1998) and Iannacchione (2003). However, this model is limited to the relationship between the response and background variables. It does not take into account the relationship with the survey items.

For discrete bivariate variables, the bivariate probit model with sample selection was used to adjust for the nonresponse bias. Standard statistical software is available to analyze situations where there is one selection equation and a bivariate survey item. The sample selection model proposed by Heckman (1979) describes a model that can be used for continuous survey items. This model was shown to work well for the survey situation where nonresponse is based on one selection equation.

12.5 KEY TERMS

Bivariate probit model with sample selection: A model that can be used to introduce a correlation between the different response types, and that accounts for the censoring of unobserved households.

Censoring: The situation that occurs when some variables are only observed for a part of the sample. This is the case with nonresponse in that the target variable is not observed for the nonrespondents, but the background characteristics are known for all persons in the sample.

Heckman model or sample selection model: A (simplified) sample selection model for nonresponse that consists of two equations. The first equation models the response propensity; the outcome of the first equation determines whether the survey target variable is observed. This equation is therefore referred to as the **selection equation**. The second equation accounts for the censoring of sample elements that do not respond, and models the target variable, therefore the second equation is referred to as the **regression equation**.

Interviewer effect: An effect that occurs when the interviewers systematically affect the probability that selected persons will participate or be contacted.

Inverse Mills ratio: The inverse Mills ratio is a monotone decreasing function of the probability that a selected person responds to the survey request.

Maximum likelihood estimation (MLE): A method that is used for fitting a statistical model to the data and finding estimates for the model's parameters. MLE amounts to finding specific parameter values that optimize the probability of finding the data that have actually been observed. Hence MLE produces values for the model parameters that make the data more likely than other values for the parameters.

Multilevel model: A model that can be used to analyze hierarchical data, for instance, the clustering of respondents within interviewers.

Nested logit model: A model that consists of separate logistic regression models for each of the stages in the response process. The different stages are analyzed on the subset of observations available for each process. The assumption underlying this model is that the stages in the response process are independent.

Sequential weight adjustment method: A method that consists of sequentially fitting a nested logit model. The estimated nested logit model is then used to construct nonresponse adjustment weights.

Truncation: The situation that occurs when, as opposed to censoring, there are no observations at all available for some part of the sample. In this situation, in addition to the missing target variable for nonrespondents, there are no background characteristics for the nonrespondents available either.

12.6 REFERENCES

Agresti, A. (2002), *Categorical Data Analysis.* Wiley, New York.

Anderson, D.A., and Aitkin, M. (1985), Variance Component Models with Binary Response: Interviewer Variability. *Journal of the Royal Statistical Society, Series B*, 47, pp. 203–210.

Bethlehem, J., and Schouten, B. (2004), *Nonresponse Analysis of the Integrated Survey on Living Conditions (POLS).* Discussion paper 04004. Statistics Netherlands, Voorburg.

Biemer, P.P., and Lyberg, L.E. (2003), *Introduction to Survey Quality.* Wiley, New York.

Cobben, F. (2009), *Nonresponse in Sample Surveys: Methods for Analysis and Adjustment.* PhD thesis. University of Amsterdam, Amsterdam.

Daalmans, J., Israëls, A., Loeve, A., and Schouten, B. (2006), *Tussenrapportage Responscijfers EBB 2002-2004.* Report TMO-R&D-2006-03-10-JDAS. Statistics Netherlands, Voorburg.

Das, M., Newey, W.K., and Vella, F. (2003), Nonparametric Estimation of Sample Selection Models. *Review of Economic Studies* 70, pp. 33–58.

Dubin, J.A., and Rivers, D. (1989), Selection Bias in Linear Regression, Logit and Probit Models. *Sociological Methods and Research* 18, pp. 360–390.

Durrant, G.B., and Steele, F. (2007), Multilevel Modelling of Refusal and Noncontact Nonrespone in Household Surveys: Evidence from Six UK Government Surveys. S3RI Working Paper M07-11. University of Southampton, Southampton, UK.

Gallant, A.R., and Nychka, D.W. (1987), Semi-nonparametric Maximum Likelihood Estimation. *Econometrica* 55, pp. 363–390.

Greene, W.H. (1981), Sample Selection Bias as a Specification Error: A Comment. *Econometrica* 39 pp. 795–798.

Greene, W.H. (2002), *Econometric Analysis*, 5th ed. Prentice-Hall, Upper Saddle River, NJ.

Gronau, R. (1974), Wage Comparisons: A Selectivity Bias. *Journal of Political Economy* 82, pp. 1119–1144.

Groves, R.M., and Couper, M.P. (1998), *Nonresponse in Household Interview Surveys*. Wiley, New York.

Groves, R.M., Dillman, D.A., Eltinge, J.L., and Little, R.J.A. (2002), *Survey Nonresponse*. Wiley, New York.

Heckman, J.J. (1974), Effects of Child-Care Programs on Women's Work Effort. *Journal of Political Economy* 82, pp. 136–163.

Heckman, J.J. (1979), Sample Selection Bias as a Specification Error. *Econometrica* 47, pp. 153–161.

Hox, J.J. (1994), Hierarchical Regression Models for Interviewer and Respondent Effects. *Sociological Methods and Research* 22, pp. 300–318.

Hox, J.J. (1995), *Applied Multilevel Analysis*, 2nd ed. TT-Publikaties, Amsterdam.

Iannacchione, V.G. (2003), Sequential Weight Adjustments for Location and Cooperation Propensity for the 1995 National Survey of Family Growth. *Journal of Official Statistics* 19, pp. 31–43.

Jenkins, S.P., Cappellari, L., Lynn, P., Jäckle, A., and Sala, E. (2006), Patterns of Consent: Evidence from a General Household Survey. *Journal of the Royal Statistical Society, Series A*, 169, pp. 701–722.

Lepkowski, J.M., and Couper, M.P. (2002), Nonresponse in the Second Wave of Longitudinal Household Surveys. In: Groves, R.M. Kalton, G. Rao, J.N.K. Schwarz, N., and Skinner, C. (Eds.) *Survey Nonresponse*. Wiley, New York, pp. 259–272.

Little, R.J., and Vartivarian, S. (2005), Does Weighting for Nonresponse Increase the Variance of Survey Means? *Survey Methodology* 31, pp. 161–168.

Lynn, P., and Clarke, P. (2002), Separating Refusal Bias and Non-contact Bias: Evidence from UK National Surveys. *Journal of the Royal Statistical Society, Series D* 51, pp. 319–333.

Maddala, G.S. (1991), A Perspective on the Use of Limited-Dependent and Qualitative Variables Models in Accounting Research. *Accounting Review* 66, pp. 788–807.

Nicoletti, C., Peracchi, F. (2005), Survey Response and Survey Characteristics: Microlevel Evidence from the European Community Household Panel. *Journal of the Royal Statistical Society, Series A* 168, pp. 763–781.

Stoop, I.A.L. (2005), *The Hunt for the Last Respondent.* PhD thesis. Faculty of Social Sciences, Utrecht University, Utrecht.

Van de Ven, W., and Van Praag, B. (1981), The Demand for Deductibles in Private Health Insurance: A Probit Model with Sample Selection. *Journal of Econometrics* 17, pp. 229–252.

Van der Klaauw, B., and Koning, R.H. (2000), Testing the Normality Assumption in Sample Selection Models with an Application to Travel Demand. *Journal of Business and Economic Statistics* 21, pp. 31–42.

Vella, F. (1998), Estimating Models with Sample Selection Bias: A Survey. *Journal of Human Resources* 33, pp. 127–169.

12.7 EXERCISES

Exercise 12.1 Which statement about interviewers is not correct?

(a) The interviewer affects the response process in the survey.

(b) The interviewer reduces the risk of a social desirability bias.

(c) Standardized interviewing may reduce the risk of measurement errors caused by the interviewer.

(d) The effect of interviewers on the response process and the survey answers is clustered within interviewers.

Exercise 12.2 Which statement about response analysis models is not correct?

(a) The multinomial model accounts for the sequential nature of the response process.

(b) The nested logit model can be used to distinguish among the different reasons for nonresponse and to account for the sequential nature of the response process.

(c) The bivariate probit model with sample selection can be used to distinguish among the different reasons for nonresponse and to account for the sequential nature of the response process.

(d) The bivariate probit model with sample selection allows for a dependency between the different stages in the response process.

Exercise 12.3 When is it not necessary to distinguish among the different reasons for nonresponse in the adjustment for a nonresponse bias?

(a) When all nonrespondents have the same answers to the survey questions.

(b) When persons refuse participation because they are not interested in the survey topic.

(c) When persons cannot be contacted because they are at work and the survey contains work-related questions.

(d) When persons are too ill to participate in a survey about healthcare.

Exercise 12.4 Which statement about the sequential weight adjustment approach is not correct?

(a) The weights that are obtained with the sequential weight adjustment approach can be used in regular nonresponse adjustment methods.

(b) The sequential weight adjustment approach accounts for the sequential nature of the response process.

(c) The sequential weight adjustment approach allows for distinctions to be made among the causes of nonresponse.

(d) The sequential weight adjustment approach allows for a correlation between the different types of nonresponse.

Exercise 12.5 Which statement about the sample selection model to adjust for nonresponse bias is not correct?

(a) The sample selection model for the simplified response process consists of a selection equation for nonresponse and an outcome equation to estimate the survey variable.

(b) The selection equation in the sample selection model for survey nonresponse determines whether the survey variable is observed.

(c) The bivariate probit model with sample selection is a modification of the sample selection model for discrete outcomes.

(d) The sample selection model can only be estimated with a maximum likelihood estimation.

Exercise 12.6 Which statement about the inverse Mills ratio is not correct?

(a) The inverse Mills ratio is a monotone decreasing function of the probability that a sample element responds to the survey.

(b) The rationale behind the inverse Mills ratio is that observations with a higher response probability have a higher probability of introducing a bias.

(c) The inverse Mills ratio can be interpreted as an omitted variable in the outcome equation that should be included to adjust for a nonresponse bias.

(d) When the coefficient for the inverse Mills ratio in the outcome equation is not significant, there is no nonresponse bias.

CHAPTER THIRTEEN

Adaptive Survey Designs

13.1 Introduction

In Chapter 3, various survey design features were described that help *reduce nonresponse* and increase response rates. In Chapter 4, the survey mode was discussed in detail. In both chapters it was assumed that all sample units receive the same treatment and that the same design features apply to all selected persons and households. In this chapter we see how survey designs can be tailored to optimize response rates and to reduce nonresponse selectivity. Although a general terminology is lacking in the literature such designs are sometimes referred to as *adaptive survey designs*.

Adaptive survey designs assume that different persons or households can receive different treatments. These treatments may be defined before the survey starts but may also depend on data that is observed during data collection. In other words, allocation of treatments is based on data that are linked to the survey sample and also based on *paradata*. Paradata are data about the survey data collection process, such as observations of interviewers about the neighborhood, the dwelling and respondents, or the performance of interviewers themselves. In this chapter paradata are used in the widest sense as data observed during data collection about the response behaviors of sampled persons and households. A general introduction to adaptive survey designs is given by Wagner (2008).

Adaptive survey designs find their origin in the literature on medical statistics where treatments are varied beforehand over patient groups but also depend on the responses of patients, namely measurements during data collection (e.g., see Heyd

Handbook of Nonresponse in Household Surveys, By Jelke Bethlehem, Fannie Cobben, and Barry Schouten

and Carlin, 1999; Murphy, 2003). Generally, in the treatments a number of stages are identified at which the patients status is evaluated for any need to change the treatment regime. Such designs are referred to as *dynamic* designs. In this chapter, *static* and *dynamic* survey designs are distinguished. Static survey designs are set before data collection starts and do not depend on paradata obtained during data collection. Dynamic survey designs choose between different design features based on observations made by interviewers or data collection staff. The observations have to be made for all sampled persons. To illustrate the difference, offering different levels of pre-paid incentives would constitute a static design, whereas offering different maximum levels of incentives, given an interviewer assessment of a person's propensity to respond, is a dynamic design.

In medical statistics a surrogate or marker of the target quantity of interest, such as blood cell count or blood oxygen content, is usually employed to determine the effectiveness of treatments. Such surrogates are needed because the target quantity, such as the cure for a disease, cannot be measured directly or in timely manner. There is again a close parallel to survey designs where the quality or representativeness of response is of interest. Representative response (see Chapter 7) cannot be measured exactly but can be approximated. Hence adaptive survey designs need indicators of survey quality.

There is an extensive literature on assumptions about treatments and treatment responses that enable statistical inferences to be made from the treatment data. One important assumption is about unobserved covariates in the allocation of treatments; the observed data must be complete and no unobserved data can interfere with the decision to assign treatments. If such covariates did exist, then conclusions drawn would not be valid. Treatment must not be confounded with the final outcome.

In interviewer-administered surveys as well, the interviewer is taught to tailor the approach to the respondent and to maintain interaction in order to continue to do so. Interviewers often base their timing of subsequent visits on the result of the first visit. Based on the appearance of the dwelling, the apartment building, and the mail box, or based on contact with neighbors, they may decide to adjust their contact strategy.

Although the strategy of the interviewer may be effective and efficient, from the point of view of the survey organization such changes employ unobserved data and assumptions about the contacts. Timing protocols are violated when data underlying the decisions of interviewers are missing or incomplete. For this reason most survey organizations ask interviewers to keep records of the sampled units that are assigned to them. These records should enable data collection staff to recreate the history of the sample. Adaptive survey designs therefore are restricted to design features that are fully under control and that employ only observed data.

A special case of an adaptive survey design is the *responsive survey design*. Responsive survey designs were introduced by Groves and Heeringa (2006). Like general adaptive survey designs, responsive survey designs may apply differential design features to sample units. However, the main restriction is that responsive survey designs must identify promising and effective treatments or design features during data collection. To do so, the data collection is divided into at least two *design phases*. The first phase is the *initial design phase*. This phase deploys the outcomes of *randomized contrasts* between sample units to distinguish effective from ineffective treatments and to identify costs associated with the treatments.

Randomized contrasts are differences from response rates between subpopulations for randomly assigned design features. Design features of subsequent phases are defined based on these findings (e.g., see Mohl and Laflamme, 2007). The allocation of design features must be done in such a way that each phase reaches its *phase capacity*, which is the optimal trade-off between quality and costs.

Responsive designs are motivated by survey settings where little is known about the sample beforehand and/or little information about the effectiveness of treatments is available from historic data. In these settings an initial phase is needed in order to first gain experience. If the second and higher design phases of responsive designs are considered, however, then the starting point is similar to survey settings where substantial prior information about sample units is available or where a survey is repeated many times. The only distinction is that in the initial design phase part of the sample has already responded.

Although survey literature has devoted considerable attention to trade-offs in survey designs to lessen the various surveys errors (e.g., Lyberg et al., 1997; Dillman, 2007), in survey practice, surprisingly few cases have investigated and implemented differential design features. Some literature is devoted to the timing of call attempts, visits, and call-backs, such as Greenberg and Stokes (1990), Kalsbeek et al. (1994), and Kulka and Weeks (1988). Optimal timing protocols focus on two causes of nonresponse, unknown noneligibility and noncontact. Such protocols may be regarded as dynamic adaptive survey designs as they change timing of call attempts or visits based on the contact history. However, in most cases the protocol does not depend on person or household characteristics that are known prior to the start of data collection and that are available from linked data. Research into adaptive survey designs is still in its infancy.

In this chapter an introduction is given to adaptive survey designs. It is assumed that historic data are available and that an effective protocol was identified before data collection is begun. Section 13.2 describes theory and concepts behind such designs. Section 13.3 presents an example. Section 13.4 summarizes the chapter.

13.2 Theory

13.2.1 WHAT ARE ADAPTIVE SURVEY DESIGNS?

In this section a mathematical framework is set out for adaptive survey designs. In subsequent sections, components of this framework are highlighted and elaborated.

Let the population consist of units $k = 1, 2, \ldots, N$. The population of interest may consist of all units in a population but also of only respondents to a previous wave in a panel study. To each unit will be assigned a strategy s from the set of candidate strategies $S = \{\phi, s_1, s_2, \ldots, s_M\}$. In the *survey strategy set* S an explicit empty strategy ϕ is included. The empty strategy means that no action is undertaken, that the selected unit is not sampled. This is the most general framework. In practice, one will often separate the sampling design from the strategy allocation and view the sample as given and fixed. However, one may include an explicit decision to sample a unit in the overall allocation of resources.

In general, a strategy is a specified set of design features. Some of those features may be sequential like the type of contact mode and the type of survey mode, but the features may also describe different aspect of a survey design. Examples of strategies are

$$
\begin{aligned}
s_1 &= (\text{advance letter 1, web questionnaire, one reminder}),\\
s_2 &= (\text{advance letter 1, web questionnaire, no reminder}),\\
s_3 &= (\text{advance letter 2, CATI administered, maximum of six call attempts}),\\
s_4 &= (\text{advance letter 2, CATI administered, maximum of 15 call attempts}).
\end{aligned}
$$

Section 13.2.2 discusses strategies and design features more fully.

In responsive survey designs the initial design phase is when the candidate strategies are selected for later design phases in data collection (Groves and Heeringa, 2006). It is assumed in this chapter that the set of strategies S is known and fixed when strategy allocation is started.

With each population unit is associated a vector of covariates $X_k = (X_{k1}, X_{k2}, \ldots, X_{kp})'$. X_k contains characteristics that are known before data collection starts and before strategies are allocated. The covariates must therefore be available in public registers or administrative data that can be connected to the sampling frame or are already in the sampling frame. Next to these general characteristics a second vector of covariates $\tilde{X}_k = (\tilde{X}_{k1}, \tilde{X}_{k2}, \ldots, \tilde{X}_{kq})'$ may be present that reflects characteristics observed during data collection for all sampled population units. These characteristics are termed *paradata* or process data because they are accumulated during the process survey by interviewers and data collection staff. However, other than the more traditional view on paradata as information about the process, in the adaptive survey design context $\tilde{X}_k$ contains observations about the sampled persons or households. Examples of X_k are gender, age, type of household, or educational level. Examples of $\tilde{X}_k$ are the interviewer's assessment of the sampled person's propensity to respond or to be contacted, the state of the dwelling or the neighborhood, and the presence of a gated entrance. $\tilde{X}_k$ is deliberately restricted to observations about the sample that allow for a differentiation of survey design features. It does not contain the values of the design features like the interviewer that was assigned to the address.

The important distinction between X_k and $\tilde{X}_k$ is the level of availability. $\tilde{X}_k$ is known only for those units that are sampled and cannot be used in distinguishing subpopulations a priori. Let $q(x)$ represent the distribution of X_k in the population and $\tilde{q}(\tilde{x}, x)$ the joint distribution of X_k and $\tilde{X}_k$ in the sample. Furthermore $\tilde{q}(\tilde{x})$ denotes the sample marginal distribution of $\tilde{X}_k$, and $\tilde{q}(\tilde{x}|x)$ the conditional sample distribution. It is assumed that $q(x)$ is known in advance and that $\tilde{q}(\tilde{x}, x)$ is assessed for the complete sample during data collection. In settings where no or little data can be linked, strategy allocation must be based fully on observations made during data collection. In responsive designs paradata about the effectiveness and costs of strategies are motivated by survey settings where little or nothing is known in advance.

Adaptive survey designs that allocate strategies beforehand based on population characteristics are called *static*, while adaptive survey designs that allocate strategies that depend (also) on paradata are termed *dynamic*.

EXAMPLE 13.1 Static and Dynamic Adaptive Survey Designs

A CATI survey is conducted, with the total length of data collection taking four weeks. After two weeks of data collection, for each sample unit the response status is evaluated. Sample units can either be respondents or noncontacts or refusals. For the remaining two weeks three survey modes are considered: paper, Web, and CATI.

The three candidate strategies in the design are $s_1 = (\text{CATI}, \text{CATI})$, $s_2 = (\text{CATI, web})$, and $s_3 = (\text{CATI, paper})$.

Assume that beforehand gender and age are known, $X = (\text{Gender}, \text{Age})'$, where variable *Age* has three categories young, middle age, and elderly. Let the type of response be denoted by paradata vector $\tilde{X} = (\text{response after 2 weeks})$.

If the allocation of strategies is based purely on X, then the design is static. If additionally strategies s_1, s_2, s_3 are optional for all types of nonresponse, then the design is dynamic.

A design where s_1 is always assigned to refusals and s_2 to noncontacts is the outcome of a dynamic survey design allocation.

A design where s_1 is assigned to middle and older age persons, s_2 to young males, and s_3 to young females, is the outcome of a static survey design allocation.

A design where s_1 is assigned to persons who refused, s_2 to young persons who were not contacted, and s_3 to older persons who were not contacted is again the outcome of a dynamic survey design allocation.

Let $\rho(x, s)$ be the response propensity of a unit carrying characteristic $X_k = x$ and assigned strategy s. It is assumed that $\rho(x, s)$ is available from historic data, meaning from previous versions of the same survey, from surveys with similar topics and designs, or from initial design phases. Obviously the anticipated response propensity must be a close estimate of the true propensity. Section 13.2.5 returns to this essential component of adaptive survey designs.

The expected costs of the assignment of strategy s to a unit with $X_k = x$ is denoted as $c(x, s)$. It is an individual cost component. The literature tells us that survey costs consist of many components of which some are overhead and others are individual (e.g., Groves, 1989). Section 13.2.3 discusses cost functions.

In responsive survey designs individual response propensities and cost functions are to some extent identified and assessed during data collection. Groves and Heeringa (2006) refer to *randomized contrasts* in order to assess such functions. Randomized contrasts are differences in response behavior among subpopulations for design features that are randomly assigned.

Let $p(s|x)$ be the *allocation probability* of a population unit with characteristics x for strategy s, and let $p(s|x,\tilde{x})$ be the allocation probability to that strategy given that also paradata $\tilde{x}$ is observed. The following must hold:

$$\begin{aligned} &0 \le p(s|x) \le 1, \quad 0 \le p(s|x,\tilde{x}) \le 1, \\ &\sum_j p(s|x) = 1, \quad \sum_j p(s|x,\tilde{x}) = 1, \end{aligned} \tag{13.1}$$

$$\sum\nolimits_{\tilde{x}} q(\tilde{x}|x)p(s|x,\tilde{x}) = p(s|x), \tag{13.2}$$

So all units are assigned a strategy, and dynamic allocation probabilities must match static allocation probabilities. In general, allocation probabilities may have values between 0 and 1. In other words, subpopulations with the same scores on x and $\tilde{x}$ may be (randomly) assigned to different strategies. For instance, only part of the nonrespondents may be re-approached in a follow-up. Allowing for allocation probabilities between 0 and 1 increases the flexibility in meeting quality levels or cost constraints. In the following discussion p denotes the matrix of allocation probabilities, meaning $p = \{p(s_j|x,\tilde{x})\}_{1\le j\le M,x,\tilde{x}}$, and contains the decision variables or the parameters in the optimization.

The response propensities ρ_X can be derived from the strategy response propensities, and the allocation probabilities by $\rho_X(x) = \sum_{s\in S} p(s|x)\rho(s,x)$.

The strategies, covariates, response propensities, cost functions, and allocation probabilities form the ingredients to adaptive survey designs. With these building blocks the adaptive survey design optimization problem can be formulated. Two ingredients are still missing, however, a quality function and an overall cost function. Let $Q(p)$ be some indicator of quality and $C(p)$ be an evaluation of total costs. The dependence on the allocation probabilities in both functions is stressed as the probabilities are the decision variables in the optimization.

The optimization can now be formulated as

$$\max_p Q(p) \quad \text{subject to} \quad C(p) \le C_M \tag{13.3}$$

or as

$$\min_p C(p) \quad \text{subject to} \quad Q(p) \ge Q_m, \tag{13.4}$$

where C_M represents the budget for a survey and Q_m minimum quality constraints. Problems (13.3) and (13.4) are called dual optimization problems, although the solutions to both problems may be different depending on the quality and cost constraints. Sections 13.2.3 describes a number of quality functions.

EXAMPLE 13.2 Allocation Probabilities

In Example one static and two dynamic outcomes of adaptive survey designs were presented. The corresponding allocation probabilities are

s_1 is always assigned to refusals and s_2 to noncontacts:

$$p(s_1|x, \text{refusal}) = 1, \quad p(s_2|x, \text{noncontact}) = 1.$$

s_1 is assigned to middle and older age persons, s_2 to young males and s_3 to young females:

$$p(s_1|\text{age} \geq 2, \tilde{x}) = 1, \quad p(s_2|\text{age} = 1, \text{gender} = \text{M}, \tilde{x}) = 1,$$
$$p(s_3|\text{age} = 1, \text{gender} = \text{F}, \tilde{x}) = 1.$$

s_1 is assigned to persons who refused, s_2 to young persons who were not contacted, and s_3 to older persons who were not contacted:

$$p(s_1|x, \text{refusal}) = 1, \quad p(s_2|\text{age} = 1, \text{noncontact}) = 1,$$
$$p(s_3|\text{age} \geq 2, \text{noncontact}) = 1$$

13.2.2 SURVEY STRATEGIES AND SURVEY DESIGN FEATURES

A survey strategy is essentially a list of survey design features. In the literature many such design features are proposed and evaluated for effectiveness (e.g., Groves and Couper, 1998; Groves et al., 2002). Here, only a brief summary of commonly used features is presented:

- Length of the period of data collection
- Survey mode or sequence of survey modes in mixed mode surveys
- Contact mode
- Invitation or advance letter
- Type of letter
- Type of reminder
- Number of reminders
- Contact protocol
- Number of attempts per address
- Timing of attempts per address
- Incentives
- Type of incentive
- Level of incentive
- Questionnaire type
- Full questionnaire
- Basic (short) questions
- Proxy or self-report
- Assignment and re-assignment to interviewers

The length of the data collection can be varied for different persons or households, such as in cases where groups are hard to reach or follow-ups need to be made for specific groups. The main time constraint may be the publication date for the survey results. It makes a considerable difference whether questions are asked with respect to a fixed date or with respect to a fixed time lag.

The usual survey modes are paper, Web, telephone, and face-to-face interview. These survey modes do lead to different survey errors pertaining to coverage, nonresponse, and measurement, see Chapter 4. Also different costs are associated with the different modes. Interviewer-administered modes are generally more expensive than self-administered modes, depending on the number of contact or call attempts and traveling distances. Hence it may be cost-effective to mix the modes according to the expectations held about the different groups, thus moving to a mixed mode survey design. For example, elderly people would not be assigned a Web survey. Also nonrespondents could be considered for follow-up by another mode after a specified time period (see De Leeuw, 2008).

The contact mode is the mode used to inform a sample unit about the survey and to invite participation. In some surveys the contact and survey mode may coincide as the sample unit is invited to participate right away. The set of contact modes is the same as the set of survey modes. However, in surveys where samples of addresses are drawn, the most common contact mode is by paper. These surveys use advance or invitation letters. In surveys based on random digit dialing, the telephone naturally is the form of contact. In many surveys there is little opportunity to vary contact modes. This, however, changes for subsequent waves in multiple-wave surveys or panels. For instance, the Dutch Labor Force Survey asks respondents to the first wave for their telephone numbers for the other waves. Alternatively, one could ask for email addresses or even poll respondents for their favorite survey mode.

When advance letters or advance emails are sent, the content could be personalized. Considerations could be to use a different language or to tailor wording and style to the sampled individual. Also reminders could be sent. Both the number and form of reminders may be varied over subpopulations. For instance, groups with low response rates may receive more reminders or a higher frequency of reminders (see Dillman, 2007; De Leeuw et al., 2007).

One design feature that is important for the interviewer-administered survey mode is the contact protocol. Usually to some extent protocol is defined by survey organizations. They may require that a minimum number of contact attempts be made, that visits and calls be spread over the week or on weekend days, and that these be at times during the day. Generally, telephone surveys allow for more detailed contact protocols than face-to-face surveys that involve traveling distances. It is common practice for telephone surveys to be operated by a management system that makes the calls and appointments for calls. Contact protocols can take different forms for different groups in the population, in terms of different numbers of calls or visits or tailoring of timing based on linked data and paradata (e.g., see Wagner, 2008).

Some surveys employ incentives to increase response rates, especially for persons with no intrinsic interest in the survey. Apart from the type of incentive (voucher, money, coupons), also the form (pre-paid, promised, or conditional) may be varied. In rare cases higher incentives have been offered to persons or households that have high refusal rates (e.g., see Barón et al., 2009).

Besides the advance letter, the questionnaire could be differentiated in content and length. Different versions of the questionnaire could be made available in split questionnaire designs or shorter, with shorter versions of the questionnaire to be used for follow-up. Examples were discussed in Chapter 10 for the basic question approach (Kersten and Bethlehem 1984, Cobben 2009) and the PEDAKSI approach (Lynn 2003) used in re-approaching nonrespondents. Groups that have high refusal rates may be provided with short questionnaires.

A design feature that is often used to increase response rates, especially for household surveys, is proxy reporting. Proxy reporting as opposed to self-reporting may be accepted to reduce the number of calls and visits to a household and thus to reduce traveling costs. Proxy reporting may, however, lead to an increase in measurement error. It may therefore be wise to keep the option of proxy reporting to certain subpopulations only (e.g., see Moore, 1988).

Finally, for interviewer-assisted contact and survey modes the interviewers may be assigned to different sample units based on their historic performance or their match to the person or household. Examples are specialized market research companies that have multilingual interviewers or interviewers who are trained for city districts that are generally considered as difficult. Apart from the prior assignment of interviewers, certain nonresponse cases could be re-assigned to highly experienced interviewers; an example is provided in Cobben (2009).

Theoretically all or a number of the design features in this section and other design features not mentioned here may be varied. However, the more design features are treated as optional, the more decision variables need to be constructed in the optimization and the more dependence must be attached to historic data or initial design phases. It is paramount that the number of features selected conform to the information contained in historic data.

EXAMPLE 13.3 Design Features

In Example only one design feature is considered optional: the survey mode in the second half of the data collection period. The three options are Web, paper, and telephone (CATI).

Hence it is assumed that all other features have standard default implementations; for instance, the standard telephone contact timing protocols are used in the first half of the data collection and the second half of the data collection when telephone is selected as the survey mode. The same holds true for refusal conversion techniques, advance letters, and so forth.

13.2.3 QUALITY OBJECTIVE FUNCTIONS

It is beyond the scope of this book to elaborate on the quality of surveys. Because we are interested in this book in survey nonresponse, we will instead consider in this section the quality of response to a survey. Still it is important to keep in mind that survey errors other than nonresponse can sometimes override the results and that design features that have a high risk of such errors should be avoided. Examples are survey modes with low coverage of some subpopulations or a high risk of response bias, increased item nonresponse in proxy reporting, and increased response bias in follow-ups.

When quality is optimized according to equation (13.3), then quality functions map the survey sample with linked data, paradata, and answers to survey items to a scalar value. This means that the outcome of the quality function is a single number that can be optimized. When costs are minimized subject to constraints on quality as in equation (13.4), then quality may be multidimensional (but the cost functions are one-dimensional). In the following paragraphs, quality functions that allow for an optimization of quality are discussed.

In general, two types of quality functions can be distinguished; quality functions that deploy covariates from linked data and paradata only, and quality functions that also deploy the answers to the survey target variables. We refer to these as *covariate-based* and *item-based* functions, respectively. The main distinction between covariate-based and item-based quality functions is that item-based quality can be assessed only for respondents. Obviously the answers of nonrespondents are missing. Hence the quality evaluation is limited to relations between target variables and covariates as observed in the response. As a consequence there is an imminent risk attached to item-based quality functions that originates directly from the phenomenon it attempts to measure.

Relations between target variables and covariates may be different for nonrespondents, so item-based quality can create a false image. Furthermore, in surveys with many survey target variables, different target variables may lead to different decisions and optimal survey designs. However, contrary to covariate-based quality functions, item-based quality functions tailor survey designs to a specific topic. For two different surveys with similar response patterns, covariate-based quality functions will lead to the same optimal survey designs.

The most commonly used quality function is the response rate. Since the 0–1 response indicator may be viewed as the simplest form of paradata, it is termed a covariate-based quality function. There are various definitions of response rates, and one would have to pick a suitable one for the optimization. The response rate is represented as the mean response propensity

$$Q(p) = \hat{\bar{\rho}}_X \qquad \text{(Response rate)}. \tag{13.5}$$

Other candidate covariate-based quality functions are given in Särndal and Lundström (2008), Schouten et al. (2009), and in Chapter 7:

$$Q(p) = \hat{B}_m(\hat{\rho}_X) = \frac{\hat{S}(\hat{\rho}_X)}{\hat{\bar{\rho}}_X} \qquad \text{(Maximal bias)}, \tag{13.6}$$

$$Q(p) = q^2 = \left[\sum_{k=1}^{N} a_k R_k d_k\right]^{-1} \left[\sum_{k=1}^{N} a_k R_k d_k (\hat{\phi}_k - \hat{\bar{\phi}})^2\right]$$

(Variance of calibration weights). (13.7)

In equations (13.5) and (13.6), $\hat{S}(\hat{\rho}_X)$ is the standard deviation of response propensities estimated with covariates X. In equation (13.7), d_k is the inclusion weight, a_k is the 0–1 sample indicator, R_k is the 0–1 response indicator, and $\hat{\phi}_k = \hat{\rho}_{X_k}^{-1}$ is the inverse response propensity, also referred to as *response influence*.

The maximal bias attempts to estimate the standardized maximal absolute bias of response means. It assumes maximal correlation between response and survey target variable. The variance of calibration weights was originally designed to serve as a tool in the selection of weighting variables. However, when fixing X, it can also serve as a quality function. A smaller q^2 is preferred as it implies that one has to rely less on the nonresponse adjustment.

Quality functions (13.5), (13.6), and (13.7) do not account for the impact of nonresponse on precision. An alternative that does may be the extension of (13.6) to the maximal mean square error, namely the sum of square of the maximal bias plus the standard error:

$$Q(p) = MSE_m(\hat{\rho}_X) = \left(\frac{\hat{S}(\hat{\rho}_X)}{\hat{\bar{\rho}}_X}\right)^2 + \lambda \frac{1}{n\hat{\bar{\rho}}_X} \quad \text{(Maximal mean square error).}$$

(13.8)

The quality function MSE_m represents the standardized maximal mean square error in a simple random sample. The standard error in such a sampling design is $S^2(y)/(n\bar{\rho}_X)$, if we ignore the finite sample correction. Standardization by $S^2(y)$ leads to the second term on the right-hand side of (13.8) except for λ. Parameter λ is included so that the weight can be adjusted to the desired precision value in opposition to maximal bias. The larger the λ, the more importance is given to small standard errors. Since the standardized variance is proportional to the inverse of the response rate, a large λ corresponds to maximizing the response rate.

Examples of item-based quality functions are presented by Groves and Heeringa (2006), Wagner (2008), and Rao et al. (2008). For a specific target variable Y, Groves and Heeringa (2006) suggest the estimated nonresponse bias of the response mean

$$Q(p) = \hat{B}(\bar{y}_R) = \frac{\mathrm{cov}(Y, \hat{\rho}_x)}{\hat{\bar{\rho}}} \quad \text{(Estimated nonresponse bias),} \qquad (13.9)$$

with $\mathrm{cov}(Y, \hat{\rho}_X)$ the response covariance between the target variable and the estimated response propensities given covariates X. Wagner (2008) proposes the fraction of missing information as a quality criterion

$$Q(p) = (1 + \frac{1}{D}) \frac{DB_D}{DW_D + (D+1)B_D} \quad \text{(Fraction of missing information),}$$

(13.10)

where B_D and W_D are, respectively, the between-imputation variance and the within-imputation variance of the estimated population parameter in D multiply imputed data sets. The obvious population parameter is the population mean of a survey target variable, which may be estimated by the sample mean of the imputed target variable. Multiple imputation means that the nonresponse for the target variable is fully imputed in D independent runs. For relatively modest D, equation (13.10) is equal to the relative size of the between-imputation variance. A small fraction of missing information is preferable as it implies that the imputed data sets lead to stable estimates for the population mean.

Quality functions (13.5) through (13.10) indicate a dependence on the allocation probabilities p, although this is not directly apparent from their definitions. The dependence is illustrated for the response rate. The response rate can be written in terms of a weighted response rate:

$$Q(p) = \bar{\rho}_X = \sum_{x,j} q(x)p(s|x)\rho(x,s). \tag{13.11}$$

Hence (13.11) indicates that the response rate is a linear function of the allocation probabilities, the decision variables in the optimization. Similar derivations can be made for the other quality functions, but in general, the quality functions lead to nonlinear objective functions in the optimization.

EXAMPLE 13.4 Quality Functions

Suppose that in the setting of Example 13.1 the following population distributions and response propensities apply with X representing *Age* × *Gender* and $\tilde{X}$ the response after two weeks:

$q(\text{refusal}) = 0.25,\ q(\text{noncontact}) = 0.40,\ q(\text{response}) = 0.35$

$q(\text{Age} \geq 2) = 0.60,\ q(\text{Age} = 1, \text{Gender} = \text{M}) = 0.20,$
$q(\text{Age} = 1, \text{Gender} = \text{F}) = 0.20$

$q(\text{Age} = 1, \text{noncontact}) = 0.30,\ q(\text{Age} \geq 2, \text{noncontact}) = 0.10$

$\rho(s_1, \text{refusal}) = 0.25,\ \rho(s_2, \text{noncontact}) = 0.30,\ \rho(s_1, \text{Age} \geq 2) = 0.70$

$\rho(s_2, \text{Age} = 1, \text{Gender} = \text{M}) = 0.50, \rho(s_3, \text{Age} = 1, \text{Gender} = \text{F}) = 0.55$

$\rho(s_2, \text{Age} = 1, \text{noncontact}) = 0.40,\ \rho(s_3, \text{Age} \geq 2, \text{noncontact}) = 0.50$

In sum, 25% of the sample units has refused after two weeks, and 60% is contacted. The 40% noncontacted consist of 30% younger persons and 10% elderly persons, so 75% of the noncontacted is young. Furthermore 25% of the refusals is converted to respondents when telephone is used in the second half of the data collection, while 30% of the noncontacted fills in a Web questionnaire.

The response rates can be derived for each of the three designs in Example 13.1.

s_1 is always assigned to refusals and s_2 to noncontacts:

$$Q(p) = \bar{\rho}_X = 0.35 + 0.25 \times 0.25 + 0.40 \times 0.30 = 0.53.$$

s_1 is assigned to middle and older age persons, s_2 to young males, and s_3 to young females:

$$Q(p) = \bar{\rho}_X = 0.60 \times 0.70 + 0.20 \times 0.50 + 0.20 \times 0.55 = 0.63.$$

s_1 is assigned to persons who refused, s_2 to young persons who were not contacted, and s_3 to older persons who were not contacted:

$$Q(p) = \bar{\rho}_X = 0.35 + 0.25 \times 0.25 + 0.30 \times 0.40 + 0.10 \times 0.50 = 0.58.$$

Hence the second survey design ensures the highest response value.

13.2.4 COST FUNCTIONS

Cost functions are the counterpart of the quality functions. As for the quality functions in the previous section, they should be one-dimensional when they are minimized subject to constraints on quality. However, again the optimization of quality is the starting point. As a consequence the cost function is allowed to be multidimensional.

There are several components to cost functions. It is important to restrict specification of costs relative to the design features that are varied in the adaptive survey design. For example, when it is the incentive that is differentiated with respect to different subpopulations, the costs do not need to be detailed for interviewer traveling times. When it is the contact timing protocol that is the design feature to be tailored, traveling times and costs play a dominant role.

If all features of Section 13.2.2 are optional, then the cost functions take more complex form with many overhead cost components and other variable cost components:

- Overhead costs of data collection staff.
- Overhead costs associated with sampling and processing of sample.
- Variable costs associated with training and instruction of interviewers.
- Variable costs associated with development of questionnaires and advance letters.
- Variable costs from mail and print of questionnaires and reminders.
- Variable costs associated with processing of paper questionnaires.
- Variable costs following from interviewer hour rates and travel expenses.
- Variable costs from incentives.

- Variable costs associated with telephone number linkage, telephone usage and computer servers.

The list of costs is not exhaustive.

Before an optimization can be begun, a distinction must be made between permanent cost components and start-up cost components. For example, once survey questionnaires are developed for all survey modes, the cost for questionnaire design does not have to be accounted for as long as the questionnaires remain unchanged. Start-up cost components can be ignored or spread over time when it is anticipated that the survey will be conducted frequently in the future without major redesigns.

In the optimization two cost components should be identified: fixed and variable costs. Variable costs depend on the allocation of population units to strategies, whereas fixed costs apply to all remaining costs. It should be pointed out that the fixed component is different for adaptive survey designs that focus on different design features. For example, when pre-paid incentives are of varied value during face-to-face surveys, the variable component would consist of the incentives and the variable interviewer costs in conducting the interviews (provided that contact do not depend on the incentives). All other costs, such as travel expenses of interviewers, are fixed. Hence the cost function $C(p)$ is the sum of two components:

$$C(p) = C_F + C_V(p), \tag{13.12}$$

of which only the second, the variable component, depends on the allocation probabilities. Clearly, in the optimization the fixed costs are subtracted from the overall budget to derive a budget for the variable costs.

In general, with a survey strategy $s$costs $c(x, s)$ are associated for population units from group x. The individual cost function may be a function of response propensities, or even more specifically of contact and participation propensities. For instance, the interviewer costs in different contact timing protocols depend on the contact rates of the selected subpopulations. The cost function $c(x, s)$ is a relative cost function as it describes only the contribution of the strategy to the variable cost component $C_V(p)$:

$$C_V(p) = \sum_{x,j} q(x)p(s|x)c(x, s). \tag{13.13}$$

Three additional remarks are in place.

First, the derivation of fixed and variable cost components is complicated when a survey organization runs many surveys in parallel. Usually interviewers work on different surveys at the same time, thus allowing for efficiency in traveling costs by clustering sample units. On the one hand, the interaction between surveys makes it hard to separate costs per survey, especially when strategies are tailored. Simple heuristics must be used to disentangle costs. On the other hand, when multiple surveys are conducted some of the variable costs components may be labeled as fixed. For example, when only a relatively small number of population units are assigned to the face-to-face survey mode, then traveling costs may be assumed to remain unchanged as the addresses are clustered with addresses from other surveys.

The second remark concerns the multidimensional aspect of costs. Apart from the overall budget, it may be requested that interviewers have a full workload throughout time or that none of the interviewers work overtime more than a fixed amount of time. Such constraints can be incorporated in the cost function as well. As a consequence the cost function becomes a vector and the constraint a vector of constraints. For the optimization this only affects the range of feasible solutions.

The third remark concerns the validity of the cost functions. Since cost functions are hard to construct in practice, it may turn out that an optimization was too idealistic. This will result in incomplete or aborted strategies during data collection, like a smaller number of interviewer visits. It is important therefore to monitor data collection closely and to build in indicators for strategies, preferably for every decision variable in the survey design. This form of paradata is needed to check the validity of the optimization and the adaptation of the design if necessary.

EXAMPLE 13.5 Cost Functions

In Example 13.1 the design of the first half of the data collection was fixed, so the corresponding costs may be viewed as fixed as well. Now suppose that the following conditions hold:

- The interviewer's hourly rate is 25 euros.
- An average of 2.1 call attempts is needed to contact persons who refused in the first two weeks, and an average of 1.5 call attempts is used for persons in the two highest age groups for the second half of data collection.
- An interview lasts 20 minutes and noncontact and refusals take up 2 minutes of the interviewers time on average.
- Postage for an advance letter plus the paper questionnaire costs 2 euros, the same cost as an advance letter with a reference to the website where the Web questionnaire can be found.
- The processing of paper questionnaires is automated and costs on average 5 minutes per questionnaire for manual checks and corrections

The costs per strategy are as follows:

$c(s_1, \text{refusal}) = 25 \times (2.1 \times 2 \times 0.75 + 2.1 \times 21 \times 0.25)/60 = 5.9,$
$c(s_2, \text{noncontact}) = 2$

$c(s_1, \text{Age} \geq 2) = 25 \times (1.5 \times 2 \times 0.30 + 1.5 \times 21 \times 0.70)/60 = 9.6$

$c(s_2, \text{Age} = 1, \text{Gender} = \text{M}) = 2,\ c(s_3, \text{Age} = 1, \text{Gender} = \text{F}) = 2+ (0.55 \times 5 \times 25)/60 = 3.1,$

$c(s_2, \text{Age} = 1, \text{noncontact}) = 2,\ c(s_3, \text{Age} \geq 2, \text{noncontact}) = 2+ (0.50 \times 5 \times 25)/60 = 3.0$

13.2.5 ESTIMATING RESPONSE PROBABILITIES

Next to cost functions, the other important ingredient of adaptive survey designs is the set of response propensities for the various strategies. These propensities may be known from past surveys, and if not the same survey, then a similar survey. Alternatively, Groves and Heeringa (2006) suggest using the first phase of the data collection to learn and derive propensities. This will be at the expense of efficiency because part of the survey has already been conducted. Nonetheless, gathered information directly feeds back to a current survey. In either case, it is imperative that response propensities be estimated from randomized contrasts, meaning that historic data must have randomized assignments to different design features. For example, from a face-to-face survey with a maximum of three visits, it cannot be inferred what the response propensities will be if the maximum is raised to 10 visits. Similarly, in telephone surveys with evening calls only, it cannot be deduced what contact propensities are for daytime shifts. Randomization over design features is, however, not sufficient. Historic data also needs randomization over subpopulations or more generally over sample units with different scores on the covariate vectors X from linked data and $\tilde{X}$ from paradata. A well-known example is the face-to-face interview. Interviewers usually work in regions spread over a country. It is impossible or at least risky to predict what the response propensities will be the same interviewers were to work in other regions. The successes of interviewers may at least partially be due to their familiarity with customs, dialect, and lifestyles of persons in their regions.

As a consequence care must be exercised in selecting the number and type of design features to be optimized. Historic survey data may not have the appropriate randomizations to support the estimation of corresponding propensities without strong assumptions. Responsive survey designs to some extent account for this problem by explicitly allowing for data mining of design features in the initial design phase. Still informative randomization is indispensable, and obviously only some design features can be randomized.

The literature on household surveys gives an extensive number of models for response propensities that include design features. A detailed account of the various models was given in Chapter 12 of this book. The classic models are logistic regression models that, apart from covariates, contain a number of design features. Since some design features affect only certain types of nonresponses, in that incentives would have an impact mostly on refusals, the regression models can be extended to nested or sequential regression models. For example, the model may take the form of an equation for obtaining contact and an equation for obtaining cooperation. In the cooperation equation the incentive type and level may be included as independent variables. When design features are only partially randomized over subpopulations, then often multilevel models are used. The first level then corresponds to the design feature, such as the interview type or the survey mode, and the second level to sample units.

The common denominator in all models is that response propensities are estimated using a number of assumptions about the true nature of the nonresponse missing-data mechanism. In general, such models are simplifications. Consequently anticipated response propensities $\rho(x, s)$ have a standard error. In the optimization this uncertainty can be accounted for by treating the response propensities as random variables rather than fixed quantities. The randomness allows for sensitivity analyses and evaluations of the robustness of the optimization. It should be understood that the optimization leads to an expected quality level and to expected costs. Sensitivity analyses provide insight into the variation of quality and costs when the survey is conducted multiple times (under the same circumstances).

EXAMPLE 13.6 Estimating Response Propensities

In Example 13.1 three adaptive survey designs were presented. There were six subpopulations based on gender and three age categories. Furthermore paradata were identified as the outcome of the first two weeks of data collection. Three results were possible: response, refusal, and noncontact. Finally, there were three candidate strategies.

If all options are still open, namely all three strategies are applicable for all six subpopulations and for each response type except, of course, response, then in total 54 probabilities need to be estimated. First, there are the 18 covariate distributions $q(x, \tilde{x})$ for each group and result. Next, there are the 12 response probabilities $\rho(s, x, \tilde{x})$for refusal and noncontact and each of the age-gender groups per strategy.

The covariate distributions may be estimated by simply computing the percentages from previous versions of the survey. The response probabilities can only be estimated when the three strategies are randomly assigned to each of the age-gender groups in the second two weeks of data collection in a pilot study. Alternatively, one could estimate the probabilities based on similar surveys that followed at least one of the strategies.

13.3 Application

Numerous settings for adaptive survey design can be given. In this section one simple example is presented: the interviewer's assignment. The example is taken from project RISQ, which was funded by the Seventh EU Research Framework Programme. The website www.risq-project.eu gives background on the project as well as the research papers.

The Dutch Survey of Consumer Confidence (SCC) is a monthly telephone survey of households to learn about their economic situations and expenditures. The survey provides insight into short-term economic growth, and early indicators for differences in consumer trends. Each month 1500 households are sampled. The two most influential causes of nonresponse in the SCC are noncontact and refusal. Of the sample, 95% is contacted, and of the contacted, 71% of the households participate. The response rate is 67%.

An important influence on the participation rate is the interviewer. Interviewers' performances can vary greatly when it comes to obtaining responses. In total 60 interviewers worked on the SCC in 2005. So each interviewer had, on average, contact with 280 households. Interviewer participation rates ranged from 50% to 79%. The lowest rate of 50% was exceptional, but the lowest participation rate was 61%. The mean interviewer participation rate was 67%. Households were randomly assigned to interviewers in the CATI management system. Hence it can be concluded that with respect to the interviewer the data are randomized. In the following the interviewer will be the design feature of interest, for a survey strategy set S that consists of 60 strategies, $S = \{I_1, I_2, \ldots, I_{60}\}$.

The available covariate vector X contains ethnicity, gender composition of the household core (male, female, or mix), average age of the household core in five-year classes, type of household, extent of urbanization of the neigborhood, and average value of houses in the neighborhood. No paradata is used, so the adaptive survey design is static. In the optimization the allocation probabilities $p(I_k|x)$ need to be chosen, namely there needs to be decided to which interviewers the subpopulations should be assigned.

The maximal bias

$$Q(p) = \frac{\hat{S}(\hat{\rho}_X)}{\hat{\bar{\rho}}_X} \tag{13.14}$$

is selected as quality function. For the SCC the maximal bias was 0.14.

To estimate the response propensities $\rho(I_k, x)$ for interviewers, two models are employed: a logistic regression for the contact propensities and a multilevel model for the participation propensities. The interviewers form the first level of the model and the households the second level. The multilevel model is used to separate individual participation propensities and interviewer participation propensities. By separating interviewer and individual, the interviewer effect can be isolated, and the interviewer assignment can be optimized and evaluated in a simulation study.

The logistic regression model for contact does not include the interviewer as the interviewer is not responsible for the timing of call attempts. This is done within the CATI management system by data collection staff.

For the interviewer effect, it was first investigated whether it was sufficient to use a fixed slope multilevel model, whereby the interviewer is added as a main effect only and there are no interactions with auxiliary variables. All covariates were selected for the multilevel model, but interactions with interviewer were not significant at the 5% level.

TABLE 13.1 Response rate and maximal bias of the SCC and the SCC after re-assignment of interviewer

	$\bar{\rho}_X$	$Q(\rho)$
SCC	67%	0.14
Re-assignment	68%	0.07

The resulting model is

$$\log\left(\frac{\lambda_i}{1-\lambda_i}\right) = \alpha_0 + \alpha_X X_i, \tag{13.15}$$

$$\log\left(\frac{\kappa_{ik}}{1-\kappa_{ik}}\right) = \beta_0 + \beta_X X_i + I_k, \tag{13.16}$$

where λ_i is the contact propensity of household i, κ_{ik} is the participation propensity of household i given interviewer k was assigned to the household, X_i is the covariate vector containing the six auxiliary variables, α_0 and β_0 are constant terms or intercepts, and α_X and β_X are the slope parameters. The response propensity now follows as $\rho(I_k, X_i) = \lambda_i \kappa_{ik}$.

Models (13.14) and (13.15) were fitted to the SCC data set. Next the estimated interviewer effect I_k was used to smooth the participation propensities. This was done using the following steps:

1. Remove I_k from the estimated propensities.
2. Sort the amputated propensities in ascending order.
3. Sort the removed I_k in descending order.
4. Match the interviewers to the households.
5. Add I_k back to the amputated propensities.

The fifth step ensures that interviewers have the same workload when the sample units are artificially re-assigned, and that the response rate stays unchanged. As a consequence costs are also the same after re-assignment. However, because strong interviewers are assigned to difficult addresses and weak interviewers to easy addresses, the individual interviewer response rates change, and differences between interviewers are leveled out.

Table 13.1 gives the response rate and maximal bias before and after re-assignment of interviewers. The re-assignment leads to a smaller maximal bias and hence higher perceived quality for the same costs. The predicted response rate after re-assignment is slightly higher than in the SCC. The slight increase is probably caused by small interactions between the interviewers and the covariates.

13.4 Summary

This chapter described survey designs in which different population units receive different treatment or survey strategies. Differences between population units are reflected by covariates from either linked data from registrations or paradata. Survey strategies are defined as different specifications of survey design features. Such designs are termed adaptive survey designs as they adapt or tailor data collection to the population of interest. Adaptive survey designs resemble responsive survey designs, the main difference being the presence of an initial design phase in responsive designs that is used to learn how effective and costly different design features are.

The basic ingredients to adaptive survey designs are survey strategies, population covariates, response propensities, cost functions, and strategy allocation probabilities. Adaptive survey designs attempt to optimize response quality by assigning different strategies to different population units. The strategy allocation probabilities represent the parameters in the optimization.

Adaptive and responsive survey designs are beginning to emerge in survey practice. Although tailoring of design features is done for several years, it has rarely been done in a systematic, mathematical way by optimizing quality given costs, and vice versa. The terms adaptive and responsive designs have been suggested only very recently. As a result research into such designs is still young.

Response quality and survey costs are known to be complex and hard to quantify. However, they determine the optimization problem. In this chapter a number of quality functions were presented that have come up in the recent literature. Furthermore the optimization depends on the quality of anticipated response propensities and costs. Both must either be estimated in an initial phase of data collection as is done in responsive designs or based on historic data on the same survey or similar surveys. Given the complexity of quality—cost trade-offs and the robustness of estimates of response and costs—it is imperative that adaptive survey designs are kept simple. In other words, the set of candidate survey strategies and the vector of covariates should be manageable.

13.5 KEY TERMS

Adaptive survey design: A survey design where the population units with different characteristics receive different treatments.

Design phase: A separate part of data collection in responsive survey designs where the design features are varied. The initial design phase is different as it is used to identify and evaluate the effectiveness and costs associated with design features.

Dynamic adaptive designs: An adaptive survey design where the strategy allocation depends on paradata and possibly also on linked data.

Paradata: Data collected during the data collection process by interviewers and data collection staff other than the survey questionnaire items.

Quality function: The function that is optimized in the adaptive survey design. Quality functions can either be covariate or item based. Covariate-based functions depend only on linked data and paradata while item-based functions employ also the relation to survey target variables.

Responsive survey design: A survey design consisting of multiple design phases where the design features are varied. Per phase the objective is to reach phase capacity, meaning optimal quality given the available budget.

Static adaptive designs: An adaptive survey design where the strategy allocation depends on linked data only.

Strategy allocation probabilities: The probability that a population unit with a specific set of characteristics is assigned a specific strategy.

Survey strategy: A specified sequence of survey design features.

Survey strategy set: The set of possible survey strategies in the allocation.

Survey design feature: A single feature or aspect of a survey design such as contact mode, survey mode, and timing protocol.

13.6 REFERENCES

Barón, J.D., Breunig, R.V., Cobb-Clark, D., Gørgens, T., and Sartbayeva, A. (2009), Does the Effect of Incentive Payments on Survey Response Rates Differ by Income Support History? *Journal of Official Statistics* 25, pp. 483–507.

Cobben, F. (2009), *Nonresponse in Sample Surveys. Methods for Analysis and Adjustment.* PhD Thesis. University of Amsterdam, Amsterdam.

De Leeuw, E.D. (2008), Choosing the Method of Data Collection. In: De Leeuw, E.D., Hox, J.J., and Dillman, D.A. (Eds.), *International Handbook of Survey Methodology,* Erlbaum Associates, New York, pp. 113–135.

De Leeuw, E., Callegaro, M, Hox, J., Korendijk, E., and Lensvelt-Mulders, G. (2007), The Influence of Advance Letters on Response in Telephone Surveys. *Public Opinion Quarterly* 71, pp. 413–443.

Dillman, D. A. (2007), *Mail and Internet Surveys: The Tailored Design Method.* Wiley, Hoboken, NJ.

Greenberg, B.S., and Stokes, S.L. (1990), Developing an Optimal Call Scheduling Strategy for a Telephone Survey. *Journal of Official Statistics* 6, pp. 421–435.

Groves, R.M. (1989), *Survey Errors and Survey Costs.* Wiley, New York.

Groves, R.M., and Couper, M.P. (1998), *Nonresponse in Household Interview Surveys.* Wiley, New York.

Groves, R.M., Dillman, D., Eltinge, J., and Little, R. (2002), *Survey Nonresponse.* Wiley, New York.

Groves, R.M., and Heeringa, S.G. (2006), Responsive Design for Household Surveys: Tools for Actively Controlling Survey Errors and Costs. *Journal of the Royal Statistical Society, Series A* 169, pp. 439–457.

Heyd, J.M., and Carlin, B.P. (1999), Adaptive Design Improvements in the Continual Reassessment Method for Phase I Studies. *Statistics in Medicine* 18, pp. 1307–1321.

Kalsbeek, W.D., Botman, S.L., Massey, J.T., and Liu, P.W. (1994), Cost-efficiency and the Number of Allowable Call Attempts in the National Health Interview Survey. *Journal of Official Statistics* 10, pp. 133–152.

Kersten, H.M.P., and Bethlehem, J.G. (1984), *Exploring and Reducing the Non-response Bias by Asking the Basic Question.* Report BPA 627-84-M1. Statistics Netherlands, Voorburg.

Kulka, R.A., and Weeks, M.F. (1988), Towards the Development of Optimal Calling Protocols for Telephone Surveys: A Conditional Probabilities Approach. *Journal of Official Statistics* 4, pp. 319–322.

Lyberg, L.E., Biemer, P., Collins, M., De Leeuw, E.D., Dippo, C., Schwarz, N., and Trewin, D. (1997), *Survey Measurement and Process Quality.* Wiley, Hoboken, NJ.

Lynn, P. (2003), PEDAKSI: Methodology for Collecting Data about Survey Non-respondents. *Quality and Quantity* 37, pp. 239–261.

Mohl, C., and Laflamme, F (2007), Research and Responsive Design Options for Survey Data Collection at Statistics Canada. *Proceedings of ASA Joint Statistical Meeting,* Section 293, July 29–August 2, Salt Lake City.

Moore, J. (1988), Self/Proxy Response Status and Survey Response Quality: A Review of Literature. *Journal of Official Statistics* 4, pp. 155–172.

Murphy, S.A. (2003), Optimal Dynamic Treatment Regimes. *Journal of the Royal Statistical Society, Series B* 65, pp. 331–355.

Rao, R.S., Glickman, M.E., and Glynn, R.J. (2008), Stopping Rules for Surveys with Multiple Waves of Nonrespondent Follow Up. *Statistics in Medicine* 27, pp. 2196–2213.

Särndal, C.E., and Lundström, S. (2008), Assessing Auxiliary Vectors for Control of Nonresponse Bias in the Calibration Estimator. *Journal of Official Statistics* 24, pp. 167–191.

Schouten, B., Cobben, F., and Bethlehem, J.G. (2009), Indicators for the Representativeness of Survey Response. *Survey Methodology* 35, pp. 101–113.

Wagner, J. (2008), *Adaptive Survey Design to Reduce Nonresponse Bias.* PhD thesis. University of Michigan, Ann Arbor.

Wagner, J., and Raghunathan, T. (2007), Bayesian Approaches to Sequential Selection of Survey Design Protocols. *Proceedings of ASA Joint Statistical Meeting,* Section 501, July 29–August 2, Salt Lake City.

13.7 EXERCISES

Exercise 13.1 For each of the following settings derive whether the survey design is static or dynamic

(a) A survey design where the respondent is allowed to choose the survey mode.

(b) A survey design where proxy reporting is allowed for persons under 12 years of age, and age is included in the sampling frame.

(c) A survey design where persons that refused participation are re-approached using short, condensed questionnaires with a few basic questions.

(d) A survey design where single person households receive a maximum of 10 call attempts, multiple person households receive a maximum number of 5 call attempts, and the size of the household is linked to the sample file.

(e) A survey design where persons younger than 25 years of age are offered an incentive of 10 euros, persons older than 25 years of age receive an incentive of 5 euros, and age is included in the sampling frame.

(f) A survey design where nonrespondents to a Web survey are invited to participate in a telephone survey if a phone number is registered and to a face-to-face survey if no number is available.

Exercise 13.2 Compute the quality function $Q(\rho) = S(\rho_X)/\rho_X$ for each of the three designs in Example 13.3.

Exercise 13.3 For Example 13.4 compute the variable cost component $C_V(p)$ for each of the three designs. What is the cheapest strategy?

Exercise 13.4 In Examples 13.1 through 13.5 the size of the sample is not mentioned. Suppose that in total 10,000 euros are available and that the required response rate is 50%. What does this mean for each of the three designs in terms of sample size? What if the required response rate is 60%?

Chapter Fourteen

Item Nonresponse

14.1 Introduction

Nonresponse is the phenomenon where persons in the sample from the population do not provide the requested information or provide information that is not usable.

Two types of nonresponses can be distinguished. First is *unit nonresponse*. This occurs when a selected person does not answer any question. The questionnaire form remains completely empty. Second is *item nonresponse*. This occurs when some questions have been answered, but no answer is given to other, possibly sensitive, questions. So, the questionnaire form has been partially completed.

Item nonresponse may occur when persons refuse to answer a question (e.g., because they do not want to answer a sensitive question) or when they do not know the answer. If a paper questionnaire is used for data collection, some persons could also have accidentally skipped a question.

Item nonresponse can also occur as a consequence of a data-editing procedure. Data editing is the process of checking questionnaire forms for errors in the answers, and correcting detected errors. Particularly large survey organizations spend substantial resources on data editing, as it may improve data quality considerably. How to correct detected errors in questionnaire forms? One way to accomplish this is to contact respondents again and to mention the errors to them. They may then offer to correct their answers. Unfortunately, this approach is not feasible in daily survey practice. Respondents consider completing a questionnaire already a burden. Having to reconsider their answers would in many cases lead to refusal to do so. Moreover this approach is time-consuming and expensive. Therefore survey agencies have to depend on other approaches to dealing with

Handbook of Nonresponse in Household Surveys, By Jelke Bethlehem, Fannie Cobben, and Barry Schouten

errors in the collected data. One such approach is to treat a wrong answer as unknown. This comes down to item nonresponse.

Because of item nonresponse the resulting survey data file will contain missing data. This may hamper proper analysis of the data. For example, how to compute the correlation between two variables *X* and *Y* if sometimes values of *X* are missing and sometimes values of *Y*?

One way around this problem is to apply *list-wise deletion*, so that the analysis is restricted to records that contain no missing values. Then the response data set is reduced to a subset of records. If values are missing completely at random (MCAR), this will not affect the properties of estimators. They will just have a larger variance. If values are not missing at random (NMAR), estimators may turn out to be biased. For surveys with large and complex questionnaires there may be some data missing in many questionnaires. Case-wise deletion would then mean throwing away many records that could contain valuable information with respect to other variables.

A less rigorous approach is to apply *pair-wise deletion*. To compute an estimate, only those cases are used for which all variables needed for the computation have been answered. Other variables may have missing values. So, to compute the correlation between *X* and *Y*, the cases are selected for which both *X* and *Y* have values. And the correlation between *Y* and *Z* will be based on cases for which values of *Y* and *Z* are both available. An advantage of pair-wise deletion is that more data can be used to compute estimates. A disadvantage is that computation of different estimates may be based on different data sets, reducing comparability.

EXAMPLE 14.1 List-wise and Pair-wise Deletion

The effects of list-wise deletion and pair-wise deletion are shown using the very small data in the table below.

Record	*X*	*Y*	*Z*
1	4	4	5
2	5	-	6
3	-	5	4
4	6	6	-

Notice that the data set contains only 4 records and 3 variables (*X*, *Y*, and *Z*). There is item nonresponse in the last three records. Suppose that the correlation coefficient R_{XY} between variables *X* and *Y* has to be computed. Application of list-wise deletion would result in removing records 2, 3, and 4 from the data set. Only one record will remain, making it impossible to compute the correlation coefficient. Notice that even record 4 will be removed, which contains values for both *X* and *Y*.

The correlation coefficient R_{XY} is defined as $R_{XY} = S_{XY}/(S_X \times S_Y)$, where S_{XY} is the covariance between *X* and *Y*, and where S_X and S_Y are the respective standard deviations of *X* and *Y*. Computation of the covariance requires records with values for both X and Y. Application of pair-wise deletion means that records 1 and 4 can be used for this. This results in $S_{XY}=2$. Computation of S_X requires records with a value of X. There are 3 such records (1, 2, and 4), resulting in $S_X=1$. Likewise records 1, 3, and 4 can be used for the computing S_Y. This gives $S_Y=1$. So the correlation coefficient is equal to $R_{XY}=2/(1 \times 1)=2$. Apparently there is something wrong because the value of the correlation coefficient must be between -1 and $+1$. A value of 2 is impossible. This problem is created by the fact that computation of the various quantities is based on different sets of records.

To avoid missing data problems, often some kind of imputation technique is applied. Imputation means that each missing value is replaced by a synthetic value. This synthetic value is obtained by a technique that attempts to estimate the missing values.

After applying that imputation technique, there should be no more "holes" in the data set. So the analysis can proceed without worry about missing values. However, in the analysis results the presence of synthetic values needs to be accounted for by adjustments to significance tests and standard errors of estimators.

Various imputation techniques are described in this chapter. They all have their advantages and disadvantages. So the choice of a particular technique should be made carefully. It is also important that users of a data set with imputed values be aware of the fact that some values are not real but synthetic. This may help them explain why some analysis results are unexpected.

14.2 Theory

There is no guarantee that an imputation technique will reduce a bias caused by item nonresponse. Bias reduction depends on the properties of the missing data mechanism and the specific imputation technique that is applied. The three types of missing data mechanisms are shown in Figure 14.1.

In the figure X represents a set of auxiliary variables that are completely observed. Y is a target variable that is partly missing. Z represents causes of missingness unrelated to X and Y, and R the missingness.

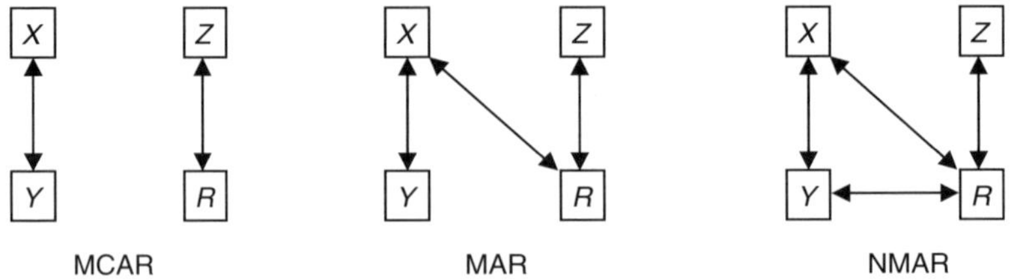

FIGURE 14.1 Missing data mechanisms.

In the case of *missing completely at random* (MCAR), missingness is caused by a phenomenon Z that is completely unrelated to X and Y. Estimates for parameters involving Y will be unbiased. The imputation techniques will not change this.

In the case of *missing at random* (MAR), missingness is caused partly by an independent phenomenon Z and partly by the auxiliary variable X. So there is an indirect relationship between Y and R. This may lead to biased estimates for Y. Fortunately, it is possible to correct for such a bias by using an imputation technique that takes advantage of the availability of all values of X, both for respondents and nonrespondents.

In case of *not missing at random* (NMAR), there may be a relationship between Z and R and between X and R, but there is also a direct relationship between Y and R that cannot be accounted for by X. This situation also may lead to biased estimates for Y. Unfortunately, imputation techniques using X will not be able to remove the bias in all cases.

14.2.1 SINGLE IMPUTATION TECHNIQUES

Among the many imputation techniques available those described in this section are all single imputation techniques. This means that a missing value is replaced by one synthetic value. Section 14.2.3 is about multiple imputation. There a missing value is replaced by a set of synthetic values.

To keep things simple, it is assumed that the sample has been selected with equal probabilities and without replacement. The sample can be represented by means of the set of indicators $a_1, a_2, \ldots, a_N$. The value of the indicator a_k is equal to 1 if element k is selected in the sample, and otherwise, it is equal to 0.

Let Y be the target variable for which some values are missing in the sample. Missingness is denoted by the set of indicators $R_1, R_2, \ldots, R_N$. Of course, $R_k = 0$ if $a_k = 0$. Item nonresponse for a sampled element k is indicated by $a_k = 1$ (sampled) and $R_k = 0$ (missing). Let X be an auxiliary variable for which no missing values occur in the sample. So the value X_k is always available if $a_k = 1$.

The number of available observations is denoted by

$$m = \sum_{k=1}^{N} a_k R_k. \tag{14.1}$$

The mean of the available observations is equal to

$$\bar{y}_R = \frac{1}{m} \sum_{k=1}^{N} a_k R_k Y_k. \tag{14.2}$$

This would be the obvious estimator if no imputation is carried out. It can be shown that the expected value of this estimator is approximately equal to

$$E(\bar{y}_R) \approx \frac{1}{N} \sum_{k=1}^{N} \frac{\rho_k}{\bar{\rho}} Y_k, \tag{14.3}$$

where $\rho_k = E(R_k) = P(R_k = 1)$ is the probability that the value for element k is missing. The bias of this estimator is equal to

$$B(\bar{y}_R) = E(\bar{y}_R) - \bar{Y} = \frac{R_{\rho Y} S_\rho S_Y}{\bar{\rho}}, \tag{14.4}$$

where $R_{\rho Y}$ is the correlation between the target variable and the probabilities ρ_k, S_Y is the standard deviation of the variable Y, and S_ρ is the standard deviation of the probabilities. From this expression of the bias a number of conclusions can be drawn:

- The bias vanishes if there is no relationship between the target variable and the probability of a missing value. Then $R_{\rho Y} = 0$. The stronger the relationship is between target variable and the probabilities, the larger the bias will be.
- The bias vanishes if all probabilities are equal. Then $S_\rho = 0$. In this situation the nonresponse is not selective; it just leads to a reduced sample size.
- The magnitude of the bias increases as the mean of the response probabilities decreases. Translated in practical terms, this means that more missing values lead to larger biases.

14.2.1.1 Deductive Imputation Sometimes the value of a missing item can be logically deduced from the non-missing values of other variables. If strict rules of logic are followed, this technique has no impact on the properties of the distribution of estimators.

EXAMPLE 14.2 Deductive Imputation

A number of questions are asked of every household member in a household. Of one member of a household, gender is female and age is 5 years. The question is unanswered about the number of children this persons has given birth to. In this case it is certain that answer must be equal to 0.

In some business and household surveys questions are asked about the various income components. Also total income is asked. By definition, the income components should add up to the total income. Although the total income question is in fact redundant, it is often included to have an extra check on the answers. It may happen that a respondent only provides the income components but not total income. The imputed value of total income is then equal to the sum of the income components.

Although deductive imputation is the ideal form of imputation, it is frequently not possible to apply it. In some cases relationships between variables are not so straightforward.

14.2.1.2 Imputation of the Mean Imputation of the mean implies that a missing value of a variable is replaced by the mean of the available values of this variable.

Let k be an element in the sample ($a_k = 1$) for which the value Y_k is missing ($R_k = 0$). The imputed value for element k is defined by

$$\hat{Y}_k = \bar{y}_R = \frac{\sum_{k=1}^{N} a_k R_k Y_k}{\sum_{k=1}^{N} a_k R_k}. \tag{14.5}$$

Since all imputed values will be equal to the same mean, the distribution of this variable in the completed data set will be affected. It will have a peak at the mean of the distribution.

14.2.1.3 Imputation of the Mean within Groups On the basis of one or more qualitative auxiliary variables the sample is divided into a number of nonoverlapping groups. Within a group a missing value is replaced by the mean of the available observations in that group.

Imputation of the mean within groups will perform better than imputation of the mean if the groups are homogeneous with respect to the variable being imputed. Since all values are close to each other, the imputed group mean will be a good approximation of the true, but unknown, value.

14.2.1.4 Random Imputation Random imputation means that a missing value is replaced by a value that is randomly chosen from the set of available values for the variable.

The set of available values is equal to

$$\{Y_k | a_k = 1 \wedge R_k = 1\}. \tag{14.6}$$

This imputation is also called *hot deck imputation*. It is a form of *donor imputation*: a value is taken from an existing record where the value is not missing.

The distribution of the values of the variable for the complete data set will not contain peaks. However, this distribution does not necessarily resemble the true distribution of the variable. Both distributions may differ if the missing values are not randomly missing.

14.2.1.5 Random Imputation within Groups On the basis of one or more qualitative auxiliary variables, the sample is divided into a number of nonoverlapping groups. Within a group a missing value is replaced by a randomly chosen value from the set of available values in that group.

Random imputation within groups will perform better than random imputation if the groups are homogeneous with respect to the variable being imputed. Since all values are close to each other, the randomly selected value will be a good approximation of the true, but unknown, value.

14.2.1.6 Nearest Neighbor Imputation The idea of nearest neighbor imputation is to search for a record that resembles as much as possible the record in which a value is missing. Some kind of distance measure is defined to compare records on the basis of the values of auxiliary variables that are available for all records.

If all auxiliary variables are of a quantitative nature, some kind of Euclidean distance may be used. Suppose that there are p such variables. Let X_{kj} be the value of variable X_j, for $k = 1, 2, \ldots, N$ and $j = 1, 2, \ldots, p$. Then the distance between the records of two elements i and k could be defined by

$$D_{ik} = \sqrt{\sum_{j=1}^{p} (X_{ij} - X_{kj})^2}. \tag{14.7}$$

Let k be an element in the sample for which the value Y_k is missing. The imputed value is copied from the record of a sampled element i with the smallest distance D_{ik}, and for which the value of Y is available.

14.2.1.7 Ratio Imputation Ratio imputation assumes a relationship between the target variable Y (with missing values) and an auxiliary variable X (without missing values). If this relationship is (approximately) of the form $Y_k = BX_k$, for some constant B, then a missing value of Y for element k can be estimated by BX_k.

If the value of B is not known, it can estimate using the available data:

$$b = \frac{\sum_{k=1}^{N} a_k R_k Y_k}{\sum_{k=1}^{N} a_k R_k X_k}. \tag{14.8}$$

Let k be an element in the sample for which the value Y_k is missing. The imputed value is defined by

$$\hat{Y}_k = bX_k. \tag{14.9}$$

Ratio estimation is often used when the same variable is measured at two different moments in time in a repeated survey.

14.2.1.8 Regression Imputation Regression imputation assumes a relationship between the target variable Y (with missing values) and an auxiliary variable X (without missing values). If this relationship is (approximately) of the form $Y_k = A + BX_k$, for some constants A and B, then a missing value of Y for element k can be estimated by $A + BX_k$.

If the values of A and B are not known, they can be estimated by applying ordinary least squares on the available data:

$$b = \frac{\sum_{k=1}^{N} a_k R_k (Y_k - \bar{y}_r)(X_k - \bar{x}_r)}{\sum_{k=1}^{N} a_k R_k (X_k - \bar{x}_r)^2} \tag{14.10}$$

and

$$a = \bar{y}_r - b\bar{x}_r. \tag{14.11}$$

Let k be an element in the sample for which the value Y_k is missing. The imputed value is defined by

$$\hat{Y}_k = a + bX_k. \tag{14.12}$$

Only one auxiliary variable was used in the regression model described here. Of course, it is possible to put more variables in the regression model. This will often increase the explanatory power of the model, and therefore imputed values will be closer to the true, but unknown, values.

14.2.2 A GENERAL IMPUTATION MODEL

At first sight, all the imputation techniques mentioned above seem rather different. Nevertheless, almost all of them fit in a general model.

Let k be an element in the sample for which the value Y_k is missing. In general, the imputed value can be written as

$$\hat{Y}_k = B_0 + \sum_{j=1}^{p} B_j X_{kj} + E_k, \tag{14.13}$$

where X_{kj} denotes the value of auxiliary variable X_j for element k, B_0, B_1, . . ., B_P are regression coefficients, and E_k is a random term, whose nature is determined by the specific imputation technique.

By taking B_0 equal to the mean of the available values of Y, and setting the other coefficients B_j and E_k equal to 0, the model reduces to imputation of the mean.

If the auxiliary variable $X_1, X_2, \ldots, X_p$ are taken to be dummy variables that indicate to which group an element belongs ($X_{kj} = 1$ if element k is in group j, and otherwise, $X_{kj} = 0$), $B_0 = 0$ and $E_k = 0$, then imputation of the group mean is obtained.

Model (14.13) reduces to random imputation if the starting point is the model for imputation of the mean, but a random term E_k is added. Its value is obtained by a random drawing from the set of values

$$\left\{ \bar{y}_R - Y_k | a_k = 1 \wedge R_k = 1 \right\}. \tag{14.14}$$

Random imputation within groups is obtained if the starting point is the model for imputation of the group mean, but a random term E_k is added whose value is a random drawing from a set of values. These values are obtained by subtracting the available values from their respective group means.

It is clear that also ratio imputation and regression imputation are special cases of model (14.13). Nearest neighbor imputation, however, does not fit in this model.

EXAMPLE 14.3 Single Imputation Techniques

The use of single imputation techniques is shown using an artificial example. The table below contains a small data set. Three variables have been measured: income, level of education (low or high), and number of years of work experience. There is a clear pattern in the data: income is higher for highly educated people. Also income increases with years of work experience. Income is missing for one person.

Person	Income	Education	Years
1	2041	Low	1
2	2110	Low	2
3	2142	Low	3
4	2201	Low	4
5	2247	Low	5
6	?	Low	6
7	4099	High	1
8	4204	High	2
9	4298	High	3
10	4401	High	4
11	4497	High	5

Several imputation methods are now applied to see how well they perform. Note that the income data have been generated such that the missing value for person 6 should be around 2300.

The mean value of the available incomes is 3224. So imputation of the mean would replace the missing value by 3224. This is clearly not a good value. There are no persons with approximately this income.

The person with a missing income has a low level of education. So one could try imputation of the mean within the group of low educated. The average income of the low educated is equal to 2148. This is a better value than 3224, but it disregards the linear relationship with years of work experience.

Random imputation could produce a value close to the true value, for example, if person 5 were selected as a donor. It could, however, also produce a very bad value, for example, if person 11 is the chosen donor.

Random imputation within the group of low educated would always result in a more reasonable value, but also here the relationship with years of experience is not taken into account.

Nearest neighbor imputation attempts to find a donor resembling the person with the missing value. Using the two variables education and works experience, person 5 is the best candidate. The imputed value of 2247 is not bad, but it does not take into account that the person with the missing value has one additional year of work experience.

If ratio imputation is applied, with work experience as an explanatory variable, it is assumed that income is proportional to years of experience. The constant b is obtained by dividing the total income of all persons (32,240) by the total years of experience (30). The result is 1075. So the income for person 6 is predicted by $6 \times 1075 = 6450$. This imputed value is very wrong. Apparently the model does not fit.

If ratio imputation is applied within the group of low educated, the value of b is equal to $10741/15 = 716$. The missing income is predicted by $6 \times 716 = 4296$. This is still a bad value. Also within the group, the model does not fit.

Regression imputation could be applied, where number of years of work experience is an explanatory variable. This assumes a linear relationship between income and years of experience. The regression coefficients turn out to be equal to $a = 3000$ and $b = 75$. Therefore the missing value is predicted by $3000 + 6 \times 75 = 3450$. This is not a good value. Apparently the regression model does not fit.

Regression imputation could also be applied within the group of low educated. The regression coefficients turns out to be equal to $a = 1997$ and $b = 50$. Therefore the missing value is predicted by $1997 + 6 \times 50 = 2297$. This is the best guess for the missing income.

14.2.3 PROPERTIES OF SINGLE IMPUTATION

When choosing among the many single imputation techniques for a practical situation, there are several considerations to be made: the measurement level of the variable, the adjustment for bias, and the impact on standard errors.

The first is the type of variable for which missing values have to be imputed. In principle, all mentioned imputation techniques can be applied for quantitative variables. However, not every single imputation technique can be used for qualitative variables. A problem is that the synthetic value produced by the imputation technique does not necessarily belong to the domain of valid values of the variable. For example, if the variable gender has to be imputed, mean imputation produces an impossible value (what is the mean gender?). Therefore one should stick to some form of "donor imputation" for qualitative variables. These techniques always produce "real" values.

The single imputation techniques described in Section 14.2.1 can be divided in two groups. One contains *deterministic* imputation techniques, and the other *random* imputation techniques. The random term E_k in model (14.13) is zero for deterministic techniques, and not zero for random techniques.

For some deterministic imputation techniques (e.g., imputation of the mean), the mean of a variable before imputation is equal to the mean after imputation. This shows that not every imputation technique is capable of reducing a bias due to missingness. For random imputation techniques the mean before imputation is never equal to the mean after imputation. However, expected values before and after imputation may be equal.

Deterministic imputation may affect the distribution of a variable. It tends to produce synthetic values that are close to the centre of the original distribution. The imputed distribution is more "peaked." This may have unwanted consequences. Estimates of standard errors may turn out to be too small. A researcher using the imputed data (not knowing that the data set contains imputed values) may get the impression that his estimates are very precise, while in reality this is not the case.

The possible effects of imputation on bias and variance are shown by analyzing two single imputation techniques in some more detail: imputation of the mean and random imputation.

14.2.4 EFFECTS OF IMPUTATION OF THE MEAN ON BIAS AND VARIANCE

Imputation of the mean replaces missing values by the mean of the available values. Let $Y_1, Y_2, \ldots, Y_N$ be the values of the variable to be imputed. A simple random sample without replacement of size n is denoted by the set of indicators $a_1, a_2, \ldots, a_N$. Missingness of sample values is denoted by the set of indicators $R_1, R_2, \ldots, R_N$, where $R_k = 1$ only if k is in the sample ($a_k = 1$) and the value Y_k is available. By definition $R_k = 0$ if $a_k = 0$. Item nonresponse for a sampled element k is indicated by $a_k = 1$ and $R_k = 0$. The number of available observations is denoted by

$$m = \sum_{k=1}^{N} a_k R_k. \tag{14.15}$$

The mean of the available observations is equal to

$$\bar{y}_R = \frac{1}{m} \sum_{k=1}^{N} a_k R_k Y_k. \tag{14.16}$$

In case of imputation of the mean, the missing observation is replaced by the mean of the available values. So, if $a_k = 1$ and $R_k = 0$ for some element k, the imputed value

$$\hat{Y}_k = \bar{y}_R \tag{14.17}$$

is used. Now a new estimator for the population mean is obtained by taking the average of all "real" and all "synthetic" values of Y. Let

$$\bar{y}_I = \frac{1}{n-m} \sum_{k=1}^{N} a_k (1-R_k) \hat{Y}_k \tag{14.18}$$

denote the mean of all imputed values. Then the mean after imputation can be written as

$$\bar{y}_{IMP} = \frac{m\bar{y}_R + (n-m)\bar{y}_I}{n}. \tag{14.19}$$

If the response mean is imputed, expression (14.18) reduces to

$$\bar{y}_I = \frac{1}{n-m} \sum_{k=1}^{N} a_k (1-R_k) \hat{Y}_k = \frac{1}{n-m} \sum_{k=1}^{N} a_k (1-R_k) \bar{y}_R = \bar{y}_R, \tag{14.20}$$

and therefore

$$\bar{y}_{IMP} = \frac{m\bar{y}_R + (n-m)\bar{y}_I}{n} = \frac{m\bar{y}_R + (n-m)\bar{y}_R}{n} = \bar{y}_R. \tag{14.21}$$

So the conclusion can be drawn that the mean after imputation is equal to the mean before imputation. Imputation does not affect the mean.

To determine the characteristics of an estimator after imputation, it is important to understand the effects of two different types of probability mechanisms. While there always is the probability mechanism of the sample selection, an extra source of randomness may come from the introduced imputation technique, for example, if the imputed value is randomly selected. To take this into account, the expected value of an estimator (after imputation) is determined using the expression

$$E(\bar{y}_{IMP}) = E_S E_I(\bar{y}_{IMP}|S). \tag{14.22}$$

E_I denotes the expectation over the imputation distribution and E_S the expectation over the sampling distribution S. Expression (14.22) is now applied to the case of imputation of the mean. Given the sample, the estimator is a constant. So taking the expectation over the imputation distribution, results in the same constant. The conclusion can be drawn that the expected value of the estimator before and after imputation are the same. Hence imputation of the mean will not be able to reduce a possibly existing bias due to missingness.

To compute the variance of the estimator after imputation of the mean, the expression

$$V(\bar{y}_{IMP}) = V_S E_I(\bar{y}_{IMP}|S) + E_S V_I(\bar{y}_{IMP}|S). \tag{14.23}$$

is used. Given the sample, estimator (14.21) is a constant. This means the second term in expression (14.23) is equal to 0. The first term is equal to the variance of the estimator before imputation. So it can be concluded that imputation of the mean does not change the variance of the estimator.

Problems may arise when an uninformed researcher attempts to estimate the variance of estimators, for example, in constructing a confidence interval. To keep things simple, it is assumed the available observations can be treated as a simple random sample without replacement, namely missingness does not cause a bias. Then the variance after imputation is equal to

$$V(\bar{y}_{IMP}) = V(\bar{y}_R) = \frac{1-(m/N)}{m} S^2, \tag{14.24}$$

where S^2 is the population variance.

It is known from sampling theory that in a simple random sample without replacement, the sample variance s^2 is an unbiased estimator of the population variance S^2. Because the response is still assumed to be a random sample, the s^2 computed using the m available observations is an unbiased estimator of S^2.

What would happen if a researcher attempts to estimate S^2 using the complete data set, without knowing that some values have been imputed? The researcher would compute the sample variance, assuming this is an unbiased estimator of the population variance, though this is not the case. For the sample variance of the imputed data set, the expression would be

$$\begin{aligned} s_{IMP}^2 &= \frac{1}{n-1}\left(\sum_{k=1}^{N} a_k R_k (Y_k - \bar{y}_R)^2 + \sum_{k=1}^{N} a_k (1-R_k)(\hat{Y}_k - \bar{y}_R)^2\right) \\ &= \frac{1}{n-1}\left(\sum_{k=1}^{N} a_k R_k (Y_k - \bar{y}_R)^2 + 0\right) = \frac{m-1}{n-1} s^2. \end{aligned} \tag{14.25}$$

Hence

$$E(s_{IMP}^2) = \frac{m-1}{n-1} S^2. \tag{14.26}$$

This is not an unbiased estimator of the population variance. The population variance is underestimated. The impression is created that the estimator is very precise, whereas in reality this is not the case. So there is a substantial risk of drawing wrong conclusions from the data. This risk is larger if there are more imputed values.

EXAMPLE 14.4 Imputation of the Mean

An Income Survey is carried out in a small town. The target population consists of 19,000 people. A sample of size $n = 1000$ is selected. From earlier research it has become clear that the population variance S^2 remains constant over time at a value of 360,000. So this value can be used for the current situation.

Suppose that all persons selected in the sample cooperate in the survey and state their income. Then the variance of the sample mean would be equal to

$$V(\bar{y}) = \frac{1-f}{n} S^2 = \frac{1-(1000/19,000)}{1000} 360,000 = 341.053.$$

Now suppose that 10% of the sample persons refuse to state their income, and imputation of the mean is applied. The number of available "real" observations has been reduced to $m = 900$ (90% of 1000). Therefore, from expression (14.24), the variance of the sample mean after imputation becomes

$$V(\bar{y}_{IMP}) = \frac{1-(m/N)}{m} S^2 = \frac{1-(900/19,000)}{900} 360,000 = 381.053.$$

Of course, this variance is larger because it is based on fewer observations.

Now, to keep things easy, suppose that the sample variance in the data set before imputation happens to be exactly equal to the population variance (360,000). If the data set is imputed, and made available to a researcher without informing him that an imputation had taken place, what would his estimate of the variance of the sample mean be? Applying expression (14.26) results in

$$s_{IMP}^2 = \frac{m-1}{n-1}s^2 = \frac{900-1}{800-1}360,000 = 319,955.506,$$

and therefore the estimate of the variance of the sample mean becomes

$$v(\bar{y}) = \frac{1-f}{n}s_{IMP}^2 = \frac{1-(1000/19,000)}{1000}319,955.506 = 303.116.$$

This estimate is much lower than the population value (381.053).

Imputation also has an impact on the correlation between variables. Suppose that the variable Y is imputed using an imputation of the mean, and suppose that a variable X is completely observed for the sample. It can be shown that in this case the correlation after imputation is equal to

$$r_{IMP,X,Y} = \sqrt{\frac{m-1}{n-1}}r_{XY}, \tag{14.27}$$

where r_{XY} is the correlation in the data set before imputation. So the more observations are missing for Y, the smaller the correlation coefficient will be. Researchers not aware of their data set having been imputed will get the impression that relationships between variables are weaker than they really are. Also there is a risk of drawing wrong conclusions as more relationships appear to be significant.

14.2.5 EFFECTS OF RANDOM IMPUTATION

The effects of imputation as mentioned in the previous section are less severe when random imputation is applied. Random imputation means that a missing value is replaced by a value that is randomly chosen from the available values for the variable.

Usually synthetic values are selected by means of drawing values without replacement. This is, of course, impossible if there are more missing values than non-missing values.

The estimator

$$\bar{y}_{IMP} = \frac{m\bar{y}_R + (n-m)\bar{y}_I}{n} \tag{14.28}$$

is now composed of two random terms: the mean of the m “real” observations and the mean of the n - m “synthetic” observations.

Given the sample, the expected value of the mean of the synthetic values over the imputation distribution is equal to the mean of the real values. If this is applied to expression (14.28), the conclusion can be drawn that the expected value of the mean after imputation is equal to the expected value of the mean before imputation. Random imputation does not change the expected value of the estimator.

The computation of the variance of the estimator is now a more complex, because the second term in expression (14.23) is not equal to zero any more. The variance turns out to be equal to

$$V(\bar{y}_{IMP}) = \frac{1-\dfrac{m}{N}}{m} S^2 + \frac{(n-m)(2m-n)}{n^2 m} S^2 \tag{14.29}$$

Apparently, random imputation increases the variance of the estimator. The variance consists of two components. The first is contributed by the sampling design, and the second by the imputation mechanism.

What happens if a researcher is unaware of the fact that random imputation has been carried out. He computes the sample variance s^2 using the complete data set, and assumes this is an unbiased estimator of the population variance S^2. This assumption is wrong in case of imputation of the mean. In case of random imputation it can be shown that

$$E(s^2) = S^2\left(1 + \frac{2m-n}{n(n-1)}\right). \tag{14.30}$$

So s^2 is not an unbiased estimator of S^2, but for large samples the bias will be small. So s^2 is an asymptotically unbiased estimator.

It should be noted that random imputation affects the values of correlation coefficients. These values will generally be too low when computed using the imputed data set. This is because imputed values are randomly selected, and existing relationships with other variables are not taken into account. The same phenomenon is encountered when other single imputation techniques are applied.

14.2.6 EM IMPUTATION

There are also estimation techniques that attempt to correct estimators for a bias due to item nonresponse without computing imputed values. EM imputation is an example of such a technique.

EM imputation (or expectation-maximization imputation) uses an iterative maximum likelihood procedure to provide estimates of the mean and variance-covariance matrix based on all available data for each respondent. The algorithm assumes that the data are from a multivariate normal distribution (or possibly some other specified distribution) and that, conditional on the reported data, the missing data are missing at random (MAR).

EM imputation is a widely used technique that derives likelihood-based inferences from incomplete data (see Little and Rubin, 2002). Actually it is not a pure imputation technique because no missing values themselves are estimated.

Rather, functions of the missing values (sufficient statistics) are substituted from the likelihood function.

In each cycle of the iterative process, there is an E-step (or expectation step) followed by an M-step (or maximization step). The E-step computes expected values based on all available data. This is followed by an M-step in which missing values are replaced by synthetic values computed in the E-step. This process is continued until convergence. Under certain conditions the algorithm converges to the maximum likelihood estimate.

14.2.7 MULTIPLE IMPUTATION

Single imputation can be seen as a technique that solves the missing data problem by filling the holes in the data set with plausible values. When analyzing the data set, a researcher is no longer bothered by missing data. This is clearly an advantage. However, there are also disadvantages. When a single imputation technique is applied in a naive way, more problems may be created than solved. As was shown in the previous section, single imputation can distort the distribution of estimates. Therefore there is a serious risk of drawing the wrong conclusions from the data set. More details about this aspect of imputation can be found, for example, in Little and Rubin (1987).

To address the problems caused by single imputation techniques, Rubin (1987) proposed *multiple imputation*. In this technique each missing value is replaced by $m > 1$ synthetic values. Typically m is small, say 3 to 10. This leads to m complete data sets. Each data set is analyzed using standard analysis techniques. For each data set an estimate of a population parameter of interest is obtained. The m estimates for a parameter are combined to produce estimates and confidence intervals that incorporate missing-data uncertainty.

Rubin (1987) developed his multiple imputation technique primarily solving the missing data problem in large public-use sample survey data files and censuses files. With the advent of new computational methods and software for multiple imputation, this technique has become increasingly attractive for researchers in other sciences where researchers are confronted by missing data (see also Schafer, 1997).

The first step in multiple imputation is the construction of a model. This model is used to generate synthetic values. Let Y be the variable with some missing values, and let $X_1, X_2, \ldots, X_p$ be variables that were completely observed. The imputation model for a quantitative variable Y will often take the form of a regression model such as

$$\hat{Y}_k = B_0 + \sum_{j=1}^{p} B_j X_{kj} + E_k. \qquad (14.31)$$

A loglinear model could be used for qualitative variables. For the sake of convenience, only quantitative variables will be considered here.

The effects of imputation depend on the missing data mechanism that has generated the missing values. The most convenient situation is *missing completely at random*. This means that missingness happens for no observed reason. It is not related to any variable, known or unknown. Missingness does not cause a bias in estimates for Y. In this case multiple synthetic drawings can be produced by applying the random imputation a number of times. It is also possible to use the imputation of the mean if the variation is properly modeled. This can, for example, be done by adding a random component to the mean that has been drawn from a normal distribution with the proper variance.

MCAR is usually not a very realistic assumption. A next best assumption is that the data are *missing at random* (MAR). This means that missingness depends on one or more auxiliary variables, and these variables have been observed completely. A model like (14.31) can be used in this case. Sets of synthetic values are generated from the regression model. To give the imputed values the proper variance, usually a random component is added to the predicted value. This component is drawn from a distribution with a specific variance.

The worst case is the situation where the data are *not missing at random* (NMAR). Then missingness depends on unobserved variables, and therefore no valid imputation model can be constructed using the available data. The distribution of the estimators cannot usually be repaired by applying multiple imputation. There is still a risk of drawing wrong conclusions from the analysis.

Rubin (1987) describes how to combine estimates for multiple data sets into one proper estimate. This procedure is summarized here, thereby concentrating on estimating the population mean.

Let $\bar{y}_j$ denote the estimator of data set j (for $j = 1, 2, \ldots, m$), and let $S(\bar{y}_j)$ be the associated standard error. The overall estimator for the population mean of Y is now defined by

$$\bar{y}_{MI} = \frac{1}{m}\sum_{j=1}^{m}\bar{y}_j. \tag{14.32}$$

The variance of this estimator is equal to

$$V(\bar{y}_{MI}) = \frac{1}{m}\sum_{j=1}^{m} S^2(\bar{y}_j) + \left(1+\frac{1}{m}\right)\frac{1}{m-1}\sum_{j=1}^{m}(\bar{y}_{MI}-\bar{y}_j)^2. \tag{14.33}$$

The first term in expression (14.33) can be seen as the within imputation variance (the variation within the data sets), and the second as the between imputation variance (the variation caused by differences in imputed values).

Rubin (1987) claims that the number of imputations per missing value should not exceed $m = 10$. He shows that the relative increase in variance of an estimator based on m imputations to an estimator based on an infinite number of imputations is approximately equal to

$$\left(1+\frac{\lambda}{m}\right), \tag{14.34}$$

where λ is the rate of missing information. For example, with 50% missing information ($\lambda = 0.5$), the relative increase in variance of an estimator based on $m = 5$ imputations is equal to 1.1. This means that the standard error will only be 5% larger.

Multiple imputation can be a useful tool for handling the problems caused by missing data, but it has to be done with care as this has the potential to be misleading. If an imputation does not model the missing data mechanism properly, analysis of the imputed data sets can be seriously flawed. This means that the models used should always be checked.

14.3 Application

Some of the concepts introduced in this chapter were applied to a real survey data set. This data set was based on a Dutch survey that was carried out by Statistics Netherlands. It is called the General Population Survey (GPS).

Persons were selected by means of an equal probability sample. The sample of the GPS was selected form the population register in The Netherlands. The sample consisted of 32,019 persons. The number of respondents was 18,792.

Statistics Netherlands has an integrated system of social statistics. This system is called the *Social Statistics Database* (SSD). The SSD contains a wide range of characteristics on each individual in The Netherlands. There are data on demography, geography, income, labor, education, health, and social protection. SSD records have been linked to the survey data records using a personal identification number. This has been done for both respondents and nonrespondents. Thus demographic variables like gender, age, province of residence, and ethnicity are available for all sampled persons, and also socioeconomic variables like employment and various types of social security benefits.

One of these socioeconomic variables (*Socall*) will be treated as a target variable in this application. This variable records whether or not a person receives a social allowance. Its value is 1 if a person receives a social allowance, and otherwise, its value is 0. Table 14.1 summarizes the sample results for this variable.

Nonresponse causes a bias. There are more people with a social allowance among the nonrespondents than among the respondents. Using only response data, some imputation techniques are applied. Estimates based on imputed data sets can then be compared with the sample percentage.

The GPS data set only contains qualitative variables. Hence some imputation techniques cannot be applied. For example, imputation of the mean would result in a value between 0 and 1 (0.029). This value is outside the domain of valid answers (0 or 1). To avoid these problems, only two random imputation techniques will be applied.

TABLE 14.1 Having a social allowance in the GPS

Data set	Percentage having social allowance
Response	2.9%
Nonresponse	6.0%
Sample	4,2%

TABLE 14.2 Having a social allowance in the GPS after random imputation

Data set	Percentage with social allowance
Response	2.9%
Nonresponse	2.8%
Sample	2,9%

The first imputation technique is simple, straightforward random imputation. For every person k with a missing value of social allowance, a random value $\hat{Y}_k$ is drawn from a distribution with $P(\hat{Y}_k = 1) = 0.029$ and $P(\hat{Y}_k = 0) = 1-0.029 = 0.971$. By using these probabilities, the distribution of the variable in the response is not disturbed. The expected value remains the same. The result is summarized in Table 14.2. Note that random imputation did not reduce the bias. The sample based estimate is now 2.9%, and this is far away from the true sample value 4.2%.

Random imputation may affect relationships between variables. This is shown in Table 14.3. The relation between the variables social allowance (yes, no) and job (yes, no) is explored. The difference between percentages before imputation (4.1%) is larger than after imputation (2.6%). Hence the relationship between the two variables is stronger before than after imputation. This a typical effect of random imputation. Since the imputation technique does not take into account relationships between variables, relationships are always weaker after imputation. Note that the true percentages of the complete sample are even more apart. It is clear that random imputation does not help to get a better idea of the relationship between the two variables.

A better imputation approach is to construct a model able of predicting social allowance using the available auxiliary variables. If such a model can be found, it can be used to predict the missing values of social allowance. Since the variable social allowance can only assume the values 1 and 0, a logit model is appropriate. It turns out that seven variables have sufficient explanatory power: age, degree of urbanization, average house value in the neighborhood, job (yes, no), household type, married (yes, no), and percentage of nonnatives in the neighborhood. The value of Nagelkerke R^2 is then equal to 0.404. This means that the model has a reasonable amount of explanatory power.

TABLE 14.3 Relation between having a social allowance and having a job before and after random imputation

	Percentage with social allowance		
	Before imputation	After imputation	True values
Job	1.0%	1.6%	1.3%
No job	5.1%	4.2%	7.2%

TABLE 14.4 Having a social allowance after imputation with a logit model

Data set	Percentage with social allowance
Response	2.9%
Nonresponse	5.4%
Sample	3.9%

TABLE 14.5 Relation between having a social allowance and having a job before and after random imputation

	Percentage with social allowance		
	Before imputation	After imputation	True values
Job	1.0%	1.2%	1.3%
No job	5.1%	6.7%	7.2%

For every person k with a missing value of social allowance the probability of having a social allowance was estimated using the logit model. This estimate is denoted by $\hat{\rho}_{SA,k}$. Next, a random value $\hat{Y}_k$ was drawn from a distribution with $P(\hat{Y}_k = 1) = \hat{\rho}_{SA,k}$ en $P(\hat{Y}_k = 0) = 1 - \hat{\rho}_{SA,k}$. Table 14.4 shows the results for the estimator after imputation. The situation has improved with respect to random imputation. The estimate based on the complete (imputed) sample is now 3.9 and it is closer to the true value of 4.2.

Table 14.5 shows the relationship between the two variables social allowance and job before and after imputation with the logit model. It is clear that this imputation technique does a much better job than random imputation. The percentages of people with a social allowance are now much closer to the true values.

14.4 Summary

Nonresponse occurs when persons in the sample from the population and who are eligible for the survey do not provide the requested information, or provide information that is not usable.

Item nonresponse occurs when some questions have been answered but no answer is obtained for some other, possibly sensitive, questions. So the questionnaire form is only partially completed. Item nonresponse can occur when persons refuse to answer a question (e.g., because they do not want to answer a sensitive question) or when persons do not know the answer. If a paper questionnaire is used for data collection, same persons could have accidentally skipped a question. Item nonresponse can also occur as a consequence of a data editing procedure, when an incorrect value is detected and no correct value is available.

Imputation means that each missing value is replaced by a synthetic value. This synthetic value is obtained as the result of some technique that estimates the missing values. There are many imputation techniques, so selecting a technique depends on a number of factors:

- Some techniques, like imputation of the mean, can only be used for quantitative variables, while others, like random imputation, can be used both for quantitative and qualitative variables.
- Some techniques, like imputation of the mean, affect the distribution of the variable. As a consequence estimates of standard errors may be wrong.
- Imputation may lead to underestimation of the strength of relationships between variables. This happens when the imputation technique does not take into account such relations.

Imputation is most effective when values are missing at random (MAR) and the relevant auxiliary variables are used in the imputation model.

If values are not missing at random (NMAR), a bias due to nonresponse may not be reduced. In this case, Imputation is a technique that just fills the holes in the data set with values that are as close as possible to the true values.

14.5 KEY TERMS

Deductive imputation An estimation technique by which a missing value is replaced by the true value, which is computed exactly using all known information.

Donor imputation Any form of imputation where the imputed value is copied from another record in which the value is not missing.

EM imputation An estimation technique that directly estimates means and variances of estimators, without first computing individual synthetic values. The EM algorithm consists of repeated E (expectation) and M (maximization) steps that converge to a maximum likelihood estimate.

Imputation A group of techniques that replace missing values by synthetic values. The synthetic value is obtained as the result of some technique that attempts to estimate the missing values.

Imputation of the mean The mean of the available values of the variable used to replace a missing value.

Imputation of the mean within groups Imputation of the mean as applied to a population divided into groups, in each group separately.

Item nonresponse Nonresponse that occurs when some questions have been answered but no answer is obtained for some other, possibly sensitive, questions. So the questionnaire form has been partially completed.

Multiple imputation A missing value is replaced by a (small) set of synthetic values, resulting in a set of imputed data sets. The properties of the estimators are determined by repeating the analysis for all the data sets.

Nearest neighbor imputation The imputed value obtained by taking the value from another record that resembles the record with the missing value as much as possible.

Random imputation A missing value replaced by a value that is selected at random from the set of available values of the variable.

Random imputation within groups The population is divided into groups, and random imputation is applied in each group separately.

Ratio imputation A missing value is replaced by a value that is predicted by a simple regression model without a constant term and with one quantitative explanatory variable.

Regression imputation A missing value is replaced by a value that is predicted by a regression model.

14.6 REFERENCES

Little, R.J.A., and Rubin, D.B. (1987), *Statistical Analysis with Missing Data.* Wiley, New York.

Little, R.J.A., and Rubin, D. (2002), *Statistical Analysis with Missing Data*, 2nd ed. Wiley, New York.

Rubin, D.B. (1987), *Multiple Imputation for Nonresponse in Surveys.* Wiley, New York.

Schafer, J.L. (1997), *Analysis of Incomplete Multivariate Data.* Chapman and Hall, London.

14.7 EXERCISES

Exercise 14.1 Why should pair-wise deletion not be applied?

(a) Too much information is thrown away.

(b) Estimates for different parameters may not be comparable.

(c) It will lead to biased estimates.

(d) Adjustment weighting works better.

Exercise 14.2 What is a disadvantage of imputation of the mean?

(a) An extra source of variation is introduced.

(b) Variances often turn out to be too small.

(c) Variance estimates often turn out to be too small.

(d) The bias of estimates is increased.

Exercise 14.3 If there is a linear relationship between an auxiliary variable and a target variable, which imputation technique should be used?

(a) Ratio imputation.

(b) Regression imputation.

(c) Both ratio imputation and regression imputation work equally well.

(d) Neither of these techniques should be used.

Exercise 14.4 What happens to the correlation between target variabele Y and auxiliary variable X in case of imputation of the mean?

(a) The correlation will be smaller.

(b) The correlation will be larger.

(c) The correlation will remain unchanged.

(d) A positive correlation will become negative, and vice versa.

Exercise 14.5 Which of the four statements about multiple imputation is true?

(a) The data set will become at least three times bigger.

(b) The distribution of estimators after imputation will be better than before imputation.

(c) The number of imputations per missing value should not exceed 10.

(d) It does not affect the variance of estimates.

Exercise 14.6 The general imputation model can be written as

$$\hat{Y}_k = B_0 + \sum_{j=1}^{p} B_j X_{kj} + E_k.$$

Which values of the parameters have to be used in order to obtain imputation of the mean as special case?

(a) Take all B_j and E_j equal to 0, except B_0.

(b) Take all B_j equal to 0, except B_0.

(c) Take all B_j equal to 0.

(d) Just take all X_{kj} equal to 0.

Exercise 14.7 The ministry of agriculture in a country wants to have more information about the manure production by pigs on pig farms. Target variable is the yearly manure production per farm. Among other variables recorded are the number of pigs per farm and the region of the country (north or south). The table below contains part of the data:

Farm	Manure production	Region	Number of pigs
1	295,260	North	220
2	259,935	North	195
3	294,593	North	221
4	253,604	North	188
5	?	North	208
6	520,534	South	398
7	?	South	435
8	559,855	South	375
9	574,528	South	416
10	561,865	South	405

The value of manure production is missing in two records due to item nonresponse. Describe six imputation techniques for replacing the missing value by a synthetic value. Compute for each technique which values are obtained. Explain the consequences for estimating the mean manure production per country. Indicate whether these consequences are acceptable.

Exercise 14.8 It is assumed there is a relationship between the energy consumption of a house and its total floor space. A simple random sample (with equal probabilities and without replacement) of four houses has been selected. The table below contains the collected data:

House	1	2	3	4
Floor space (m^2)	116	81	73	99
Gas consumption (m^3)	1200	950	650	1050
Electricity consumption (kWh)	1715	1465	1020	—

It is a known fact that the mean floor space of all houses in the target population is equal to 103.7 m^2.

(a) Apply ratio imputation to impute the missing value of electricity consumption of house 4.

(b) Apply regression imputation to imputed the missing value of electricity consumption of house 4

Exercise 14.9 A town council has carried out an income survey among its inhabitants. A simple random sample of 1000 persons was selected. The total size of the population is 19,000. All selected persons were asked to state their net monthly income. From previous research it has become clear that the standard deviation of the net monthly income is always equal to 600.

(a) Assuming all sampled persons cooperate and state their income data, compute the standard error of the sample mean.

Suppose that 10% of the sampled persons do not provide their income data. The researcher solves this problem by imputation of the mean.

(b) Compute the standard error of the mean after imputation.

Suppose the sample standard deviation before imputation happens to be equal to the population standard deviation (600). The imputed survey data set is made available to a researcher. He does not know that imputation has been carried out.

(c) If he computes the estimated standard error, what value would he get?

(d) Which conclusion can be drawn from comparing the results of parts a and c?

CHAPTER FIFTEEN

Miscellaneous Topics

15.1 Introduction

The previous chapters devoted attention to several topics concerning nonresponse in household surveys. The focus lies on unit nonresponse as the only source of error in surveys and the description of analysis and adjustment is restricted to cross-sectional research using information from auxiliary sources. In this chapter a number of situations are discussed that require a different approach to nonresponse.

In practice, a survey will suffer from both item and unit nonresponse. Therefore in Section 15.2.1 methods to deal with a combination of item and unit nonresponse are presented.

Panel surveys have become a popular way of collecting data with the introduction of the Internet as a data collection mode. A special type of nonresponse occurs when sample elements drop out a panel survey. This is the topic of Section 15.2.2.

A key element of the analysis of nonresponse and successful adjustment for nonresponse bias is the availability of good auxiliary information about respondents and nonrespondents. This information can come from external administrative sources. But also information about the data collection process can provide useful insights into the response process. This type of data is the topic of Section 15.2.3.

The weighting technique that is most commonly used to adjust for nonresponse bias is the generalized regression estimator, or GREG estimator. In Section 15.2.4 it is described how this method can be used to estimate a consistent

Handbook of Nonresponse in Household Surveys, By Jelke Bethlehem, Fannie Cobben, and Barry Schouten

set of tables. This method of repeated weighting has been developed for use in the Dutch Virtual Census. This is not a real census in the way that every element of the Dutch population is surveyed, but information on the entire Dutch population is obtained by combining data from registers and surveys.

15.2 Theory

15.2.1 COMBINED TREATMENT OF UNIT AND ITEM NONRESPONSE

Most of the chapters in this book deal with unit nonresponse. The methods described in Chapters 8, 11, and 12 can be applied to adjust for unit nonresponse. They deal uniquely with unit nonresponse. Chapter 14 covered methods used to adjust for item nonresponse, and exclusively for item nonresponse. In practice, however, surveys typically have items missing due to both unit and item nonresponse. Therefore this section shows how the treatment of unit and item nonresponse can be combined.

In general, to correct for a unit nonresponse, a weighting technique is applied. Respondents are assigned weights in a way that represents the sample or the population. For item nonresponse, it is common practice to impute the missing values. The missing values are replaced by estimates.

Item nonresponse can occur for the target variables of a survey. Sensitive questions are known to run a higher risk of item nonresponse than nonsensitive questions. However, also the auxiliary variables can have missing items. The weighting techniques used to adjust for a unit nonresponse bias rely heavily on the availability of good auxiliary information. The perfect auxiliary information has three properties (Särndal and Lundström, 2008): (1) it is related to the response indicators R, (2) it is related to the survey target variables Y, and (3) it identifies the domains on which population characteristics are published. In addition the auxiliary information should not have any missing items; hence it should be free of item nonresponse since in that case the auxiliary variable cannot be used for any unit nonresponse adjustment.

There are three possible ways to combine the treatment of item and unit nonresponse. The first is to apply a weighting method for each of the missing target variables, regardless of whether the missingness is caused by item or unit nonresponse. This approach results in a different set of weights for each target variable. The second approach is to fully impute the missing values. Hence, not just items missing due to item nonresponse, but also the items missing due to unit nonresponse, are replaced by estimates. This approach results in a rectangular dataset, without missing values. The third approach consists of a combination of weighting and imputation. First, the items that are missing due to item nonresponse are imputed. Then, a weighting technique is applied to adjust for unit nonresponse.

Särndal and Lundström (2005) describe the different approaches. First of all, some notation is introduced. Let the target population of the survey, denoted U, consist of N elements. Then select a probability sample s of size n from the population. Assume that the inclusion probabilities π_k for $k = 1, 2, \ldots, n$ are known and nonzero. The sampling weight than is $d_k = 1/\pi_k$. Among target variables in the survey, consider a target variable Y. Suppose that the objective of the survey is to estimate population characteristics, measured by target variable Y. For instance, to estimate the population mean $\bar{Y}$, the sample mean $\bar{y} = (y_1, y_2, \ldots, y_n)/n$ could be used (in the case of simple random sampling), where $y_1, y_2, \ldots, y_n$ denote the n observations that have become available in the sample.

If y_i is observed for every sample element $i \in s$, there is complete response for variable Y. In general, however, there will be missing values for Y due to both item and unit nonresponse. Let r denote the subset of respondents to the survey. This subset consists of all the sample elements that have responded to at least one of the target variables of the survey. The unit nonrespondents belong to the subset of the sample that did not participate in the survey, namely $s - r$.

Now denote the subset of sample elements that have responded to the particular target variable Y by r_y. This is the *item response subset*. Hence Y_i is observed for every sample element i in r_y, and $r_y \subseteq r \subseteq s$.

If y_i is missing and imputed, the imputed value is denoted by $\hat{y}_i$. Imputation methods that can be applied to obtain $\hat{y}_i$ are described in Chapter 14. Here any of the imputation methods can be used. In the following sections, the three approaches as identified by Särndal and Lundström (2005) are presented.

15.2.1.1 Full Imputation The full imputation approach to the combined treatment of item and unit nonresponse is to impute all the missing values regardless of whether they are missing due to item nonresponse or unit nonresponse. This approach results in a full rectangular data set that can be described by

$$y_{\bullet i} = \begin{cases} y_i & \text{for } i \in r_y, \\ \hat{y}_i & \text{for } i \in s - r_y. \end{cases} \tag{15.1}$$

Hence $y_{\bullet i}$ is either the observed value y_i when sample element i responded, or $y_{\bullet i}$ is the imputed value $\hat{y}_i$ when sample element i did not respond.

Because $y_{\bullet i}$ is observed for every sample element, estimates for population characteristics of interest, such as the population mean $\bar{y}$, can now be computed by applying estimators for the full response case.

The weighting estimator for the full response situation, described in Chapter 2, is

$$\bar{y}_W = \frac{1}{N} \sum_{i \in s} w_i y_i, \tag{15.2}$$

where the adjustment weights are $w_i = d_i \times g_i$ with d_i the design weights and g_i a correction weight, often referred to as g-weight, produced by a weighting adjustment method. For instance, the Horvitz–Thompson estimator is obtained by setting the correction weights g to 1 for all elements in the sample:

$$\bar{y}_{HT} = \frac{1}{N}\sum_{i=1}^{n} d_i y_i. \tag{15.3}$$

Based on the imputed data, the Horvitz–Thompson estimator (15.3) can be described as

$$\hat{\bar{y}}_{IHT} = \frac{1}{N}\sum_{i=1}^{n} d_i y_{\bullet i} = \frac{1}{N}\sum_{i\in r_{ir}} d_i y_i + \frac{1}{N}\sum_{i\in s-r_{ir}} d_i \hat{y}_i. \tag{15.4}$$

To indicate the difference from the full response case, the estimator based on the imputed data $y_{\bullet i}$ is referred to as the *imputed Horvitz–Thompson estimator,* hence the subscript *IHT.*

The GREG-estimator is obtained by

$$g_i = 1 + (\bar{X} - \bar{x}_{HT})' \left(\sum_{i\in s} d_i X_i X'_i \right)^{-1} X_i. \tag{15.5}$$

In the expression for the g-weights, $\bar{X}$ is a vector of population means for the auxiliary variables and $\bar{x}_{HT}$ is the Horvitz–Thompson estimator of the population means based on the sample.

The *imputed GREG-estimator* can be described as

$$\hat{\bar{y}}_{IGREG} = \frac{1}{N}\sum_{i=1}^{n} d_i g_i y_{\bullet i} = \frac{1}{N}\sum_{i\in r_{ir}} d_i g_i y_i + \frac{1}{N}\sum_{i\in s-r_{ir}} d_i g_i \hat{y}_i, \tag{15.6}$$

with g_i given by (15.5).

Sometimes the y-variables are used in imputation methods. This may complicate the full imputation approach considerably because that means that for every target variable a different number of variables is used.

In the full imputation approach, the weighting model of the full response estimator is used for the imputed estimators. This means that element k receives the same weight regardless of whether it is imputed or observed. It is arguable whether imputed and observed values should receive different treatments in the weighting process. Point estimation is simple since the weights remain unchanged; however, variance estimation is more difficult (see Särndal and Lundström, 2005).

15.2.1.2 Combined Approach The combined approach is more widely used than the full imputation approach. In the combined approach, first the item nonresponse is imputed. This first step results in a full, rectangular data set for the subset of respondents that can be described by

$$y_{\bullet i} = \begin{cases} y_i & \text{for } i \in r_y, \\ \hat{y}_i & \text{for } i \in r - r_y. \end{cases} \tag{15.7}$$

For every element i in the response subset r the values for the target variables are known. $y_{\bullet i}$ is either the observed value y_i, when sample element i is responding, or $y_{\bullet i}$ is the imputed value $\hat{y}_i$, when there is item nonresponse for sample element i. The item nonresponse method that is used to compute $\hat{y}_i$ can be any of the methods described in Chapter 14.

Next weighting is used to adjust for unit nonresponse. The weighting estimator for the full response situation given by (15.2) can be adapted to the situation of nonresponse (see also Chapter 8). Equation (15.2) is modified to

$$\bar{y}_W^r = \frac{1}{N}\sum_{i\in r} w_i y_i. \tag{15.8}$$

Choosing $g_i = 1/\rho_i$, where ρ_i is the (unknown) response probability for sample element i, leads to a modified version of the Horvitz–Thompson estimator (15.3):

$$\bar{y}_{HT}^r = \frac{1}{N}\sum_{i\in r} \frac{d_i y_i}{\rho_i}. \tag{15.9}$$

Based on the imputed data, the modified Horvitz–Thompson estimator (15.9) can be described as

$$\hat{\bar{y}}_{IHT}^r = \frac{1}{N}\sum_{i\in r} \frac{d_i y_{\bullet i}}{\rho_i} = \frac{1}{N}\sum_{i\in r_{ir}} \frac{d_i y_i}{\rho_i} + \frac{1}{N}\sum_{i\in r - r_{ir}} \frac{d_i \hat{y}_i}{\rho_i}. \tag{15.10}$$

This estimator is referred to as the *imputed modified Horvitz–Thompson estimator*. Choosing the g-weights to be

$$g_i = 1 + (\bar{X} - \bar{x}_{HT}^r)'\left(\sum_{i\in r} d_i X_i X'_i\right)^{-1} X_i \tag{15.11}$$

leads to the modified GREG estimator. Based on the imputed data, the modified GREG estimator can be described as

$$\hat{\bar{y}}_{IGREG}^r = \frac{1}{N}\sum_{i\in r} d_i g_i y_{\bullet i} = \frac{1}{N}\sum_{i\in r_{ir}} d_i g_i y_i + \frac{1}{N}\sum_{i\in r - r_{ir}} d_i g_i \hat{y}_i, \tag{15.12}$$

with the g-weights as given by (15.11). Estimator (15.12) is referred to as the *imputed modified GREG-estimator.*

The combined approach has wider use than the full imputation approach. Särndal and Lundström (2005) describe the combined approach in more detail.

15.2.1.3 Full Weighting An alternative to full imputation and the combined approach of imputation and weighting discussed in the previous sections is full weighting. In full weighting all the missing target variables are weighted and no imputation is used.

In the case of the weighting estimator for nonresponse given by (15.8), the g-weights are now different for each item response subset, denoted by g_i^y, so that weights w_i become $w_i = d_i \times g_i^y$. Consequently (15.8) can be described as

$$\bar{y}_W^{r_y} = \frac{1}{N} \sum_{i \in r_{ir}} d_i g_i^y Y_i. \tag{15.13}$$

In the case of the Horvitz–Thompson estimator, g_i^y equals 1 for all sample elements and the *full weighted modified Horvitz–Thompson estimator* can be described as

$$\bar{y}_{FHT}^{r_y} = \frac{1}{N} \sum_{i \in r_{ir}} d_i y_i. \tag{15.14}$$

In the case of the GREG estimator, the g-weights can be expressed as

$$g_i^y = 1 + (\bar{X} - \bar{x}_{HT}^{r_y})' \left(\sum_{i \in r_y} d_i X_i X'_i \right)^{-1} X_i, \tag{15.15}$$

where $\bar{x}_{HT}^{r_y}$ is the Horvitz–Thompson estimator for the mean of the auxiliary variables based on the item response subset. The *full weighted modified GREG estimator* is given by

$$\bar{y}_{FGREG}^{r_y} = \frac{1}{N} \sum_{i \in r_{ir}} d_i g_i^y y_i \tag{15.16}$$

For $r_y = r$ there is only unit nonresponse, and weighting estimator (15.8) can be used.

For every item response subset r_{ir}, the responses y_i are weighted. This results in a different set of weights for each target variable when either the response subsets for the y's are different or when different weighting models are used. In practice, this is not desirable. In addition, if full weighting is used to estimate cross-tabulations, a new set of weights is produced for each cross-tabulation. For this reason the combined approach is often preferred over the full weighting approach.

15.2.1.4 Item Nonresponse in Auxiliary Variables So far the descriptions of the approaches have been limited to missing values for target variables. However, sometimes values for the auxiliary variables are missing too. That poses no problems for the modified Horvitz–Thompson estimator, since in this estimator no auxiliary information is used. However, the GREG estimator does make use of auxiliary information. If the auxiliary information is incomplete, it cannot be used for the bias adjustment.

From the expression for the g-weights it should be clear that the individual values for the auxiliary variables need to be known. One solution to this problem is to impute the missing values of the auxiliary variables. Weighting is not possible since for the auxiliary variables a full rectangular data set is needed for the entire sample.

Say that the value for auxiliary variable X_i is observed for a subset of the sample denoted by r_x. Hence the value for X_i is missing for $s - r_x$. Imputing this variable results in a full rectangular data set described by

$$X_{\bullet i} = \begin{cases} X_i & \text{for } i \in r_x, \\ \hat{X}_i & \text{for } i \in s - r_x. \end{cases} \tag{15.17}$$

Hence $X_{\bullet i}$ is either the observed value X_i, when sample element $i \in r_x$, or $X_{\bullet i}$ is the imputed value $\hat{X}_i$, when sample element $i \notin r_x$.

This results in imputed g-weights that can be expressed as

$$\hat{g}_i = 1 + (\bar{X} - \hat{\bar{x}}_{HT})' \left(\sum_{i \in s} d_i X_{\bullet i} X'_{\bullet i} \right)^{-1} X_{\bullet i}. \tag{15.18}$$

In the expression for the g-weights, $\bar{X}$ is a vector of population means for the auxiliary variables and $\hat{\bar{x}}_{HT}$ is the Horvitz–Thompson estimator of the population means based on $X_{\bullet i}$.

These g-weights can be inserted in the different expression for the full imputation, combined, or full weighting approach.

15.2.2 NONRESPONSE IN LONGITUDINAL STUDIES

Missing data in surveys are usually considered to be of two types: unit nonresponse, where all the survey data are missing because the sample element did not participate in the survey, or item nonresponse, where one or more survey target variables are missing because the sample element did not complete the survey. In the case of unit nonresponse, the only information available about the sample element may be information from the sampling frame. For item nonresponse, there is information about the other questions in the survey that the sampled population member did answer.

The surveys discussed in this book so far were *cross-sectional studies*. A sample is selected from the population and approached with a request to participate in a one-time survey. The sample elements are hence observed at the same point in time. A one-time cross-sectional study takes a snapshot of the population and can only be used to describe the population at that one point in time. The cross-sectional studies performed by statistical offices can best be described as *rolling cross-sectional studies*. These are cross-sectional surveys that are performed continuously so that the population can be described continuously too. A different sample is selected each time. This type of survey can hence be used to analyze changes in variables over time at an aggregated population level.

Longitudinal studies are surveys where repeated observations are made on the same sample elements over a period of time. A *panel survey* is a longitudinal study where a sample is selected from the population, and this same sample is surveyed at regular time intervals. Each sample element is asked to participate in a survey at more than one point in time. Each period of data collection is called a *wave*. As in a cross-sectional study, the data from panel surveys can also be used to analyze changes in variables over time. In addition the survey data can be used to analyze differences in variables among the sample elements.

There are two distinct situations of missing data in longitudinal surveys, as shown in Figure 15.1. *Panel attrition* or *dropout* occurs when sample elements drop out prior to the end of the study and do not return. In this case one or more waves of panel data would be missing for a sample element that participated at least once. Dropout usually refers to a monotone missing data pattern. But if a sample element drops out of the study at one point and is found willing to participate in a later wave, the data follows an arbitrary missing data pattern.

Dropout follows a *monotone* missing data pattern (Figure 15.1*a*). The target variables (or groups of target variables) $Y_1, \ldots, Y_p$ may be ordered in such a way that if Y_j is missing for an element, then $Y_{j+1}, \ldots, Y_p$ are missing as well. Wave nonresponse (Figure 15.1*b*) follows an *arbitrary* missing data pattern. There is no structure or ordering in the missingness of the values of the target variables.

Wave nonresponse can be regarded as unit nonresponse. Therefore the remainder of this section will focus on dropout. Suppose the dropout occurs at time *t*. In this case the target variables can be ordered in such a way that if Y_t is missing for a sample element, $Y_{t+1}, \ldots, Y_p$ are missing as well.

There is usually less information available for dropouts than is the case with item nonresponse, but more than for unit nonrespondents because data from a previous wave are available. As with item and unit nonresponse, the most important question is whether the dropout is selective. In other words, are sample elements that drop out of the panel survey systematically different from sample elements that remain in the panel? If they are not different, then dropout will not result in a bias. If

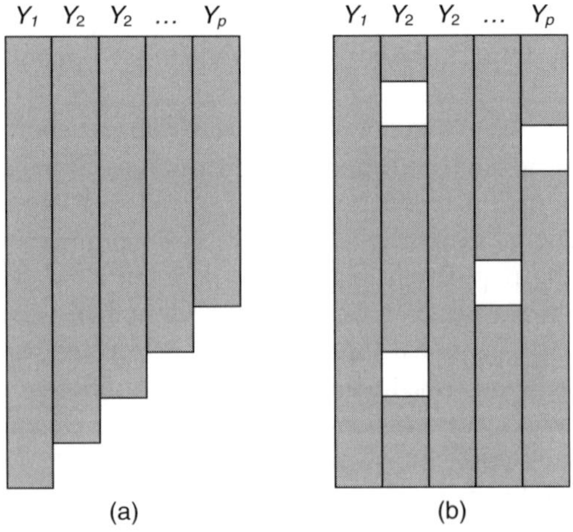

FIGURE 15.1 Missing data patterns in longitudinal studies.

they are different, there will be a bias that, depending on the mechanism of the missing data, can be adjusted for using auxiliary information.

The dropout can be missing completely at random (MCAR), missing at random (MAR), or not missing at random (NMAR). These mechanisms were described for the situation of unit nonresponse in Chapter 2. In addition to the auxiliary variables *X*, now also the target variables *Y* from previous waves, may be regarded to determine the nature of the missing data mechanism.

In the case of MCAR, dropout is caused by a phenomenon that is completely unrelated to auxiliary variables and target variables *Y* in the present wave and previous waves. The dropout is harmless. It only reduces the sample size. Estimates for parameters involving *Y* will not be biased

In the case of MAR, the missingness due to dropout is caused partly by an independent phenomenon and partly by the auxiliary variables *X* and the target variables from previous waves. So there is an indirect relationship between target variables *Y* in the present wave and the dropout. This leads to biased estimates for *Y*. It is possible to correct for such a bias by using a technique that takes advantage of the availability of all values of *X* and target variables from previous waves, both for respondents and dropouts.

In the case of NMAR, there may be a relationship between an independent phenomenon and dropout and between auxiliary variables *X* and target variables from previous waves and dropout, but there is also a direct relationship between the target variable *Y* in the present wave and dropout that cannot be accounted for by *X* or the target variables from previous waves. This situation also leads to biased estimates for *Y*. Correction techniques using *X* or target variables from previous waves will not be able to remove the bias.

Fitzmaurice, Laird, and Ware (2004) describe a number of ways to handle dropout. These will be discussed in the following sections.

15.2.2.1 Complete Case Analysis One way to handle dropout is to simply disregard all the information from a sample element that drops out of the panel survey, that is, auxiliary information as well as information from previous waves in the panel. In doing so, only those sample elements that provided all the information are analyzed. A lot of information is not used.

In this approach there is a high risk of a dropout bias. Only when the dropout is MCAR (i.e., the remaining respondents are a random subsample of the complete sample) will there be no bias. In practice, this is hardly ever the case. Besides, analysis based on completed cases alone also reduces the sample size and hence leads to an increased variance or reduced power of statistical tests. Consequently this approach is unappealing in practice.

15.2.2.2 Available Data Analysis In this approach the information that is only partially available for dropouts is also used in the analysis. Therefore this approach should be preferred over the complete case analysis. However, for the available data analysis to yield unbiased results the dropout needs to be MCAR, the same as in the complete case analysis. In practice, this property is hardly ever satisfied, which makes this approach also unappealing.

15.2.2.3 Imputation The most common way to handle dropouts in panel surveys is by imputation. Imputation of missing responses following dropout is referred to as *longitudinal imputation*. Longitudinal imputation is different from the imputation described in Chapter 14 in that it is possible to use observations on the target variables from previous waves in which the dropouts did participate.

When there is dropout in a longitudinal survey, it is profitable to use the longitudinal character of the survey instead of just the cross-sectional data. First of all, previous observations of the same sample element will be highly correlated with the target variables in the wave in which the dropout occurs and hence make good predictors for the missing value. The quality of the imputation can hence be improved compared to the situation of cross-sectional data. Most of the methods discussed in Chapter 14 can be applied for this purpose, by using target variables from the previous wave(s) as auxiliary information in the imputation models. A second reason is that the longitudinal data, in general, is used to analyze changes in target variables over time. To be able to do this in a correct way, one needs to take into account the previous values of the target variables when imputing missing values.

Fitzmaurice, Laird, and Ware (2004) discuss a number of methods that can be used for longitudinal imputation. One method that is widely used is the *last observation carried forward* (LOCF). This method imputes the missing observation using the last observed value for the dropout element. There are strong assumptions underlying this method, the strongest being that the behavior of the dropout with respect to the target variables in the survey is constant after dropout and remains unchanged. This assumption is most credible in clinical trial studies, where elements may drop out of a study after they have recovered and do not need clinical care anymore, or elements drop out because they die. This method is not used in official statistics.

When the last observation is adjusted for changes over time, the method of *ratio imputation* is applied. This method was discussed in Chapter 14, and it is frequently used when it is reasonable to assume that the value of the target variable Y at time t is proportional with the value of the target variable Y at time $t + 1$.

Another method is *hot deck imputation*. This is a form of donor-imputation. One donor is assigned for a dropout and all the missing values are imputed with the values from the donor. This method is profitable to use when more than one variable is missing. In longitudinal surveys all target variables are missing for the dropouts, so this method is particularly suitable for use in longitudinal surveys. Consistent imputed values that conserve the correlations between observations over time can be obtained by this method.

Regression imputation was also discussed in Chapter 14. Longitudinal data complicates the situation but has the advantage of producing good predictors for the missing values. In Chapter 14 the situation is discussed where there is one missing target variable Y that needs to be imputed using a number of auxiliary variables X. In a longitudinal survey there are more observations on Y for each sample element. So a bivariate model could be fitted for every target variable that is missing, with Y_t modeled with auxiliary variables X as well as previous observations on Y. This method can become very complex, and additionally it does not account for the relationship between target variables. This can be solved by fitting a multivariate model for all the target variables simultaneously. In the multivariate model, for instance a linear model, the correlation among the target variables can be included in the model specification, hence accounting for the relationship between variables (for further reading, see Verbeke and Molenberghs, 2000, Molenberghs and Verbeke, 2005).

15.2.2.4 Weighting In general, the dropout problem in longitudinal surveys can also be resolved by applying the weighting adjustment method. When the statistic of interest describes the state of the population at a certain time, the dropout can be regarded as unit nonresponse. The underrepresentation of certain groups in the observed data can then be taken into account and adjusted for by computing an appropriate weight for the respondents.

One way to do this is by inverse probability weighting, as described in Chapter 8 and Chapter 11. For this purpose the probability of remaining in the sample needs to be estimated instead of the probability of response. The idea behind weighting in the case of dropout is that the estimation is based on data from the respondents that completed the study. The data from these respondents is then weighted to account for the probability of remaining in the panel.

Hence the probability of remaining in the panel needs to be estimated. This can be done using auxiliary information X, the same as in a cross-sectional survey, and the observed target variables from previous waves of the survey can be used as well. The survey estimates are based on the sample elements that completed the entire study, and hence the probability of interest is the probability of remaining in the panel.

Suppose that there are m waves in the survey. Then the dropout can be described by a dropout indicator D_i for every sample element i. $D_i = j$ if sample element i drops out between wave $(j - 1)$ and wave j. In that case, only the first $j - 1$ responses are observed. The probability of remaining in the sample untill the end of the study is

$$P(D_i = m+1) = \delta_i = (1-\partial_{i1}) \times (1-\partial_{i2}) \times \ldots \times (1-\partial_{im}), \qquad (15.19)$$

where $\partial_{id} = P(D_i = d|D_i \geq d)$ is the probability that sample element i drops out between wave $(d - 1)$ and wave d, given that element i did not drop out before wave d. ∂_{id} can be estimated based on the sample elements remaining at wave $(d - 1)$, given the recorded data up to wave $(d - 1)$. These probabilities can be estimated using, for example, a logit or a probit model.

The weight for the adjustment is then $1/\hat{\delta}_i$. When the dropout is MAR, the weighted observations can be used for a complete case analysis.

Fitzmaurice, Laird, and Ware (2004) note that this general approach to handling dropout can be made more efficient by regarding all the available data. Denote by δ_{id} the probability that sample element i is still in the sample at the dth wave. Then

$$\begin{aligned}\delta_{i1} &= (1-\partial_{i1}),\\ \delta_{i2} &= (1-\partial_{i1}) \times (1-\partial_{i2}),\\ &\vdots\\ \delta_{im} &= (1-\partial_{i1}) \times (1-\partial_{i2}) \times \cdots \times (1-\partial_{im}).\end{aligned} \tag{15.20}$$

Now the available data at wave d are weighted by $1/\hat{\delta}_{id}$.

15.2.3 PARADATA

Couper (1998) introduced the concept of *paradata*. Paradata are defined as auxiliary data about the process of data collection in both the contact and the response phase (see also Couper and Lyberg, 2005). Therefore often the term *process data* is used to indicate the same type of data.

Paradata are a by-product of the data collection process. A number of levels on which paradata can be collected are identified by Kreuter (2006). She distinguishes between *contact protocol data* (CPD) and data about the *interview process* (IP). The contact protocol data are only available for interviewer-assisted surveys. This type of data can be divided further into two types, the first of which is collected for every element in the sample that has been allocated to an interviewer (CPD I). These data contain information about each contact attempt, such as day, time, mode, and the outcome of each attempt. Also aggregate information about the contact process fall into this category, such as the number of contact attempts or the sequence of outcomes of the contact attempts.

The second type of contact protocol data (CPD II) consists of data that were collected only for elements where contact was established. This information is recorded on the response behavior, and other interviewer observations like type of dwelling, access impediments, and so forth. Finally, the interviewer process provides data on the actual time spent during the interview on item nonresponses. Some interviewer process data are also available for Web surveys (item nonresponse, response time) and mail surveys (item nonresponse).

The CPD categories can be extended to include information about the interviewer (e.g., age, gender, years of experience) and classification of response types (unprocessed, noncontact, not-able, refusal, response). If interviewer data are used in the models to adjust for selective nonresponse, the clustering of sample elements among interviewers may be accounted for.

The value of paradata is twofold. In a number of studies these data were used to analyze response behavior. Durrant and Steele (2007) recommend collecting paradata for guidance on contact strategies and interviews and for the identification of groups in the population that allow for tailoring (see also Chapter 3).

Heerwegh (2003) collected paradata in Web surveys. These data comprise information about the number of changes made to answers, the sequence in which the questions are answered, and the time spent on each question. Paradata that describe response behavior can serve as feedback to questionnaire designers, for example, where respondents have returned to earlier questions in the questionnaire and changed their answers so that they could follow a different routing. This could indicate that the questionnaire was not designed optimally.

Another use of paradata is to assist interviewers in their fieldwork. Henly and Bates (2005) use contact history information in a panel survey to inform interviewers about previous data collection efforts. The paradata they provided contained information about the times, days, and results of previous contact attempts, the number of attempts needed to obtain a response as well as any remarks made by the respondents. They suggest that this information can be used by interviewers to tailor their approaches, thus lowering the number of attempts needed to establish contact and at the same time increasing the response rate. They, however, do not find sufficient evidence to support this idea, which may be due to seasonal effects, and possibly to interviewers not being accustomed to the use of contact history data.

Groves and Heeringa (2005) suggest that paradata could be used to form fieldwork strategies by a so-called responsive or adaptive design. (See Chapter 13 for more on adaptive survey designs.)

Beaumont (2005) explores the possibily of using paradata for the adjustment of nonresponse bias. Remember, most variables that are used for nonresponse adjustment are nonrandom; they are fixed demographic or socioeconomic characteristics of the sample elements or the geographical location. The paradata in contrast, are random; they can change if the data collection process were to be repeated for the same sample. Furthermore paradata are only available at the sample level, and not for the entire population. Despite these restrictions, Beaumont (2005) shows that paradata can be used for nonresponse adjustment without introducing additional bias or an additional variance component in the estimates for population totals.

While it may seem far-fetched to include information that at first sight has so little to do with the survey items, such information can help an interviewer grasp what can happen during the fieldwork. Furthermore response behavior can only partially be explained by socioeconomic and demographic variables. Indeed logistic models of response propensity have an explained variance of at most 20%. This implies that 80% of the variation cannot be explained by the standard information. This effect is partly caused by the fact that the response behavior is being estimated based on one observation only.

Paradata display the behavior of respondents during a data collection strategy. Hence there is a causal relationship between response behavior and paradata. The interviewer's strategy influences response behavior, paradata makes the behavior visible. Whereas the variables that have been considered in the research thus far are all fixed attributes of the sample persons, paradata are not fixed but random.

In Chapter 12 an example of the relationship between the number of contact attempts and the employment situation was provided. Any such clear relationships that arise between survey variables and paradata should be included as well in the adjustment models.

EXAMPLE 15.1 Paradata in the General Population Survey

In the GPS the number of contact attempts are recorded for each sample element. This paradata is of the contact protocol data I type, since it is available for all elements for which contact has been attempted by an interviewer. In a number of cases, the number of contact attempts equals zero. This is the case for *unprocessed cases*, when no contact attempt has been made due to the illness of the interviewer or to a lack of time.

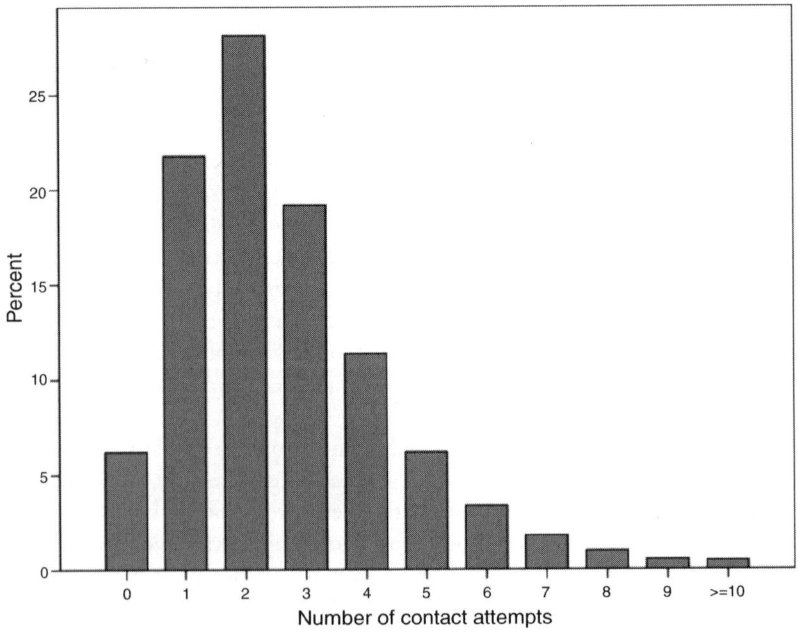

It can be seen from the figure above that the majority of the sample elements were contacted twice. This means that a final result was obtained after two contact attempts, and was recorded as either a response, or a nonresponse for some reason. In the following figure it can be seen that the paradata makes the response behavior visible. Obviously all sample elements with zero contact attempts are unprocessed in the field.

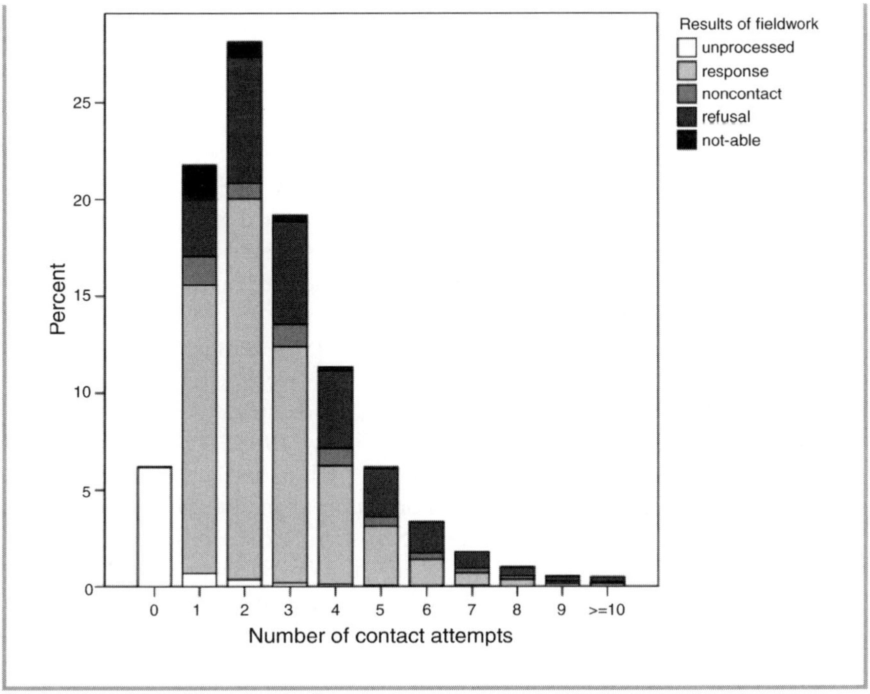

15.2.4 CONSISTENCY BETWEEN SURVEY STATISTICS

15.2.4.1 The Dutch Social Statistical Database When more than one data source is used to produce statistics, either administrative data or data from surveys, it becomes evident how well adjustment techniques work and what the effect is of measurement errors. Repeated weighting can be applied to obtain consistent estimates based on data from multiple sources. The method of repeated weighting was constructed for use in the Dutch Virtual Census in 2001. In this section a description of the census in The Netherlands is provided. Then the method of repeated weighting is outlined and illustrated with an example for the Dutch Social Statistical Database.

The last real census in The Netherlands took place in 1971. Results of pilot studies for the 1981 census predicted a nonresponse rate of 26%. Therefore the reliability of the results was at stake. The Parliament decided that participation had to be voluntary and not obligatory, and that the resulting data had to be absolutely anonymous. The government came to the conclusion that the social-psychological climate in the country would make successful completion of the census very difficult, and therefore it decided to postpone the census.

The Parliament decided to cancel the census completely in 1981, but The Netherlands still had to meet its obligations to the European Union and conduct the census. The European Commission gave permission to replace the census by compiling a set of tables based on combined data from registers and surveys. This was repeated for the 1991 census.

Demographic data for the 1981 and 1991 census were obtained from the population register. Socioeconomic data came from the Labor Force Survey. These included data on labor and education. These data sources, however, were used separately, which means that no special attention was paid to consistency. No attempt was made to link data from different sources at the micro level. Consequently tables based on data from different sources could be inconsistent. To overcome such inconsistencies, a weighting adjustment technique was applied by which tables were made to fit population register totals.

The table sets compiled for the 1981 and the 1991 census were less detailed than those of subsequent censuses. For example, there were no tables at the level of the municipalities. More details for 1981 census are described in Vliegen and Van de Stadt (1988) and for the 1991 census in Corbey (1994).

For the census of 2001, the European Union and other international organizations required more detailed information than for the census rounds of 1981 and 1991. To meet this request, Statistics Netherlands initiated a new approach. The core component was the creation of the Social Statistical Database (SSD). This database contains detailed and comprehensive information on persons and households. This information collected from various sources and linked at the micro level. The backbone of the SSD is formed by basic demographic variables that are retrieved from the population register. Socioeconomic data are added from other administrative sources, such as employment, income, benefits, and income.

Not all data included in the SSD are retrieved from registers. Some variables are measured in sample surveys. Examples are occupation and level of education, which are recorded in the Labor Force Survey.

Records from various administrative sources can be linked to records in the SSD, because all the records there contain unique social security and fiscal numbers (the so-called SoFi number). The SoFi number is not recorded in the sample surveys. Linkage for these survey data records was established using a key based on gender, data of birth, and address. More on record linkage issues can be found in Linder (2004).

Figure 15.2 gives a schematic and simplified overview of the Social Statistical Database as it was used for the virtual census of 2001.

Due to the use of sample survey data, the values of the variables are not available for every person in the population. There is no problem for tables based on variables coming from the registers and other administrative sources. This is just a matter of counting.

The situation is different for tables that contain survey variables. The population counts cannot be computed but have to be estimated. Moreover, if a variable is measured in more than one survey, there will be different estimates for its population distribution. Although the differences may be within margins of sampling error, Statistics Netherlands prefers to have consistent figures. To achieve this consistency, a new methodology was developed. This was called the method of *repeated weighting*.

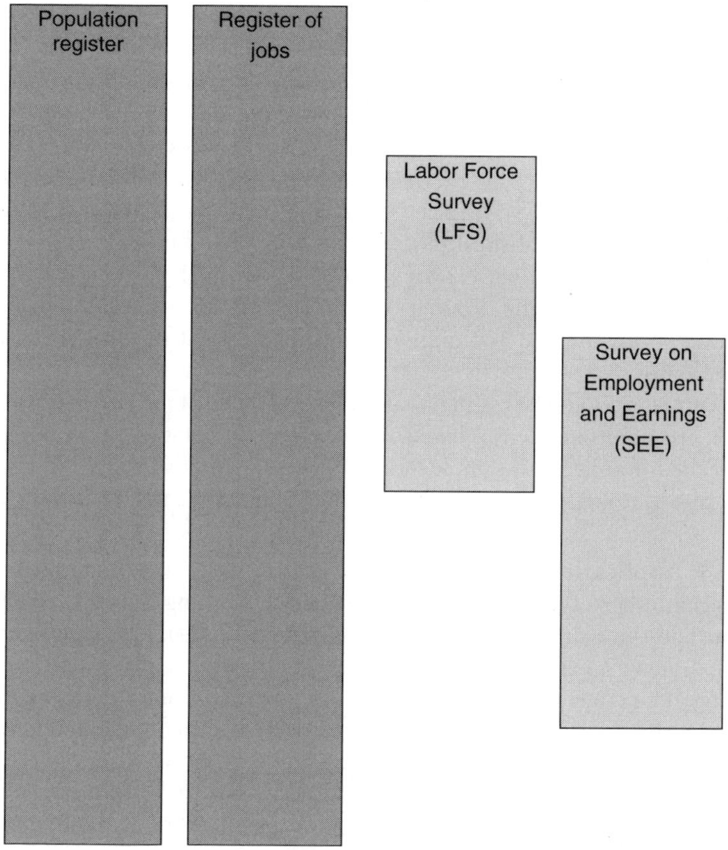

FIGURE 15.2 Schematic view of the Social Statistical Database.

15.2.4.2 Repeated Weighting Repeated weighting comes down to a repeated application of the generalized regression estimator. To produce tables that are as accurate as possible, they should be computed using the largest number of possible records. This means that tables just based on register variables use all records in the register. Tables that contain at least one survey variable are estimated using the largest possible set of records that contain all variables concerned.

To obtain a consistent set of tables, a sequential procedure is followed. For each new table are determined the marginals it has in common with tables that already have been estimated. The next step is to estimate the table while calibrating on these common marginals. The estimates, apart from being consistent, will also be more accurate, particularly if the marginals can be estimated from larger data sets or counted from register data, and as such serve as auxiliary information. In case of traditional weighting there will usually be one set of weights for the survey. Repeated weighting, however, computes a set of weights for each table. This is necessary in order to obtain consistency between tables. More details about the method of repeated weighting can be found in Houbiers et al. (2003) and Houbiers (2004).

The work on the census of 2001 started at a later moment in time than would have been the case for a traditional census. One of the reasons was that first all data sources (including registers) had to be available. Forty census tables were produced, 28 at the national level, 9 at the regional level, and 3 at the municipal level. Notwithstanding the late start, The Netherlands was one of the first to send its tables to Eurostat, the statistical office of the European Union.

The costs of the virtual census of 2001 were much lower than that of a traditional census. At the time it was estimated that a traditional census would cost approximately 300 million euros. The cost of the virtual census was 3 million euros.

Of course, it should be realized that the level of detail of the virtual census is less than that of a traditional census based on complete enumeration of the population. Some of the information is sample based. However, the advantages of lower costs and no census nonresponse problems made up for some lack of detail.

15.2.4.3 Application In this section repeated weighting is applied to the estimated number of jobs stratified to "gender × working hours × education." The data and the example are taken from Houbiers (2004). The table "gender × working hours × education" is estimated, both with and without repeated weighting. The target population is "jobs," so in the frequency table the number of jobs is estimated against some auxiliary variables, in particular gender, educational level of the person having a certain job, and the hours worked per week in that job (< 35 hours corresponds to part-time, $= 35$ hours to full-time).

In Figure 15.2 it is shown how two surveys, the Labor Force Survey and the Survey on Employment and Earnings, are linked to the population register and the register of jobs. The surveys have some overlap.

Four data blocks can be extracted from the two surveys and the registers displayed in Figure 15.2, each data block containing all records that have a maximal set of variables in common. These are:

1. A data block containing all elements in the population and all the variables in the population register and the register on jobs.
2. A data block containing all records in the largest survey, the SEE, and for each record all relevant variables from that survey and the register.
3. A data block containing all records from the smallest survey, the LFS, and for each record all relevant variables from that survey and the register.
4. A data block containing all records in the overlap from the two surveys and all variables from both surveys as well as the register variables.

The frequency table "gender × working hours × education" has to be estimated from data block 4, but the marginal "gender" can be counted from the register, the marginal "gender × working hours" can be estimated from data block 2 and the marginal "gender × education" can be estimated from data block 3.

Before estimates can be made from these rectangular data blocks, weights must be attached to the data to inflate the samples to the population. For the registers, these weights are equal to 1. For data blocks that consist of survey data (i.e., blocks 2 to 4), the weights depend on the survey design, the nonresponse, and the auxiliary information. For data blocks 2 and 3, these weights are given by the survey design weights which are calibrated to known population totals from the registers. For data block 4, the weights are given by the product of the survey design weights from each of the surveys since the two surveys are independent from each other. In addition these weights can subsequently be calibrated not only to the population totals from the registers, but also to estimated population totals from blocks 2 and 3.

Because data block 4 is the smallest data block, it may not be possible to include all known crossings from the register and estimated crossings from the larger data blocks 2 and 3 in the weighting model for block 4. This may cause inconsistencies between estimates based on data block 4 and estimates from data block 2 and 3. Cross-tabulations that cannot be estimated consistently with the block weights should be calculated with the method of repeated weighting.

Data block 4, the intersection between the two surveys, consists of approximately 50,000 records. The frequency table estimated based on data block 4 is shown in Table 15.1. The block weights of data blocks 2, 3, and 4 are all calibrated on "gender," so the estimated number of total jobs occupied by females and males is exactly equal to the register totals (3,747.2 thousand for males and 2,699.2 thousand for females).

Table 15.1 also gives the standard errors of the corresponding totals, estimated using a Taylor linearization of the generalized regression estimator. The standard errors are quite large. However, the marginal "gender × working hours" (last row in Table 15.1) can be estimated more accurately from data block 2, which contains records of approximately half of the population. Similarly the marginal "gender × education" (total male/female columns in Table 15.1) can be estimated more accurately from data block 3, which contains approximately 100,000 records. The resulting estimates are shown in Tables 15.2 and 15.3. Except for the total number of females and males with a job, the estimates of these tables clearly differ from the corresponding marginals in Table 15.1.

The standard errors for Table 15.3 are also shown. Again, they are estimated using a Taylor linearization of the generalized regression estimator. The standard errors are much smaller than the corresponding standard errors in Table 15.1. In principle, the standard errors for Table 15.2 are also smaller. Owing to the size of data block 2, they are ignored here.

Using repeated weighting, the frequency table "gender × working hours × education" can be estimated consistently with the more accurate marginals from Tables 15.2 and 15.3. This means that the estimates from Table 15.1 are calibrated to the 14 + 4 cells from Tables 15.2 and 15.3. The result is shown in Table 15.4.

Indeed the more accurate estimates of "gender × working hours" and "gender × education" from Tables 15.2 and 15.3 are reproduced in Table 15.4. The estimated standard errors of the repeated weighting estimators are also shown in Table 15.4.They are smaller than the corresponding ones in Table 15.1. This shows that repeated weighting not only leads to numerical consistency but also leads to

TABLE 15.1 Estimate of frequency table from data block 4, using the corresponding block weights (×1,000)

Gender × working hours	Male						Female					
Level of education	Part-time		Full-time		Total male		Part-time		Full-time		Total female	
Primary or less	92.2	(12.8)	277.4	(21.6)	369.6	(24.7)	173.3	(16.7)	52.1	(10.3)	225.4	(18.7)
Lower secondary general	103.5	(13.7)	144.0	(11.3)	247.5	(17.2)	193.3	(13.8)	85.6	(12.4)	278.9	(17.9)
Lower secondary vocational	87.7	(10.3)	478.0	(26.5)	565.7	(27.9)	213.1	(15.1)	61.7	(10.0)	274.8	(17.6)
Upper secondary general	74.0	(11.2)	149.7	(14.7)	223.7	(17.8)	154.8	(15.8)	76.6	(10.2)	231.5	(18.2)
Upper secondary vocational	166.4	(16.3)	1236.7	(34.1)	1403.1	(35.5)	672.6	(25.4)	363.0	(22.5)	1035.6	(29.3)
Vocational college	123.9	(11.1)	482.4	(21.5)	606.3	(23.7)	305.4	(15.2)	175.5	(9.9)	480.9	(17.8)
University or more	64.4	(6.7)	267.0	(14.4)	331.3	(15.8)	85.6	(8.20	86.5	(10.3)	172.1	(12.8)
Total	712.2	(26.5)	3035.1	(26.5)	3747.2		1798.1	(28.4)	901.1	(28.4)	2699.2	

Note: Standard errors are given in parentheses.

TABLE 15.2 Estimate of "gender × working hours" from data block 2, using the block weights (× 1,000)

Gender × working hours	Male: Part-time	Male: Full-time	Total male	Female: Part-time	Female: Full-time	Total female
Total	701.1	3,046.2	3,747.2	1,827.9	871.3	2,699.2

more accurate estimates because of better use of auxiliary information. Roughly speaking, the estimated standard errors are obtained by a repeated Taylor linearization of the generalized regression estimator (see Houbiers et al., 2003; Knottnerus, 2003).

15.3 Summary

In this chapter, a number of subjects were discussed that received little attention in the other chapters of this book but that are related to the issue of nonresponse in household surveys.

The first topic, in Section 15.2.1, concerned the combined treatment of item and unit nonresponse. Typically both types of nonresponses occur in surveys. However, the methods described in this book focus either on unit nonresponse or on item nonresponse. A combined treatment of the two types can be done by full imputation, full weighting or a combination of imputation, and weighting. The last approach is most frequently used to impute missing values for item nonresponse, followed by a weighting adjustment, using the imputed items to adjust for unit nonresponse.

The second topic, in Section 15.2.2, concerned nonresponse in longitudinal surveys, also referred to as dropout or panel attrition. The amount of available information for dropouts is usually less than is the case with item nonresponse but more than for unit nonrespondents since data from a previous wave is available. As with item and unit nonresponse, the most important question is whether the

TABLE 15.3 Estimate of "gender × education" from data block 3, using the block weights (× 1,000)

Level of education × gender	Total male		Total female	
Primary or less	357.2	(5.2)	209.8	(3.9)
Lower secondary general	267.3	(4.5)	290.5	(4.4)
Lower secondary vocational	595.8	(6.3)	320.6	(4.6)
Upper secondary general	213.2	(4.2)	205.4	(3.9)
Upper secondary vocational	1,347.0	(8.3)	1,006.9	(6.9)
Vocational college	616.0	(6.3)	500.1	(5.5)
University or more	323.8	(4.9)	166.0	(3.5)
Total	3,747.2		2,699.2	

TABLE 15.4 Estimate of frequency table "gender × working hours × education" from data block 4 using repeated weighting (× 1,000)

Gender × working hours	Male						Female					
Level of education	Part-time		Full-time		Total male		Part-time		Full-time		Total female	
Primary or less	87.3	(10.6)	269.9	(11.2)	357.2	(5.2)	163.0	(9.2)	46.8	(8.8)	209.8	(3.9)
Lower secondary general	110.2	(9.2)	157.1	(9.5)	267.3	(4.5)	203.9	(9.7)	86.6	(9.4)	290.5	(4.4)
Lower secondary vocational	90.4	(9.4)	505.3	(10.7)	595.8	(6.3)	250.6	(8.9)	70.0	(8.3)	320.6	(4.6)
Upper secondary general	69.2	(8.9)	143.9	(9.3)	213.2	(4.2)	139.6	(8.8)	65.8	(8.5)	205.4	(3.9)
Upper secondary vocational	159.0	(13.0)	1,215.0	(14.8)	1,374.0	(8.3)	664.0	(14.9)	342.8	(14.5)	1,006.9	(6.9)
Vocational college	123.3	(9.6)	492.7	(10.8)	616.0	(6.3)	322.4	(9.3)	177.7	(8.9)	500.1	(5.5)
University or more	61.5	(6.2)	262.2	(7.3)	323.8	(4.9)	84.4	(6.8)	81.6	(6.8)	166.0	(3.5)
Total	701.1		3,046.2		3,747.2		1,827.9		871.3		2,699.2	

Note: Standard errors are given in parentheses.

dropout is selective. In other words, are sample elements that drop out of the panel survey systematically different from sample elements that remain in the panel? If they are not different, the dropout will not result in a bias. If they are different, there will be a bias that, depending on the mechanism of the missing data, may be adjusted for using auxiliary information.

The dropout can be missing completely at random (MCAR), missing at random (MAR), or not missing at random (NMAR). In addition to the auxiliary variables *X*, also the target variables *Y* from previous waves may be regarded to determine the nature of the missing data mechanism. In Section 15.2.2 a number of methods were discussed that handle dropout in panel surveys.

Section 15.2.3 concerned the use of data about the process of data collection, namely paradata or process data. In a number of studies these data were used to analyze response behavior. Paradata can provide guidance for contact strategies and interviews and for the identification of groups in the population that allow for tailoring. Paradata in Web surveys comprise information about the number of changes made to answers, the sequence in which the questions are answered and the time spent on each question. These paradata serve to describe response behavior and can be helpful indicators of data quality. They can also serve as feedback to questionnaire designers, such as when respondents have returned to earlier questions in the questionnaire and changed their answers so that they can follow a different routing; this practice could indicate that the questionnaire is not designed optimally.

Another application of paradata is to inform the fieldwork about the progress of the data collection. For example, contact history information can be used to inform interviewers about the previous response process. This can be, for instance, information about the times, days, and results of previous contact attempts, the number of attempts needed to obtain a response, as well as remarks made by the respondents. The interviewers can use this information to tailor their approach, thus lowering the number of attempts needed to establish contact and at the same time increasing response.

In Section 15.2.4 it is described how the GREG estimator can be applied to estimate consistent sets of tables. This can be done by repeatedly applying the GREG estimator; therefore the method is named repeated weighting. An application of repeated weighting in the Dutch Virtual Census of 2001 is described.

15.4 KEY TERMS

A rolling cross-sectional survey: A continuous cross-sectional survey that can be used to describe the population at any given time.

Cross-sectional survey: A one-time survey for which a sample is selected from the population and approached with a request to participate in the survey.

Full weighting estimator: An estimator that can be used to treat both item and unit nonresponse based on weighting.

Imputed estimator: An estimator that can be used to treat both item and unit nonresponse and that fully imputes the missing values.

Imputed modified estimator: An estimator that can be used to treat both item and unit nonresponse and that imputes for item nonresponse and weights for unit nonresponse.

Item response subset: The subset of sample elements that have responded to a particular target variable.

Longitudinal survey: A survey where repeated observations are made on the same sample elements over a period of time.

Panel attrition: Nonresponse in a panel survey that occurs when sample elements drop out prior to the end of the study and do not return. Also referred to as dropout.

Panel dropout: Nonresponse in a panel survey that occurs when sample elements drop out prior to the end of the study and do not return. Also referred to as panel attrition.

Panel survey: A longitudinal study where a sample is selected from the population and this same sample is surveyed at regular time intervals.

Repeated weighting: A method that is used to obtain numerical consistency between tables based on a combination of surveys and registers. It comes down to repeatedly applying the generalized regression estimator.

Wave nonrespondents: Nonrespondents in one of the waves of a panel survey.

Wave: One of the surveys offered to a panel in a longitudinal study or panel survey.

15.5 REFERENCES

Beaumont, J.-F. (2005), On the Use of Data Collection Process Information for the Treatment of Unit Nonresponse through Weight Adjustment. *Survey Methodolog* 31, pp. 227–231.

Corbey, P. (1994), Exit the Population Census, *Netherlands Official Statistics* 9, pp. 41–44.

Couper, M.P., and Lyberg, L.E. (2005), *The Use of Paradata in Survey Research.* Paper presented at the 56th session of the ISI Conference, Sydney, Australia.

Couper, M.P. (1998), *Measuring Survey Quality in a CASIC Environment.* Paper presented at the Joint Statistical Meetings of the American Statistical Association, Dallas, TX.

Durrant, G.B., and Steele, F. (2007), Multilevel Modelling of Refusal and Noncontact Nonresponse in Household Surveys: Evidence from Six UK Government Surveys, S3RI Working Paper M07-11. University of Southampton, Southampton, UK.

Fitzmaurice, G.M., Laird, N.M., and Ware, J.H. (2004), *Applied Longitudinal Analysis.* Wiley, New York.

Groves, R.M., and Heeringa, S.G. (2005), Responsive Design for Household Surveys: Tools for Actively Controlling Survey Errors and Costs. *Journal of the Royal Statistical Association, Series A* 169, pp. 439–457.

Heerwegh, D. (2003), Explaining Response Latencies and Changing Answers Using Client-Side Paradata from a Web Survey. *Social Science Computer Review* 21, pp. 360–373.

Henly, M.,and Bates, N. (2005), *Using Call Records to Understand Response in Panel Surveys,* Paper presented at the Annual AAPOR meeting–ASA section on Survey Research Methods.

Houbiers, M. (2004), Towards a Social Statistical Database and Unified Estimates at Statistics Netherlands. *Journal of Official Statistics* 20, pp. 55–75.

Houbiers, M., Knottnerus, P, Kroese, A.H., Renssen, R.H., and Snijders, V. (2003), *Estimating Consistent Table Sets: Position Paper on Repeated Weighting.* Discussion Paper 03005. Statistics Netherlands, Voorburg/Heerlen.

Kreuter, F. (2006), *Paradata—Introduction to the Special Interest Section International Workshop on HH NR 2006.* Paper presented at the 17th International Workshop on Household Survey Nonresponse, Omaha, NE.

Knottnerus, P. (2003), *Sample Survey Theory: Some Pythagorean Perspectives.* Springer, New York.

Linder, F. (2004), The Use of Administrative Registers and Sample Surveys in the Dutch Census of 2001. In: Schulte Nordholt, E., Hartgers, M., and Gircour, M. (eds.), *The Dutch Virtual Census of 2001.* Statistics Netherlands, Voorburg, pp. 243–260.

Molenberghs, G., and Verbeke, G. (2005). *Models for Discrete Longitudinal Data.* Springer, New York.

Särndal, C.E., and Lundström, S. (2005), *Estimation in Surveys with Nonresponse.* Wiley, Chichester, UK.

Särndal, C.E., and Lundström, S. (2008), Assessing Auxiliary Vectors for Control of Nonresponse Bias in the Calibration Estimator. *Journal of Official Statistics* 24, pp. 167–191.

Verbeke, G., and Molenberghs, G. (2000), *Linear Mixed Models for Longitudinal Data.* Springer, New York.

Vliegen, J.M., and Van de Stadt, H. (1988), Is a Census Still Necessary? Experiences and Alternatives. *Netherlands Official Statistics* 3, pp. 27–34.

15.6 EXERCISES

Exercise 15.1 There are three possible approaches to the combined treatment of item and unit nonresponse. Which of the following statements is true?

(a) Full weighting leads to a rectangular dataset in which all the missing variables are replaced by an estimated value.

(b) Full weighting leads to a different set of weights for each target variable.

(c) Full imputation leads to a different set of weights for each target variable.

(d) The combined approach consists of weighting for item nonresponse and imputation for unit nonresponse.

Exercise 15.2 Which of the following statements about nonresponse in panel surveys is true?

(a) There is usually more information about nonrespondents in panel surveys than in cross-sectional surveys because the nonrespondents could have participated in previous waves.

(b) In a panel survey, the risk of a bias due to nonresponse is lower so that the nonresponse can be ignored and data from the respondents can be used for analysis.

(c) Available or complete case analysis in panel surveys can only be used when the dropout is MAR or NMAR.

Exercise 15.3 The last observation carried forward method can be used to adjust for panel dropout if the reason for the dropout

(a) is unrelated to the target variables of the survey.

(b) is unrelated to the target variables of the survey and does not change after the sample elements drops out of the panel survey.

(c) is unrelated to the target variables of the survey and does not change after the sample elements drops out of the panel survey and remains unchanged afterward.

Exercise 15.4 Paradata are data about the process of data collection. Which of the following statements about paradata is not true?

(a) Paradata can be used to make contact strategies for different groups in the population.

(b) Paradata can be used to analyze response behavior.

(c) Paradata can be used to inform the respondents about the interviewers.

(d) Paradata can be used for an optimal design of Web questionnaires.

Exercise 15.5 Which of the following statements about repeated weighting is true?

(a) It does not matter in which order tables are estimated with repeated weighting.

(b) Empty cells are no longer a problem when applying the method of repeated weighting.

(c) The estimates obtained by repeated weighting are consistent but less accurate than estimates based on applying the generalized regression estimator only once.

(d) The estimates obtained by repeated weighting are consistent and more accurate, particularly if the marginals can be estimated from larger data sets or counted from register data.

INDEX

Handbook of Nonresponse in Household Surveys, By Jelke Bethlehem, Fannie Cobben, and Barry Schouten

WILEY SERIES IN SURVEY METHODOLOGY

Established in Part by WALTER A. SHEWHART AND SAMUEL S. WILKS

The ***Wiley Series in Survey Methodology*** covers topics of current research and practical interests in survey methodology and sampling. While the emphasis is on application, theoretical discussion is encouraged when it supports a broader understanding of the subject matter.

The authors are leading academics and researchers in survey methodology and sampling. The readership includes professionals in, and students of, the fields of applied statistics, biostatistics, public policy, and government and corporate enterprises.

ALWIN · Margins of Error: A Study of Reliability in Survey Measurement
BETHLEHEM · Applied Survey Methods: A Statistical Perspective
BETHLEHEM, COBBEN, and SCHOUTEN · Handbook of Nonresponse in Household Surveys
*BIEMER, GROVES, LYBERG, MATHIOWETZ, and SUDMAN · Measurement Errors in Surveys
BIEMER and LYBERG · Introduction to Survey Quality
BRADBURN, SUDMAN, and WANSINK ·Asking Questions: The Definitive Guide to Questionnaire Design—For Market Research, Political Polls, and Social Health Questionnaires, *Revised Edition*
BRAVERMAN and SLATER · Advances in Survey Research: New Directions for Evaluation, No. 70
CHAMBERS and SKINNER (editors) · Analysis of Survey Data
COCHRAN · Sampling Techniques, *Third Edition*
CONRAD and SCHOBER · Envisioning the Survey Interview of the Future
COUPER, BAKER, BETHLEHEM, CLARK, MARTIN, NICHOLLS, and O'REILLY (editors) · Computer Assisted Survey Information Collection
COX, BINDER, CHINNAPPA, CHRISTIANSON, COLLEDGE, and KOTT (editors) · Business Survey Methods
*DEMING · Sample Design in Business Research
DILLMAN · Mail and Internet Surveys: The Tailored Design Method
FULLER · Sampling Statistics
GROVES and COUPER · Nonresponse in Household Interview Surveys
GROVES · Survey Errors and Survey Costs
GROVES, DILLMAN, ELTINGE, and LITTLE · Survey Nonresponse
GROVES, BIEMER, LYBERG, MASSEY, NICHOLLS, and WAKSBERG · Telephone Survey Methodology
GROVES, FOWLER, COUPER, LEPKOWSKI, SINGER, and TOURANGEAU · Survey Methodology, *Second Edition*
*HANSEN, HURWITZ, and MADOW · Sample Survey Methods and Theory, Volume 1: Methods and Applications
*HANSEN, HURWITZ, and MADOW · Sample Survey Methods and Theory, Volume II: Theory
HARKNESS, BRAUN, EDWARDS, JOHNSON, LYBERG, MOHLER, PENNELL, and SMITH (editors) · Survey Methods in Multinational, Multiregional, and Multicultural Contexts
HARKNESS, VAN DE VIJVER, and MOHLER (editors) · Cross-Cultural Survey Methods
KALTON and HEERINGA · Leslie Kish Selected Papers

*Now available in a lower priced paperback edition in the Wiley Classics Library.

KISH · Statistical Design for Research
*KISH · Survey Sampling
KORN and GRAUBARD · Analysis of Health Surveys
LEPKOWSKI, TUCKER, BRICK, DE LEEUW, JAPEC, LAVRAKAS, LINK, and SANGSTER (editors) · Advances in Telephone Survey Methodology
LESSLER and KALSBEEK · Nonsampling Error in Surveys
LEVY and LEMESHOW · Sampling of Populations: Methods and Applications, *Fourth Edition*
LUMLEY · Complex Surveys: A Guide to Analysis Using R
LYBERG, BIEMER, COLLINS, de LEEUW, DIPPO, SCHWARZ, TREWIN (editors) · Survey Measurement and Process Quality
MAYNARD, HOUTKOOP-STEENSTRA, SCHAEFFER, and VAN DER ZOUWEN · Standardization and Tacit Knowledge: Interaction and Practice in the Survey Interview
PORTER (editor) · Overcoming Survey Research Problems: New Directions for Institutional Research, No. 121
PRESSER, ROTHGEB, COUPER, LESSLER, MARTIN, MARTIN, and SINGER (editors) · Methods for Testing and Evaluating Survey Questionnaires
RAO · Small Area Estimation
REA and PARKER · Designing and Conducting Survey Research: A Comprehensive Guide, *Third Edition*
SARIS and GALLHOFER · Design, Evaluation, and Analysis of Questionnaires for Survey Research
SÄRNDAL and LUNDSTRÖM · Estimation in Surveys with Nonresponse
SCHWARZ and SUDMAN (editors) · Answering Questions: Methodology for Determining Cognitive and Communicative Processes in Survey Research
SIRKEN, HERRMANN, SCHECHTER, SCHWARZ, TANUR, and TOURANGEAU (editors) · Cognition and Survey Research
SUDMAN, BRADBURN, and SCHWARZ · Thinking about Answers: The Application of Cognitive Processes to Survey Methodology
UMBACH (editor) · Survey Research Emerging Issues: New Directions for Institutional Research No. 127
VALLIANT, DORFMAN, and ROYALL · Finite Population Sampling and Inference: A Prediction Approach